# Introductory Experiments in Digital Electronics and 8080A Microcomputer Programming and Interfacing—Book 2

by

Peter R. Rony

Also Published as

**Introductory Experiments in Digital Electronics, 8080A Microcomputer Programming, and 8080A Microcomputer Interfacing**

by E & L Instruments, Inc.

Howard W. Sams & Co., Inc.
4300 WEST 62ND ST. INDIANAPOLIS, INDIANA 46268 USA

Copyright © 1977 by Peter R. Rony

FIRST EDITION
THIRD PRINTING—1981

All rights reserved. No part of this book shall be reproduced, stored in a retrieval system, or transmitted by any means, electronic, mechanical, photocopying, recording, or otherwise, without written permission from the publisher. No patent liability is assumed with respect to the use of the information contained herein. While every precaution has been taken in the preparation of this book, the publisher assumes no responsibility for errors or omissions. Neither is any liability assumed for damages resulting from the use of the information contained herein.

International Standard Book Number: 0-672-21551-9
Library of Congress Catalog Card Number: 78-56564

*Printed in the United States of America.*

# Preface

Welcome to the new electronics revolution. In ten years, integrated-circuit technology has transformed the digital integrated-circuit chip from an expensive electronic component containing only simple logic functions and few transistors into a highly complex component containing up to ten thousand transistors. The computer-on-a-chip is here! It contains everything—central processing unit, read/write memory, read-only memory, and interface circuitry—required of a digital computer. Within several years, you will be able to purchase a handful of such chips for $10 to $20. There are now only 250,000 minicomputers and large computers in the United States. By 1982, there may be one billion microcomputers in existence. A computer revolution? Certainly.

In education, I believe that the new electronics revolution will create important opportunities and changes:

- More students, including engineers, chemists, biologists, physicists, agricultural scientists, biochemists, and experimental psychologists, will need to learn about digital technology and microcomputers.
- Theoretical courses on Boolean algebra, Karnaugh mapping, and the like will become less important for the majority of students who are interested in digital technology.
- Students of computer science will be exposed to more digital hardware, e.g., in laboratory courses on digital electronics and microcomputers. Many students will have their own microcomputers.
- Hundreds of microcomputers will be present on the typical community college or university campus. Perhaps thousands.
- Courses in digital telecommunications and digital controls will grow in importance.

In the face of these changes, one thing will remain essentially invariant: the time that a student spends in school. If anything, the number of credit hours required for graduation will decrease. Educators will be faced with the problem of incorporating the above topics into various curricula without cutting back on other important courses. How can this be done? Perhaps by integrating several courses together and covering only essential concepts.

This series of books on digital electronics, microcomputer interfacing, and microcomputer programming is an attempt to integrate these three subjects into a single unified course. This course is oriented toward laboratory experiments, for I believe that this is the best way to convey the excitement and importance of the new electronics revolution. The three subjects will be given approximately equal weight. You will learn how to program a microcomputer, how to interface a microcomputer to external digital devices, and how the external devices operate from a digital point of view. Important digital concepts will be illustrated both with integrated-circuit chips and with microcomputer programs, usually side by side in the same or adjacent units.

For the reader of these books, little or no background in digital electronics or microcomputers is assumed. You will first treat microcomputers and integrated-circuit chips as *functional modules.* With exposure to the modules, you will gradually learn their basic operational characteristics. We will not discuss how they are manufactured, since the technology is sophisticated and changes every several years.

This book is a laboratory-oriented text in a series of books that approach the field of electronics in a somewhat different manner. Rather than start you, as is customarily done in introductory electronics courses, with experiments on electronic *components,* such as *resistors, capacitors, diodes,* and *transistors,* I instead introduce you immediately to *integrated-circuit chips.* I also introduce you immediately to the concepts of *logic switches, lamp monitors, pulsers,* and *displays* and show you how to use such auxiliary functions. I provide you with many experiments that are based upon connections between integrated-circuit chips and such devices. All this is done in *Logic & Memory Experiments Using TTL Integrated Circuits, Books 1 and 2.*

Once you have mastered the basic concepts of digital electronics and are knowledgeable with the techniques of wiring digital circuits using integrated-circuit chips, you are exposed to more complicated digital chips and digital systems. *Introductory Experiments in Digital Electronics and 8080A Microcomputer Programming and Interfacing (Books 1 and 2)* is an experiment in digital electronics education. As mentioned earlier, I am attempting to integrate the subjects of digital electronics, microcomputer interfacing, and microcomputer programming into a single unified course. In effect, I am consolidating the material found in other books of the series into a single laboratory textbook. The concepts and techniques of microcomputer programming and interfacing are discussed at the same time that you learn basic digital concepts and perform experiments on TTL integrated-circuit chips. Some material in the earlier books has been omitted, and much new material added, especially in the microcomputer sections.

I believe that the pendulum of digital electronics will now move steadily toward the use of microcomputers. Such being the case, there will be considerable incentive in educational institutions to introduce microcomputers at an early stage in a student's curriculum. What is true for the college student should also be true for the professional scientist or engineer who desires to update his knowledge of digital electronics. Books 1 and 2 of *Introductory Experiments in Digital Electronics and 8080A Microcomputer Programming and Interfacing* are directed toward such individuals.

Books 1 and 2 of *Introductory Experiments in Digital Electronics and 8080A Microcomputer Programming and Interfacing* are self-instructional texts. Answers to all experimental and review questions will be found in the texts. When you perform an experiment, you will be told what you should observe. Who can use these books successfully? They are directed toward the same audience as the earlier books in the series. You need no initial background in digital electronics or microcomputers. If you have the ability to organize and grasp new concepts, to extrapolate knowledge to new situations, and to perform experiments in wiring digital circuits carefully, you should enjoy these books. Books 1 and 2 of *Introductory Experiments in Digital Electronics and 8080A Microcomputer Programming and Interfacing* lend themselves very nicely to a self-study program for professionals who desire to update their skills in digital electronics and microcomputers.

There has been wide acceptance of these books in formal classes as well as by individual users in the United States and abroad. Selected books are being translated into German, Japanese, French, Italian, Spanish and Chinese.

With the aid of material in the *Intel 8080 Microcomputer Systems User's Manual* and the NEC Microcomputers, Inc. $\mu COM$-8 *Software Manual,* we have provided a detailed description of the 8080A microprocessor chip. I am grateful to both the Intel Corporation and NEC Microcomputers, Inc. for their kind permission to let me use information in their manuals. If you are a serious user of the 8080A chip, you should have both manuals in your possession.

PETER R. RONY

# Contents

### UNIT 16

WHAT IS INTERFACING? .............................. 11
Introduction—Objectives—The Smart Machine Revolution—Microprocessor vs Microcomputer—Hardware vs Software—What Is a Controller?—Where Microcomputers Fit—Computer Hierarchies—A Typical 8080 Microcomputer—Address Bus—Bidirectional Data Bus—Control Bus—What Is Interfacing?—What Is an I/O Device?—Review Questions

### UNIT 17

DEVICE SELECT PULSES ............................. 31
Introduction—Objectives—What Is a Device Select Pulse?—Uses for Device Select Pulses—Generating Device Select Pulses—I/O Instructions—The Fetch, Input and Output Machine Cycles—First Program—Second Program—Introduction to the Experiments—Experiments—Review Questions

### UNIT 18

THE 8080A INSTRUCTION SET ....................... 81
Introduction — Objectives — Microcomputer Programming — Sources of 8080 Programming Information—8080 Instruction Set Summaries—8080 Microprocessor Registers—What Types of Operations Does the 8080A Microprocessor Perform?—8080 Mnemonic

Instructions—The Instruction Set—The 8080 Instruction Set—Data Transfer Group—Arithmetic Group—Logical Group—Rotate Instructions—Branch Group—Stack, I/O and Machine Control Group—Instruction Set—Introduction to the Experiments—Experiments—Octal/Hexadecimal Listing of the 8080 Instruction Set—8080 Instruction Set Summary—Review Questions

## UNIT 19

DATA BUS TECHNIQUES USING THREE-STATE DEVICES ....... 173

Introduction—Objectives—What Is a Bus?—Three-State Bussing—Examples of Simple Bus Systems—74125 Three-State Buffer—74126 Three-State Buffer—8095 Three-State Buffer—Other Three-State Devices—Introduction to the Experiments—Experiments—Review Questions

## UNIT 20

AN INTRODUCTION TO ACCUMULATOR
INPUT/OUTPUT TECHNIQUES ........................... 191

Introduction—Objectives—What Is Input/Output?—Microcomputer Output—Some Output Latch Circuits—Output Drive Capability—Microcomputer Input—Some Three-State Buffer Input Circuits—Accumulator I/O Instructions—First Input/Output Program—Second Program—Third Program—Fourth Program—Fifth Program—Introduction to the Experiments—Experiments—Listing of Subroutine KBRD—Listing of Subroutine TIMOUT—Experiment No. 3, Characteristics of the DAA Instruction—Review Questions

## UNIT 21

AN INTRODUCTION TO MEMORY-MAPPED
INPUT/OUTPUT TECHNIQUES ........................... 227

Introduction—Objectives—Memory-Mapped I/O vs Accumulator I/O—Generating Memory-Mapped I/O Address Select Pulses—Memory-Mapped I/O: Use of Address Bit A-15—Memory-Mapped I/O Instructions—The Memory Read and Memory Write Machine Cycles—First Program—Some Input/Output Circuits—Second Program—Third Program—Fourth Program—Fifth Program—Sixth Program—Seventh Program—Eighth Program—Introduction to the Experiments—Experiments—Review Questions

## UNIT 22

SOME EXAMPLES OF MICROCOMPUTER INPUT/OUTPUT ...... 261
Introduction—Objectives—Data Logging With an 8080 Microcomputer—First Program: Logging Sixty-Four 8-Bit Data Points—Second Program: Logging Slow Data Points—Third Program: Output From a Data Logger—Fourth Program: Detecting an ASCII Character—Other Methods of Generating Time Delays—Introduction to the Experiments—Experiments—Additional Information—Experiment No. 9, A Staircase-Ramp Comparison Analog-to-Digital Converter—Review Questions

## UNIT 23

FLAGS AND INTERRUPTS ............................. 309
Introduction—Objectives—What Is a Flag?—First Example: Interfacing a Keyboard—Second Example: Solvent-Level Control—Polled Operation—What Is an Interrupt?—Types of Interrupts—Restart: RST X—Enable and Disable Interrupt: EI and DI—Third Example: Interrupt-Driven Keyboard Interface—Priority Interrupts—Hardware Priority Interrupts—Interrupt Software—Introduction to the Experiments—Experiments—Review Questions

## APPENDIX A

REFERENCES ........................................ 385

## APPENDIX B

DEFINITIONS ....................................... 387

## APPENDIX C

OCTAL/HEX CONVERSION TABLE ....................... 391

## APPENDIX D

ANSWERS TO REVIEW QUESTIONS ...................... 393

INDEX ............................................. 401

# 16

# What Is Interfacing?

**INTRODUCTION**

This unit introduces you to a few of the objectives of interfacing and provides definitions for some of the concepts involved.

**OBJECTIVES**

At the completion of this unit, you will be able to do the following:

- Distinguish between microprocessor and microcomputer.
- Define data processor.
- Distinguish between hardware and software, and give examples of each.
- Define controller.
- Discuss the spectrum of computer-equipment complexity, from hardwired logic to the large mainframe computers.
- Describe the three important busses in an 8080A-based microcomputer.
- List five important control signal lines on the microcomputer.
- Define synchronous, I/O device, CPU, and memory.

**THE SMART MACHINE REVOLUTION**

In Book 1, we provided you with information on basic digital electronics that you will need as you *interface* an 8080-based microcomputer. We still have more subjects to cover—three-state bussing, shift registers, arithmetic/logic units, and buffers—but you have already been exposed to the logic operations—AND, OR, NAND, NOR, and Exclusive-OR; the gating characteristics of the four basic gates; decoders; latches and flip-flops; counters; monostable multivibrators; input/output devices such as logic switches, pulsers, clocks, and displays; digital codes; and the important terms, strobe, enable, disable, gate, and inhibit. Before you jump into the subject of microcomputer interfacing, we believe that it would be useful to provide you with some perspective on why you would want to interface a microcomputer in the first place.

For those of you who have access to the McGraw-Hill publication *Business Week,* we would direct your attention to the July 5, 1976 issue and the article entitled, "The Smart Machine Revolution: Providing Products with Brainpower." Some excerpts from the article are as follows:

"This is the second industrial revolution," says Sidney Webb, executive vice-president of TRW Inc. "It multiplies man's brainpower with the same force that the first industrial revolution multiplied man's muscle power."

The engine of the revolution is the microprocessor, or computer-on-a-chip, a tiny slice of silicon that is the arithmetic and logic heart of a computer. The first surge of products with microprocessor brains is just now starting to hit the marketplace, and this is demonstrating that never before has there been a more powerful tool for building "smart" machines—machines that can add decision-making, arithmetic, and memory to their usual functions. Included in the first wave of smart machines are:

> The smart watch
> The smart scale
> The smart mobile phone
> The smart can-making system
> The smart video game

A tidal wave of smart products such as these is on the way. They will dramatically change the marketplace for consumer, commercial, and industrial products. The computer-on-a-chip, powering the brains of smart products, will spawn new industries and thousands of new companies. And in the process, it will wipe out some existing companies, and even some industries.

A key to the sudden surge in sales of microprocessors and to the wave of new smart machines they will power is simply price. C. Lester Hogan, vice-chairman of Fairchild Camera & Instrument Corp., demonstrated this element dramatically at a Boston convention a few weeks ago. He pulled 18 microprocessors from his pocket and tossed them out to his audience. "That's $18 million worth of computer power—or it was 20 years ago," he said. Hogan explained that his $20 microprocessor is as powerful as International Business Machines' first commercial computer, which cost $1 million in the early 1950s. "The point I'm making," Hogan said, "is that computer power today is essentially free."

Even a year ago, those $20 microprocessors cost more than $100, and the sudden slash in price led designers to start work on the beginnings of the flood of smart products. Switching from conventional electronic parts, such as integrated circuits, to the MPU cuts design time and manufacturing costs because it replaces hundreds of ICs and other parts. Once the MPU is designed into a product, it can provide tremendous marketing advantages; a product's functions can be altered not by a costly redesign of its electronics but simply by changing the instructions, or software, stored in the MPU's memory. New features can be added with little increase in cost, and the new smart machines can handle work that could not be done economically before.

The most exciting new products to come from the computer-on-a-chip will be for the consumer. Microprocessors will go into homes, autos, appliances, and other consumer goods in far greater numbers than into other products. "Between 7 and 10 microprocessors will be in each home by 1980," predicts Andrew A. Perlowski, who heads microprocessor activities at Honeywell Inc. His company is already hard at work on energy management and security systems for the home.

In the factory, the computer-on-a-chip is dropping the cost of electronic intelligence so low that it is turning the smallest product units into smart machines. And it is speeding the day of the automated factory by linking the smart production machines, sensors, and other instruments into distributed data acquisition and control systems.

Factory automation has moved slowly, partly because manufacturers did not want an entire plant shut down because one bit in a computer failed. "The advantage of the MPU is that it chops up the control job in smaller pieces, and an individual MPU won't pull down a whole network if it fails," says Sheldon G. Lloyd, engineering vice-president at Fisher Controls Co. "The microprocessor makes it economically possible to develop and build hierarchical systems."

In a hierarchy, the microprocessors in the smart production machines are linked to supervisory minicomputers that collect and send management reports and status information to a central factory computer. At the top is the big corporate computer, which when linked to the factory system, will be able to generate up-to-the-second financial reports for the entire company.

Many jobs now being done by microprocessors were too small to automate before. Dow Chemical Co. is considering MPUs for a variety of jobs "where computations are required that aren't quite complex enough to justify a minicomputer," says Charles R. Honea, process instrument manager at Dow's Texas Div. For instance, Dow uses microprocessors to calculate the flow of ethylene piped into the plant. The information was charged manually and required half a dozen people. "And it was always a day behind," Honea says. "You had no way of knowing how much ethylene you used today."

The process industries are a conservative lot, partly because of the reliability needed in control gear to keep their plants running continuously. It usually takes five to six years for a major technology breakthrough to find widespread use in the process control industries. "Microprocessors will be no different," says Nicholas P. Scallon, vice-president for marketing at Fisher Controls. "But the microprocessor will speed up automation," he says, "by breaking up control loops into smaller segments. Instead of trying to control the whole system, we will use a dedicated microprocessor to control such things as a boiler, an evaporator, or a catalytic conversion."

Software is not only the biggest problem now for the MPU users, but it is also where most of the costs are. "Software costs are actually even more for a microcomputer than for a minicomputer," says Richard Marley, a New Hampshire consultant who has designed smart products for several small companies. He says that he spends up to $100,000 on every software design, while the cost of hardware designs is down to around $20,000.

The microprocessor is probably affecting no other single industry as much as the instruments business. In the next two years, predicts technology consultant Lynwood O. Eikrem, analytical instruments using microprocessors will rise from 2% to 50% of the market. "Companies are rushing into microprocessors, and those who don't move fast will lose the market," he says.

So far the biggest MPU effort is coming in digital test instruments, such as voltmeters, counters, and frequency synthesizers, and in such analytical instruments as spectrometers and chromatographs. "Probably 90% of digital instruments selling for $2000 or more will use microprocessors by 1980," says industry analyst Galen W. Wampler.

Over the next several years, smart products and machines will spread at an ever increasing rate. Software will become available so that anyone will be able to program a microcomputer. Schools will be turning out a flood of young people familiar with microprocessors and eager to build products with them. The semiconductor industry will continue to develop more powerful parts. "In the next 5 to 10 years, we will be able to turn out 1 million devices on a single chip," predicts Richard L. Petritz, vice-president of New Business Resources, a venture capital company. This will mean that the power either of a large mainframe computer or of a complete minicomputer with large amounts of memory will be available on a single chip.

Development time is so short for a smart product now and the entry costs are so low that there will be "myriad examples of new companies spawning, with bright, young fellows developing MPU-based products," says Fairchild's Hogan. Petritz says: "The MPU will reduce the application of electronics essentially to that of writing a computer program, and the average person can be educated to program a computer."

That spells danger for the established companies. Already, manufacturers have to be looking at microprocessors "or somebody will come along and obsolete their product," warns Donald V. Kleffman, a marketing manager at Ampex Corp. Says Kessler of NCR: "There will be many new companies coming in with MPU technology, and they will replace some of the old companies. A lot of companies will be beaten down."

When that time comes, microprocessors will be everywhere—from the smart machines of the factory and the office to the handheld, personal microcomputers costing less than $100, and the personal mobile telephone.

We shall amplify on some of the above points in the following sections.

## MICROPROCESSOR VS MICROCOMPUTER

We have had difficulty in finding a good definition for the term *digital computer*. In looking around for such a definition, we found an excellent pair of paragraphs by Donald Eadie in his book, *Introduction to the Basic Computer,* that provide some insight into what is meant by the term *processor*. Thus:

> This chapter serves as a general introduction to the field of digital devices, with particular emphasis on those devices called *computers,* or more properly, *data processors.* The name data processor is more inclusive because modern machines in this general classification not only compute in the usual sense, but also perform other functions with the data which flow to and from them. For example, data processors may gather data from various incoming sources, sort it, rearrange it, and then print it. None of these operations involve the arithmetic operations normally associated with a computing device, but the term computer is often applied anyway.
>
> Therefore, for our purpose a computer is really a data processor. Even such data processing operations as rearranging data may require simple arithmetic such as addition. This explains why a certain amount of imprecision has entered our language and why confusion exists between the terms *computer* and *data processor.* The two terms are so loosely used at present that often one has to inquire further to determine exactly what is meant.[15]

Eadie defined the term data processor as follows:

> *data processor*—A digital device that processes data. It may be a computer, but in a larger sense it may gather, distribute, digest, analyze, and perform other organization or smoothing operations on data. These operations, then, are not necessarily computational. Data processor is a more inclusive term than computer.[15]

Thus, it is tempting to define the term *microprocessor* as follows:

> *microprocessor*—An extremely small data processor.

At this time, the microprocessor does not quite have such a definition.

As semiconductor manufacturers develop the capability to manufacture an entire computer on a single chip, including memory and I/O ports, we believe that the term microprocessor will assume the meaning given above.

At the moment, there is a distinction between a *microprocessor* and a *microcomputer*. To quote the Texas Instruments Incorporated *Microprocessor Handbook:*

> This lesson begins with the word "microprocessor." To some people microprocessor means microcomputer. To other people, the words microprocessor and microcomputer are different. To them, "microprocessor" is a broader and more generic term which describes an extremely small electronic system capable of performing specific tasks. Thus, microcomputer is an application of microprocessors.[16]

At the present time, we consider a microprocessor to be a single integrated-circuit chip that contains approximately 75% of the power of a very small digital computer. It usually cannot do anything without the aid of support chips and memory. In contrast, a microcomputer is a full operational computer system based upon a microprocessor chip. Such a system contains memory, latches, counters, input/output devices, buffers, and a power supply in addition to the microprocessor chip. There may be as much cost involved in the other hardware components as there is in the microprocessor chip itself.

While on this subject, we would also like to quote from the article by Laurence Altman in the April 18, 1974 issue of *Electronics:*

> What a microprocessor is . . . but first, what it isn't. A microprocessor is not a computer but only part of one. To make a computer out of a microprocessor requires the addition of memory for its control program, plus input and output circuits to operate peripheral equipment. Also, the word is not short for microprogrammable central processing unit. For though some microprocessors are controlled by a microprogram, most are not.
> 
> What a microprocessor is, then, is the control and processing portion of a small computer or microcomputer. Moreover, it has come to mean the kind of processor that can be built with LSI MOS circuitry, usually on one chip. Like all computer processors, microprocessors can handle both arithmetic and logic data in bit-parallel fashion under control of a program. But they are distinguished both from a minicomputer processor by their use of LSI with its lower power and costs, and from other LSI devices (except calculator chips) by their programmable behavior.
> 
> In short, if a minicomputer is a 1-horsepower unit, the microprocessor plus supporting circuitry is a ¼-hp unit. But as LSI technology improves, it will become more powerful. Already single-chip bipolar and CMOS-on-sapphire processors are being developed that have almost the capability of the minicomputer.

As the complexity of MOS LSI circuitry increases, it will become much more difficult to distinguish between a microprocessor and a microcomputer. The Intel Corporation, Rockwell International Corporation and Mostek Corporation, to name just a few, have *one-chip microcomputers*. Intel's series of one-chip microcomputers have 8-bit CPUs, a 1K- or 2K-byte ROM for program storage, 64 or 128 bytes of R/W memory for data storage and either 21 or 27 programmable I/O lines. Rockwell's R6500/1 is also an 8-bit CPU with a 2K ROM for program storage, 64 bytes of R/W memory for data storage and 32 programmable I/O lines. Mostek's MK 3870 has an 8-bit CPU along with the same amount of ROM, R/W memory and I/O lines as the Rockwell R6500/1. Of course, each chip or "family" of chips has its own instruction set, so a program developed for an Intel chip will not execute on an R6500/1 or an MK 3870 chip. Some of the common characteristics of these chips is that they all require a single +5-volt power supply, that they have simple clock circuitry and some interrupt circuitry, and they all contain an internal interval counter/timer. In the near future, we will see one-chip microcomputers with more memory, "richer" instruction sets and even internal analog-to-digital and digital-to-analog converters. The point that we wish to make is that this single chip is much closer to a true *computer-on-a-chip* than most microprocessor chips that are currently on the market. The 8080A microprocessor chip discussed in this book is still a microprocessor; it contains no built-in read/write memory, ROM, or I/O capability.

## HARDWARE VS SOFTWARE

*Hardware* and *software* are important terms that will be used repeatedly in this unit. It is appropriate, therefore, to define them early:

hardware—The mechanical, magnetic, electronic, and electrical devices from which a computer is fabricated; the assembly of material forming a computer.[15]

software—The totality of programs and routines used to extend the capabilities of computers, such as compilers, assemblers, narrators, routines, and subroutines. Contrasted with hardware.[5]

The microcomputer, along with any integrated-circuit chips, wire, breadboarding aids, and peripheral devices, are all considered to be the hardware. The programs and subroutines that you use and write are the software. In time, you will observe that it requires a considerable effort to write good programs that take maximum advantage of available memory, the instruction set, and the time that is required to execute individual instructions.

## WHAT IS A CONTROLLER?

Graf has defined a *controller* as

*controller*—An instrument that holds a process or condition at a desired level or status as determined by comparison of the actual value with the desired value.[2]

Controllers can be analog or digital, and can be electronic, mechanical, electromechanical, or pneumatic, or some combination of these. A *digital controller* acquires the actual value of the condition in digital form and compares it to the desired value contained within the controller. If there is any difference between the two, a digital signal is sent out to the device, machine, or process to initiate actions to reduce this difference. The digital controller itself consists either of integrated-circuit chips and discrete components that are wired to a printed-circuit board, or else, a computer of any size with a limited number of chips to serve as an interface between the computer and the external world.

The question of cost becomes an important factor when one considers the use of computers as controllers. One would not control 100 devices, each with a value of $500, with a $1,000,000 computer. The use of such a large computer to control $50,000 worth of equipment is a form of overkill. On the other hand, such a computer would be useful in the control of a $20,000,000 chemical plant. However, with today's technology, it is doubtful that a million-dollar computer would be required; probably $200,000 would buy a very large minicomputer system that would serve the requirements of the plant. One can justify the cost of a computer/controller if it represents only a modest percentage of the cost of operating a process or producing a product. The trade-offs in cost and performance constantly change as the prices of computer systems decrease. With the advent of microcomputers, it is quite likely that the cost of controlling equipment will decrease at no sacrifice in reliability.

## WHERE MICROCOMPUTERS FIT

The *Business Week* article that we excerpted earlier in this unit should provide you with some perspective concerning the potential applications for microcomputers. We would like to discuss this subject in further detail with the aid of Fig. 16-1 and Chart 16-1.

In Fig. 16-1, we have plotted the number of microcomputer applications, on a normalized scale of 0 to 1, versus the type of application. As can be observed, we do not expect many of today's microcomputers to be used as number-crunching machines or as substitutes for simple relay logic systems. Basically, most of the exciting micro-

Chart 16-1. The Spectrum of Computer-Equipment Complexity

| WORD LENGTH | 1 | 2 | 4 | 8 | 16 | 32 | 64 |
|---|---|---|---|---|---|---|---|
| COMPLEXITY | Hardwired logic | Programmed logic array | Calculator | Microprocessor | Minicomputer | | Large computer |
| APPLICATION | | | Control | Dedicated computation | Low-cost general data processing | | High-performance general data processing |
| COST | Under $100 (1978) | | | $1000 (1978) | $10,000 (1978) | | $100,000 and up (1978) |
| MEMORY SIZE | Very small 0-4 words | | Small 2-10 words | Medium 10-1000 words | Large 1000-1 million words | | Very large More than 1 million words |
| PROGRAM | Read-only | | | | | | Reloadable |
| SPEED CONSTRAINTS | Real time | | Slow | | Medium | | Throughput-oriented |
| INPUT-OUTPUT | Integrated | | Few simple devices | | Some complex devices | | Roomful of equipment |
| DESIGN | Logic | | Logic + microprogram | Microprogram macroprogram | Macroprogram | | Macroprogram high-level language software system |
| MANUFACTURING VOLUME | Large | | | | | | Small |

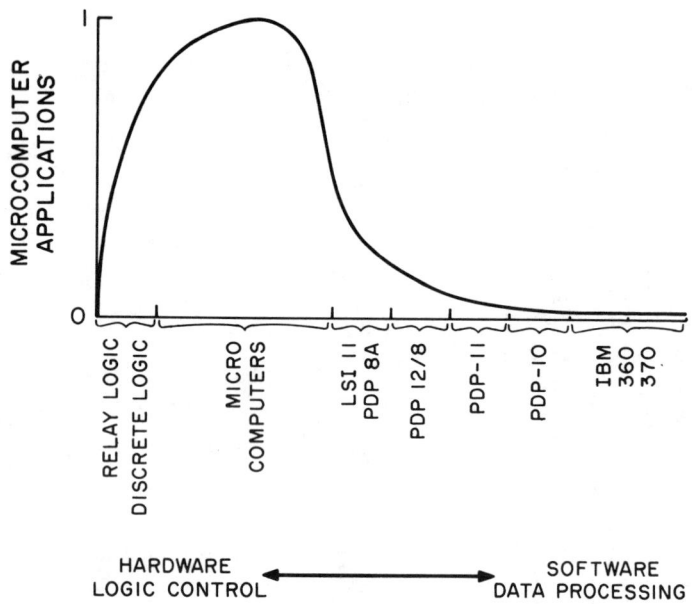

**Fig. 16-1.** Diagram showing the application potential of microcomputers and their niche between discrete logic and inexpensive minicomputers.

computer applications will fall between discrete random logic (gates and flip-flops) on one hand and inexpensive minicomputers on the other. Microcomputers are carving out an entirely new market, one that has not been previously served either by minicomputers—owing to their cost—or by complex digital circuits. This is the domain of the "smart" machine. The domain will grow at the expense of both discrete random logic and minicomputers as the cost of microcomputers decreases.

At the moment, you can even construct a minicomputer using microcomputer chips. Intersil, Incorporated manufactures the IM 6100, which executes the instruction set of Digital Equipment Corporation's PDP-8. Fairchild Semiconductor's "Microflame" 9440 executes the instruction set of the Data General Nova. Texas Instruments Incorporated probably has the most diversified set of "microprocessor/minicomputer" chips, including the TMS 9940, TMS 9980 and TMS 9900. All three of these chips execute the instruction set (or a subset of the instruction set) of the Texas Instruments Incorporated 990 series of minicomputers. Digital Equipment Corporation manufactures the LSI-11 and the LSI-11/2. However, these computers may or may not be based on microprocessor chips, simply because the CPU consists of four or five 40-pin integrated circuits. Even so, the LSI-11 chips can execute the same instructions as many of the

PDP-11s. The Data General microNova is a microcomputer, simply because it is based on a one-chip microprocessor. Of course, it executes the Nova instruction set.

At the higher end of the computer spectrum, microcomputers are not currently being used to replace large number-crunching computers of the PDP 10, IBM 360, and IBM 370 class. However, one California company has proposed the use of 256 8080A microcomputers arranged in the form of a "hypercube." According to them, such a computer would rival or exceed large computers in number-crunching capability. It is quite possible that future computer generations will take advantage of distributed computer architecture. Again, the problem is software development.

Chart 16-1 depicts the spectrum of computer-equipment complexity, from simple hardwired logic systems to high-performance data processing equipment. Costs are declining across the board. Every five years, the cost for an equivalent amount of computing capability decreases approximately ten-fold. (Chart 16-1 is adapted from an article by Wallace B. Riley in the October 17, 1974 issue of *Electronics*. The article indicates that the chart is based on Pro-Log Corporation material.)

## COMPUTER HIERARCHIES

A *hierarchy* is a series of items classified according to rank or order.[2] Microcomputers will control the behavior of individual machines or instruments. Minicomputers will collect data from groups of microcomputers and compare such data to more complex mathematical models, such as the model of a process that is being controlled by ten microcomputers. Larger computers might periodically interrogate minicomputers for the status of entire processes, and might format the received information in a manner that is easy to understand by production supervisors. In Fig. 16-2, we depict a hierarchy consisting of seven 8080-based microcomputers and a single minicomputer. The individual instruments A through G are controlled by built-in 8080 microprocessors. These microprocessors also communicate back and forth with the minicomputer, which monitors the operation of the entire system. Communication between the microcomputers and minicomputer most likely will be serial.

## A TYPICAL 8080 MICROCOMPUTER

A typical microcomputer constructed from an 8080A microprocessor chip is shown in Fig. 16-3. To properly understand the diagram of Fig. 16-3, several definitions are necessary.

*bus*—A path over which digital information is transferred, from any of several sources to any of several destinations. Only one transfer of information can take place at any one time. While such transfer of information is taking place, all other sources that are tied to the bus must be disabled.

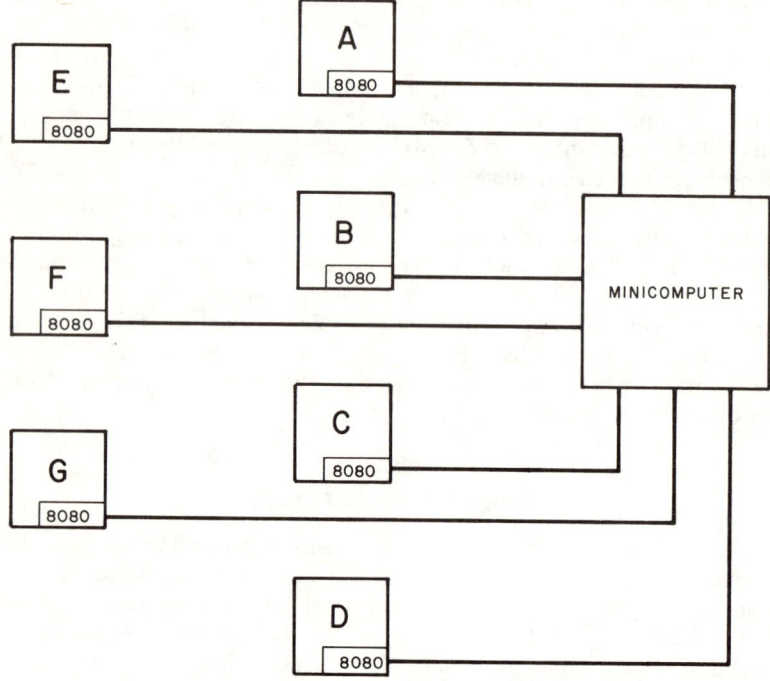

Fig. 16-2. An example of a computer hierarchy.

*bidirectional data bus*—A data bus in which digital information can be transferred in either direction. With reference to an 8080A-based microcomputer, the bidirectional data path by which data is transferred between the CPU, memory, and input-output devices.

*address bus*—A unidirectional bus over which digital information appears to identify either a particular memory location or a particular I/O device. The 8080A address bus is a group of sixteen lines.

*address*—A group of bits that identify a specific memory location or I/O device. An 8080A microcomputer uses sixteen bits to identify a specific memory location and eight bits to identify an I/O device.

*control*—Those parts of a computer which carry out instructions in proper sequence, interpret instructions, and apply proper signals.[14]

*control bus*—A set of signals that regulate the operation of a microcomputer system, including I/O devices and memory. They function much like "traffic" signals or commands. They may also originate in the I/O devices, generally to transfer to or receive signals from the CPU. According to the Intel Corporation literature, a control bus is a unidirectional set of signals that indicate the type of activity—memory read, memory write, I/O read, I/O write, or interrupt acknowledge—in progress.

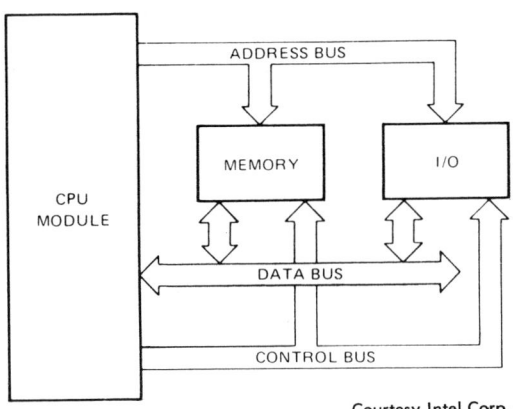

Courtesy Intel Corp.

**Fig. 16-3. Diagram for a typical microcomputer.**

*I/O*—Abbreviation for input-output.

*I/O device*—Input/output device. A card reader, magnetic tape unit, printer, or similar device that transmits data to or receives data from a computer or secondary storage device.[2] In a more general sense, any digital device, including a single integrated-circuit chip, that transmits data to or receives data or strobe pulses from a computer.

*CPU*—Abbreviation for central processing unit.

*central processing unit (large computer)*—Also called central processor. Part of a computer system which contains the main storage, arithmetic unit, and special register groups. Performs arithmetic operations, controls instruction processing, and provides timing signals and other housekeeping operations.

*central processing unit (microprocessor)*—A single integrated-circuit chip that performs data transfer, control, input-output, arithmetic, and logical operations by executing instructions obtained from memory.

*memory*—Any device that can store logic 0 and logic 1 bits in such a manner that a single bit or group of bits can be accessed and retrieved.

A typical microcomputer constructed from an 8080A chip posesses all of the minimum requirements for a digital computer:

- It is programmable, with the data and program instructions capable of being arranged in any sequence desired.
- It is digital.
- It is clocked. (In most microcomputers, the internal operations in the CPU chip proceed synchronously.)
- It contains an arithmetic/logic unit, located within the CPU chip, that performs arithmetic and logic operations.
- It can exchange data with memory or I/O devices.
- It contains "fast" memory. (Speed is an important requirement for a functional digital computer.)

## ADDRESS BUS

The Intel 8080A microprocessor chip contains a 16-bit address bus that is used for the identification of specific memory locations or specific I/O devices. It is a unidirectional bus, which means that address information can only be output from the 8080A chip. When addressing memory, $2^{16} = 65{,}536$ different memory locations can be accessed. We say that the 8080A can address "64K" where the "K" is an abbreviation for kilobyte, or 1024 bytes.

The Intel 8080A address bus is also used to supply the 8-bit device code for input and output devices. When addressing input-output devices, the address bus assumes a new identity, i.e., it is subdivided into two identical 8-bit device address busses, either of which you can use when wiring an interface circuit to I/O devices. When addressing I/O devices using the IN or OUT microcomputer instructions, you can address $2^8 = 256$ different input devices and $2^8 = 256$ different output devices.

Whenever you encounter the term *bus,* you should be alert for the possibility that different types of information appear on the bus lines at different times. In the case of the 8080A address bus, this is certainly true. Most of the time, the information that appears on the address bus is the address of a specific memory location. Occasionally, the information that appears on the address bus is a device code. The microcomputer knows when the bus is being used to access memory and when it is being used to identify I/O devices. It provides the appropriate *control pulse* that informs you what it is doing! We shall discuss these control pulses in a later section.

## BIDIRECTIONAL DATA BUS

The Intel 8080A microprocessor chip contains an 8-bit bidirectional data bus that permits eight bits of data, known as a *byte,* to be transferred between the 8080A chip and memory or I/O devices. Different types of information appear on the data bus lines at different times. Much of the time, the data that appears is an instruction byte from memory. At other times, the data that appears on the data bus is one of the following:

- A data byte that is being input from an input device.
- A data byte that is being output to an output device.
- A data byte that is being written into or read from memory.
- Control status bits used to derive some of the control bus signals.
- A HI or LO address byte that is being stored in an area of memory called the *stack.*
- A HI or LO address byte that is being retrieved from the stack.
- An instruction byte that is being output by an I/O device during an interrupt.

How do you know when these different types of data transfers are occurring? The microcomputer tells you, by providing the appropriate *control or status pulses* needed to inform you of the type of activity in progress. It should be clear now that an understanding of the control bus is essential to the understanding of the behavior of the 8080A microcomputer. Such a statement is true for any type of digital computer that you encounter.

## CONTROL BUS

Although called a control bus, the set of signals in quesion do not actually comprise a bus since different types of information *do not* appear on the individual signal lines at different times. Each signal line is unidirectional and unifunctional. With this caveat, we shall continue to call the set of control signals associated with the 8080A chip a control bus; the term is too widely used in microcomputer literature for us to suggest any reasonable alternative.

The five basic types of activities in which the 8080A microprocessor chip engages are:

1. Memory Read
2. Memory Write
3. I/O Read
4. I/O write
5. Interrupt/Interrupt Acknowledge

Some useful definitions include the following:

*read*—To transmit data from a specific memory location to some other digital device. A synonym for *retrieve*.
*write*—To transmit data from some other digital device into a specific memory location. A synonym for *store*.
*interrupt*—In a digital computer, a break in the normal execution of a computer program such that the program can be resumed from that point at a later time.

Five separate control signal lines are provided, one for each of the above activities. These lines have the following abbreviations:

1. Memory Read: $\overline{\text{MEMR}}$
2. Memory Write: $\overline{\text{MEMW}}$
3. I/O Read: $\overline{\text{I/O R}}$ or $\overline{\text{IN}}$
4. I/O write: $\overline{\text{I/O W}}$ or $\overline{\text{OUT}}$
5. Interrupt Acknowledge: $\overline{\text{INTA}}$ or $\overline{\text{I ACK}}$

Observe that in all cases, the signal is a negative pulse,

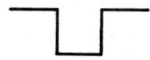

The pulse width depends on the speed of the 8080A-based microcomputer. For the microcomputer, which is clocked at 750 kHz, the pulse width is 1.333 μs.

The uniqueness of each of the control signals can be seen with the aid of the following truth table. These control signals are available from the SK-10 bus socket on the microcomputer printed-circuit board (Fig. 16-4). You will use them, especially $\overline{\text{IN}}$ and $\overline{\text{OUT}}$, to gate the transfer of data between digital integrated-circuit chips (wired on the breadboard) and the CPU of the 8080A-based microcomputer. (A0 through A7 are the eight least significant bits on the address bus; D0 through D7 are the bidirectional data bus; INTE is the interrupt enable flip-flop output; INT is the interrupt request input; and $\overline{\text{MEMR}}$, $\overline{\text{MEMW}}$, $\overline{\text{IN}}$, $\overline{\text{OUT}}$, and $\overline{\text{I ACK}}$ are output control signals. RDYIN and WAIT are used with the single-step circuit described in Unit 11, Experiment No. 5.)

| $\overline{\text{MEMR}}$ | $\overline{\text{MEMW}}$ | $\overline{\text{IN}}$ | $\overline{\text{OUT}}$ | $\overline{\text{I ACK}}$ | Operation |
|---|---|---|---|---|---|
| 0 | 1 | 1 | 1 | 1 | Read byte from memory |
| 1 | 0 | 1 | 1 | 1 | Write byte into memory |
| 1 | 1 | 0 | 1 | 1 | Read byte from I/O device |
| 1 | 1 | 1 | 0 | 1 | Write byte into I/O device |
| 1 | 1 | 1 | 1 | 0 | Strobe byte into instruction register during an interrupt (interrupt acknowledge) |

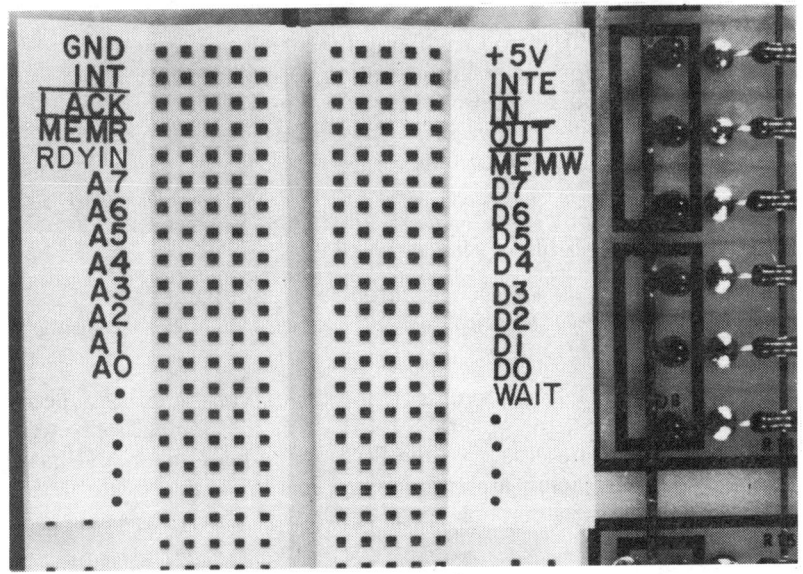

Fig. 16-4. Signals available on the Dyna-Micro microcomputer bus socket.

## WHAT IS INTERFACING?

*Interfacing* can be defined as the joining of members of a group (such as people, instruments, etc.) in such a way that they are able to function in a compatible and coordinated fashion.[17] By "compatible and coordinated fashion," we usually mean synchronized. Some important definitions include the following:

*sync*—Short for synchronous, synchronization, synchronizing, etc.

*to synchronize*—To lock one element of a system into step with another.

*synchronous*—In step or in phase, as applied to two devices or machines. A term applied to a computer, in which the performance of a sequence of operations is controlled by clock signals or pulses. At the same time.

*synchonous computer*—A digital computer in which all ordinary operations are controlled by a master clock.

*synchronous operation*—Operation of a system under the control of clock pulses.

*synchronous logic*—The type of digital logic used in a system in which logical operations take place in synchronism with clock pulses.

*synchronization pulses*—Pulses originated by the transmitting equipment and introduced into the receiving equipment to keep the equipment at both locations operating in step.

*synchronous inputs*—Those inputs of a flip-flop that do not control the output directly, as do those of a gate, but only when the clock permits and commands.

The above definitions have been obtained from Reference 2. We can thus define *computer interfacing* as:

*computer interfacing*—The synchronization of digital data transmission between a computer and external devices, including memory and I/O devices.

Although the details of computer interfacing vary with the type of computer employed, the general principles of interfacing apply to a wide variety of computers. For the 8080A microcomputer, the basic objectives of interfacing are summarized in Fig. 16-5. If you desire to interface the microcomputer, your object is to:

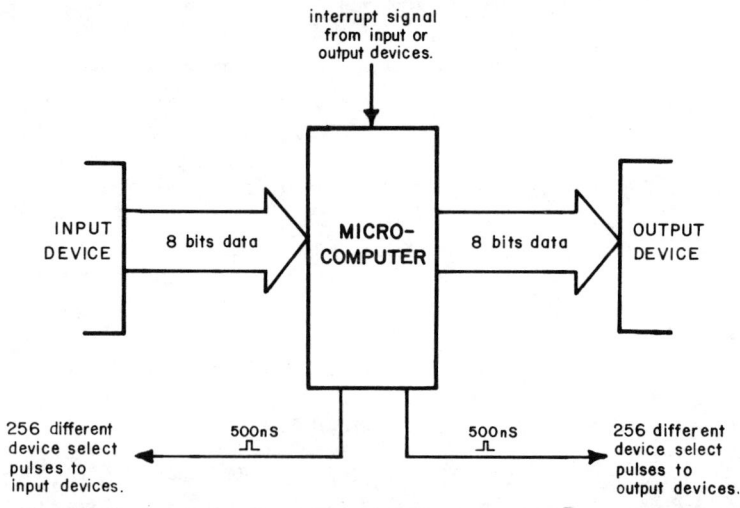

Fig. 16-5. The four principal tasks of interfacing: input, output, device select pulse generation, and interrupt servicing.

- Synchronize the transfer of 8 bits of data between the microcomputer and each output device.
- Synchronize the transfer of 8 bits of data between each input device and the microcomputer.
- Generate the appropriate input and output data transfer synchronization pulses, which are called *device select pulses*. For an 8080A-based microcomputer, you can generate 256 different

input synchronization pulses and 256 different output synchronization pulses.
- Service interrupt signals that enter the microcomputer from external I/O devices.
- Program the microcomputer to perform all input-output and interrupt servicing operations.

A better way of viewing three of the four tasks of interfacing is given in the diagram shown in Fig. 16-3. The transfer of 8 bits of data between the CPU and an I/O device occurs over the 8-bit bidirectional data bus. The specific I/O device that is involved in the data transfer is selected via the use of 8 bits on the address bus. The precise timing of the data transfer is determined by the presence of an $\overline{\text{IN}}$ or $\overline{\text{OUT}}$ pulse on the control bus. Therefore, during the transfer of data between the CPU and an I/O device, *all three busses participate!* There is much more to say about computer interfacing, but we will save it for subsequent units.

## WHAT IS AN I/O DEVICE?

Two useful definitions include:

*input-output, input/output I/O*—General term for the equipment used to communicate with a computer and for the data involved in the communication.

*I/O device*—Any digital device, including a single integrated-circuit chip, that transmits data to or receives data or strobe pulses from a computer.

The traditional view of an I/O device is that it is somewhat large or complex. Card readers, magnetic tape units, CRT displays, and teletypewriters fit such a description. However, a single integrated-circuit chip, such as a latch, three-state buffer, shift register, counter, or small memory, can be considered to be an I/O device as well. If it is digital, it usually can be an I/O device.

We have indicated previously that you must synchronize the transfer of data between a computer and an I/O device, and that this synchronization is accomplished with the aid of pulses called *device select pulses*. An important point is that several device select pulses may be required for a single I/O device. For example, the 74198 shift register has a pair of control inputs that determine whether the register shifts left, shifts right, or parallel loads eight bits of data. The chip also contains clock and clear inputs. Thus, a single 74198 chip may require three or four unique device select pulses. The fact that you can generate 256 different input device select pulses and 256 different output device select pulses does not mean that you can address 512 different "devices." A more reasonable number is of the order of 50 to 100 devices. Rarely will you require so many device select pulses. If you do, there are other tricks that you can use to generate still more such pulses.

In the next unit, you will learn how to generate device select pulses. The various uses for device select pulses are also discussed.

### REVIEW QUESTIONS

The following questions will help you review a few of the important concepts of microcomputer interfacing.

1. Which of the following constitute hardware and which constitute software?
    a. Cross-assembler
    b. Editor
    c. Integrated-circuit chip
    d. Wire
    e. Printed-circuit board
    f. Capacitors and resistors
    g. FORTRAN program
    h. Turbo-alternating grundle flusher
    i. Thermistor (temperature transducer)
2. What are the three important busses in a microcomputer?
3. Why is it useful for the data bus to be bidirectional?
4. List five important control signal lines in an 8080A-based microcomputer.

# 17

# Device Select Pulses

## INTRODUCTION

This unit will teach you how to generate input and output device select pulses. These are the microcomputer-generated synchronizing signals between the microcomputer and the external I/O devices. The I/O devices could be simple integrated-circuit chips. One of the most useful device select pulse circuits is one that is based upon the 74154 decoder chip; sixteen different device select pulses can be generated.

## OBJECTIVES

At the completion of this unit, you will be able to do the following:

- Define device select pulse.
- Explain what is meant by the statement, "the substitution of software for hardware."
- List several different uses for device select pulses.
- Give one or two schematic diagrams of circuits that can be used to generate device select pulses.
- Define state and machine cycle.
- List how many machine cycles exist for some typical 8080A instructions.
- Wire a circuit that will permit you to single step the microcomputer.

## WHAT IS A DEVICE SELECT PULSE?

A *device select pulse* is a synchronization pulse generated by a digital computer to synchronize the transfer of data between the computer and an input-output device. Associate the term *device select pulse* with the terms *to enable, to strobe, to gate, to disable, to inhibit,* and *to clock.* Basically, a device select pulse is a strobe pulse that strobes some operation in a digital circuit or chip. It can be either a gate pulse—a pulse that enables a gate circuit to pass a digital signal—or a clock pulse in a clocked logic system.

## USES FOR DEVICE SELECT PULSES

Device select pulses are inexpensive and are easy to generate. In a typical application, a 74L154 decoder chip is used. At a cost of $1.25 per chip, it will cost you approximately 8¢ for each device select pulse that you generate, in quantities of sixteen.

### The Substitution of Software for Hardware

In interfacing a microcomputer, your object is to minimize the number of external chips required, assuming that you are constructing thousands of units rather than a one-of-a-kind unit. One way of minimizing external chips is by performing much of the digital logic within the microcomputer rather than external to it. We call this process *the substitution of software for hardware.* Remember this theme: software vs hardware. There exists a tradeoff between the two, but your main objective in using microcomputers is to substitute microcomputer programs for electronic and mechanical hardware devices. Since many I/O devices are slow by microcomputer standards, you will be very successful in many applications that incorporate such devices. However, occasionally you will encounter a situation where the speed of the microcomputer is not fast enough. It takes time to execute each instruction in a microcomputer program. If the program is too long, too much time will be consumed and you may not be able to accomplish a specific task.

To demonstrate the substitution of software for hardware, we would like to discuss several different uses for device select pulses. In each example presented, please keep in mind that there exists an accompanying microcomputer program that times the generation of the device select pulses.

It is easy to write a program that generates a single device select pulse, such as the negative pulse, $\overline{\text{OUT 0}}$, shown below,

$\overline{\text{OUT 0}}$ ──────────▶ ─⊔─        OUT 0

For the MMD-1 microcomputer, the pulse width is 1.333 μs. In a subsequent unit, you will learn how to write various types of time-delay loops that repeatedly execute a small group of microcomputer instructions. By writing such a program, you will be able to generate a series of device select pulses, as shown for $\overline{OUT}$ 001,

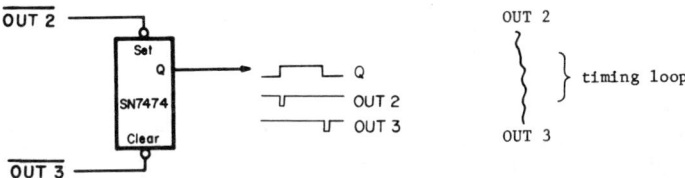

The duration between successive pulses is determined by a programmed *software time-delay loop*.

With a pair of device select pulses and a single preset-clear flip-flop such as the SN7474, you can write a time-delay loop program that generates a single clock pulse at the output of the latch.

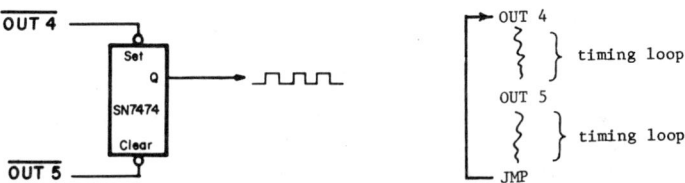

By adding a second time-delay loop to your program, you can generate a series of clock pulses with a known duty cycle that is specified by the program,

With these last two circuits, you have substituted software for either a monostable multivibrator or an astable multivibrator, such as the 555 IC timer wired as an oscillator. You have already used a number of different Outboards®, such as the Pulser Outboard and the Clock Outboard. With your ability to substitute software for hardware, you are now able to replace such Outboards with an SN7474 chip (which contains two flip-flops) and appropriate programs in the microcomputer.

In complex digital printed-circuit boards, you frequently encounter flip-flops and gates that are used to provide pulses or logic states at certain points within the circuit at certain instants of time. With the microcomputer, it is relatively easy to accomplish tasks of this type. You can also use a single bit, D0, on the bidirectional data bus and a device select pulse, OUT 006, to latch the bit and, thus, control the logic state at a particular point in an external digital circuit.

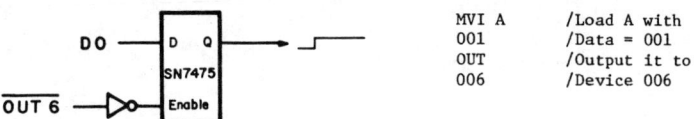

```
MVI A    /Load A with
001      /Data = 001
OUT      /Output it to
006      /Device 006
```

A much more efficient interface circuit is based around an 8-bit latch and a single device select pulse.

```
MVI A    /Load A with
013      /Control word
OUT      /Output it to
007      /Device 007
```

With device select pulse OUT 007 and an 8212, it is possible to latch all eight bits on the bidirectional data bus and use such bits as individual control lines at various points in a digital circuit. In the next unit, you will learn to wire such a latch circuit.

You will not be able to substitute software for certain types of chips, such as latches and three-state buffers. Nevertheless, as long as speed is not your requirement, you can use software to substitute for most of the important functions of MSI and many LSI integrated-circuit chips. This point is made quite well with the PACE microcomputer in the manual *Logic Designers Guide to Programmed Equivalents to TTL Functions,* which is available from National Semiconductor Corporation. All the PACE programs can be converted to operation on an 8-bit microcomputer such as the 8080A or Motorola 6800. The *hardware* for which they have provided equivalent microcomputer software includes the following:

- 7408 quad 2-input AND gate.
- 7409 quad 2-input AND gate with open-collector outputs wired together.
- 7411 triple 3-input AND gate.
- 74H21 dual 4-input AND gate.
- 7432 quad 2-input OR gate.
- 7486 quad 2-input Exclusive-OR gate.
- 7483 4-bit binary full adder.
- 74121 and 555 monostable multivibrators (time delays greater than 10 $\mu$s).
- 74150 16-line-to-1-line data selector/multiplexer.

- 74151 8-line-to-1-line data selector/multiplexer.
- 74154 4-line-to-16-line decoder/demultiplexer.
- DM8220 9-bit parity generator/checker.
- 7485 4-bit magnitude comparator.
- 74160 to 74163 synchronous 4-bit counters, BCD and Binary.
- 74185 binary-to-BCD converter.
- 74184 BCD-to-binary converter.
- 74190 up/down BCD counter.
- 74191 up/down binary counter.

They have also provided software equivalents for the following types of digital systems, each of which consists of a number of 7400-series integrated-circuit chips:

- Digital servo (74193, 7485, and various gates).
- Digital tachometer (74163, 7485, 74123, 7475, 7400, and 7404).
- Modulo-N-divider (74163, 7485, and 7402).
- Real-time clock and interval timer (modulo-N-divider, 74160, 7404, and 7400).
- Pseudo-random number generation (74C14).
- State sequencer (DM8551, 7473, and numerous gates).

Although it is possible to treat a microcomputer as a minicomputer or use it as a super programmable calculator, the dominant use of microcomputers will be in digital systems in which software is substituted for much of the hardware originally present. If you learn how to substitute software for hardware, you will have learned one of the most important aspects of microcomputer applications.

### Strobing Integrated-Circuit Chips

An important application for microcomputers is to strobe the operation of individual integrated-circuit chips in instruments and electronic devices. For example, such pulses can:

- Clear shift registers, counters, flip-flops and latches.
- Load shift registers, counters and latches.
- Enable multiplexers, demultiplexers, decoders, data selectors, counters, shift registers, memories, priority encoders, UARTs and a variety of other chips.
- Inhibit clock inputs to counters and shift registers.
- Set, clear and clock flip-flops.
- Select shift left, shift right, load and inhibit functions in shift registers.

By using device select pulses to control the operation of individual integrated-circuit chips, you are substituting software for hardware.

You are already familiar with a number of integrated-circuit chips in the MSI category. Some of them require strobing or enabling in order to perform their digital function. Thus:

| | |
|---|---|
| 7490 decade counter: | Logic 1 at pins 2 and 3 clears counter. Logic 1 at pins 6 and 7 sets counter to 9. Clock input is at pin 14. |
| 7493 binary counter: | Logic 1 at pins 2 and 3 clears counter. Clock input is at pin 14. |
| 7442 decoder: | Logic 0 at pin 12 enables octal decoder operation. |
| 74154 decoder: | Logic 0 at pins 18 and 19 enables decoder. |
| 7474 flip-flop: | Logic 0 at pin 1 clears first flip-flop. Logic 0 at pin 4 sets first flip-flop. D input to first flip-flop is at pin 2. Clock input to first flip-flop is at pin 3. |
| 7475 latch: | Logic 1 at pin 4 enables first two latches. Logic 1 at pin 13 enables second two latches. |

In microprocessor systems, the 7442 and 74154 decoders are used to help in the generation of device select pulses. However, they can also be used as general decoders that are enabled by such pulses.

## GENERATING DEVICE SELECT PULSES

To generate a device select pulse, you require two types of information from the 8080A microcomputer:

1. The 8-bit identification code, called a *device code,* of the I/O device.
2. A single-bit synchronization pulse, either $\overline{\text{IN}}$ or $\overline{\text{OUT}}$, that synchronizes the decoding of the device code.

The origin of both types of information is in software, i.e., the IN and OUT instructions that you encountered in the early modules in this book. These instructions include the 8-bit device code and cause the microcomputer to generate the appropriate $\overline{\text{IN}}$ or $\overline{\text{OUT}}$ synchronization pulse. The location of the IN and OUT instructions in the program determines the specific instant when the device select pulses are generated.

In other words, during the generation of a device select pulse, both the address bus and the control bus are active. As shown in Fig. 17-1, the 8-bit device code is obtained from the 16-bit address

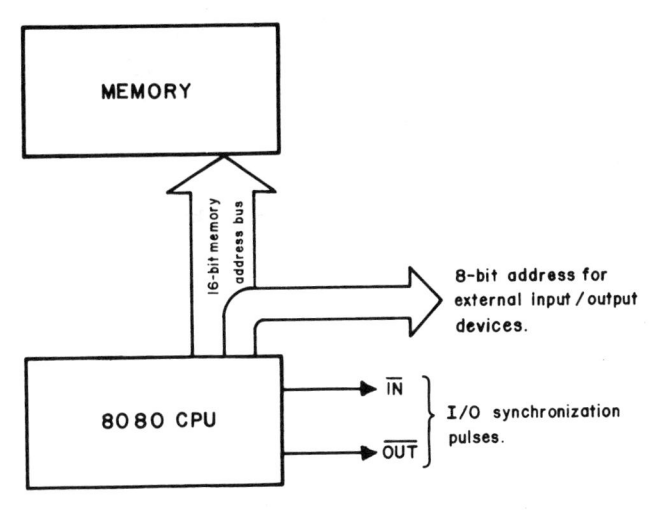

Fig. 17-1. To generate a group of device select pulses, you require eight bits from the address bus and the two synchronization pulses, $\overline{\text{IN}}$ and $\overline{\text{OUT}}$, from the control bus.

bus, and the two synchronization pulses, $\overline{\text{IN}}$ and $\overline{\text{OUT}}$, from the control bus. In the 8080A microcomputer, the 16-bit address bus is subdivided into two 8-bit device codes. For the Intel 8080A chip, both codes are identical during the execution of the third IN or OUT machine cycle.

What do you do with the 8-bit address bus and synchronization pulses? You might wire them to a 74154 4-line-to-16-line decoder as shown in Fig. 17-2. With a single 74154 decoder, you use only four of the eight address bus bits and either $\overline{\text{IN}}$ or $\overline{\text{OUT}}$. A more ambitious device select pulse decoding circuit is shown in Fig. 17-3. All eight address bits are used, and seventeen 74154 decoders pro-

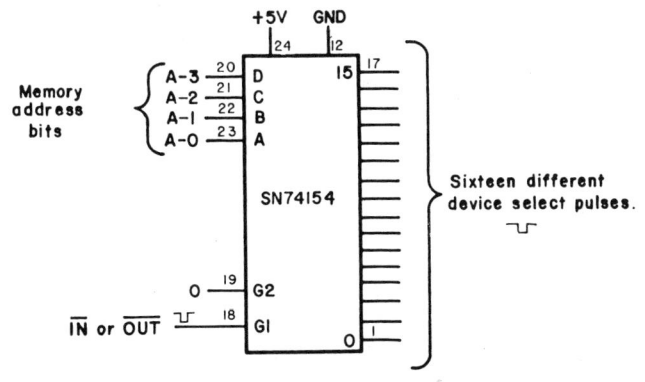

Fig. 17-2. To generate sixteen different device select pulses, you need four bits from the address bus and either the $\overline{\text{IN}}$ or $\overline{\text{OUT}}$ synchronization pulse.

37

vide you with the opportunity to generate 256 unique pulses. To generate all 512 input and output device select pulses, two circuits of the type shown in Fig. 17-3 would be required. This is rarely done in actual interfacing applications.

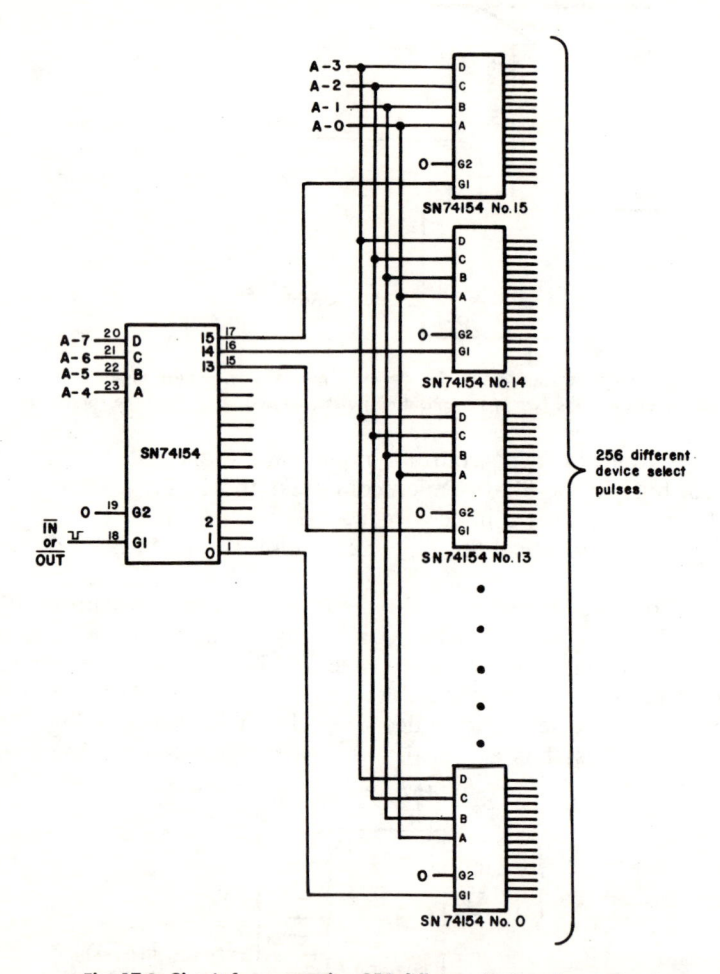

Fig. 17-3. Circuit for generating 256 different device select pulses.

It is not likely that you will need to generate 512 different device select pulses. A more limited decoding circuit, based upon Figs. 17-2 and 17-3, is shown in Fig. 17-4. The circuit includes an *absolute decoding* of the complete 8-bit device select address byte, i.e., all eight bits, not just four. This is not a widely used circuit and is mentioned for illustration only. It may be more useful to use one of the 74154

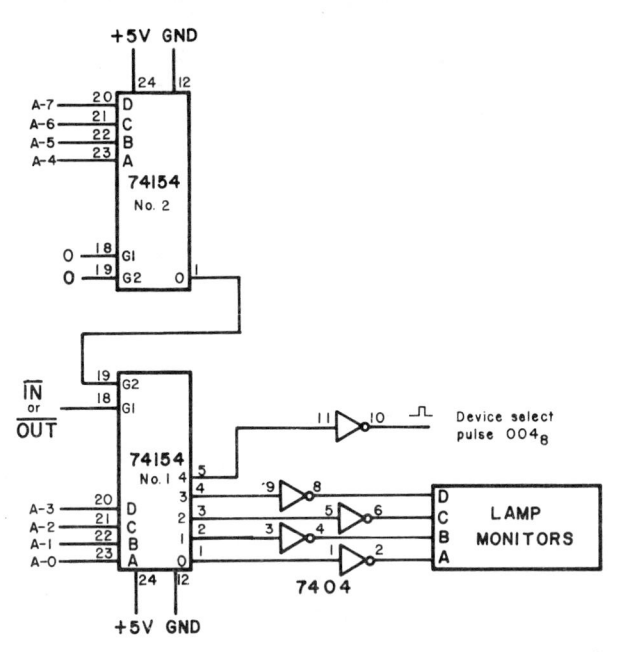

Fig. 17-4. One possible decoding circuit for generating sixteen absolute decoded device select pulses.

chips for input device select pulses, and the other one for output device select pulses.

The preceding circuits are ones that we use to generate device select pulses. However, there exist alternative schemes to decode the address and control buses and we would like to illustrate them. In Fig. 17-5, we use a pair of 74154 decoders and thus absolutely decode the 8-bit address bus byte. Each device select pulse that we generate requires a separate 7402 (or 7432) gate, as shown in Fig. 17-6. However, this circuit is useful only if the device select pulses in your system are scattered randomly in the range, 000 to $377_8$ and all functions are centrally located. Only two 74154 decoder chips and four 7402 2-input NOR gate chips—which contain a total of sixteen NOR gates—are required to obtain sixteen different device select pulses. If the device codes are sequential, then this circuit is preferred.

A related decoding technique is shown in Fig. 17-7. Rather than connect the IN or OUT control signals to a decoder, you connect them to individual 7402 (or 7432) gates. Such a decoding scheme is used in the microcomputer (Fig. 17-7). Four address bus bits are connected to a 74L42 decoder, outputs of which are then gated with 7402 2-input NOR gates to provide the necessary device select pulses for the 7475 latches on the microcomputer board. Note in Fig.

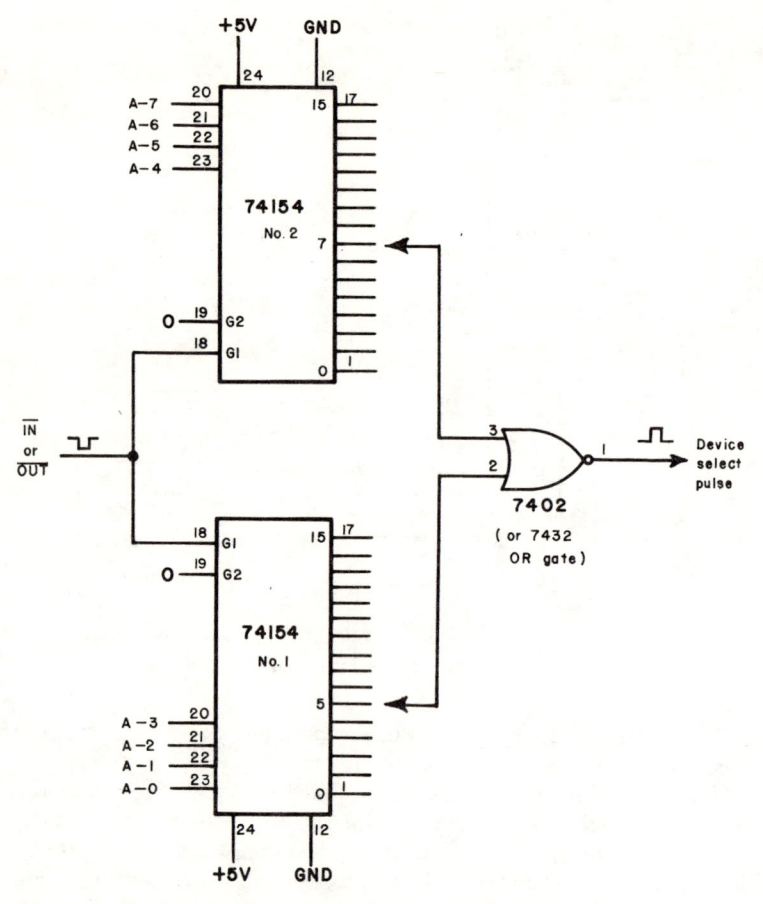

Fig. 17-5. Absolute decoding scheme for device select pulses that requires a 7402 or 7432 gate for each device select pulse desired.

17-7 that address bus bits A3 through A7 are wired to 74LS05 inverters, the outputs of which are all connected together and then tied to the D input of the 74L42 decoder. The D input must always be at logic 0 if device select pulses are to be generated. The technique employed here is the *open-collector* bussing technique of tieing the outputs of special open-collector integrated-circuit chips to a common bus line. We have essentially constructed a five-input OR gate, which enables the 74L42 decoder. We shall discuss this later.

It is also possible to decode 8-bit device code bytes using gates and comparators. This is particularly useful in real situations where only a few device select pulses are needed. For example, consider the circuit shown in Fig. 17-8. On the left, two distinct device select pulses

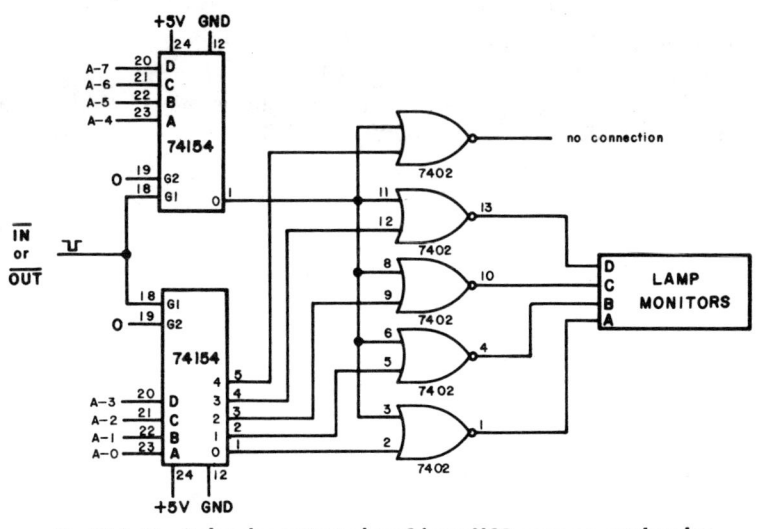

Fig. 17-6. Circuit that demonstrates how 2-input NOR gates are employed to generate individual device select pulses.

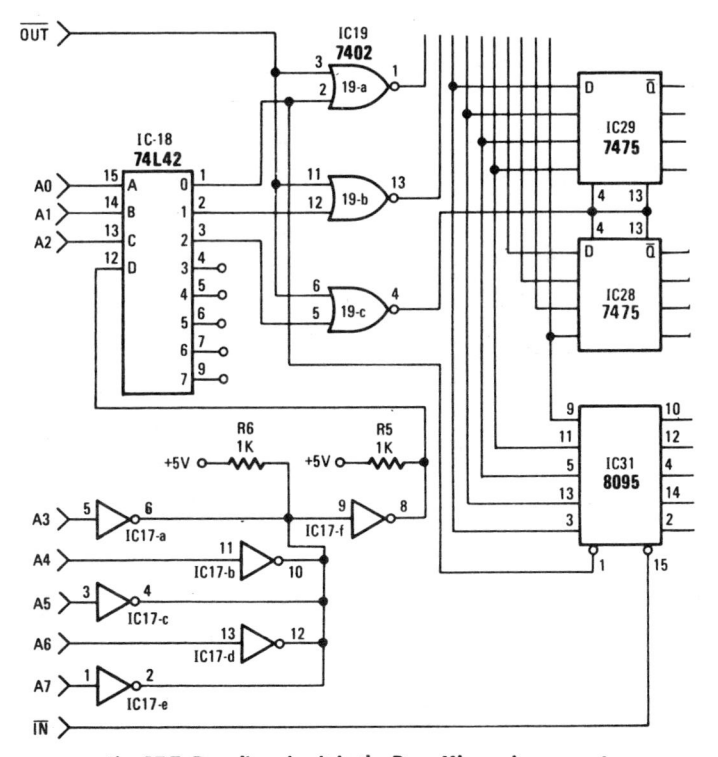

Fig. 17-7. Decoding circuit in the Dyna-Micro microcomputer.

are generated with the aid of a pair of 7430 8-input NAND gate chips. The device code for the top 7430 gate is $11000110_2$, or 306 in octal code, whereas, the device code for the bottom gate is $11001111_2$, or 317 in octal code. The unique output state from the 7430 NAND gate is logic 0, which occurs only when the proper device code has been applied to the gate. If the $\overline{\text{OUT}}$ control signal is also at logic 0, the unique output from the 7432 2-input OR gate, i.e., logic 0, either clears or presets the 7476 flip-flop.

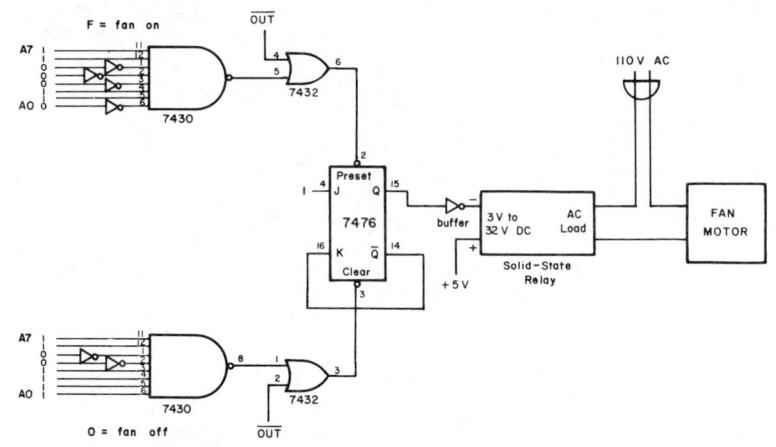

Fig. 17-8. Solid-state relay circuit that employs a pair of 7430 8-input NAND gates to absolutely decode two output device select pulses.

The circuit of Fig. 17-8 demonstrates how you can use device select pulses to control ac power. Optically-isolated solid-state relays permit you to use digital signals to turn on and off ac loads operating at either 115 volts or 220 volts. Solid-state relays can be obtained that will switch 10 amperes at these voltage levels, and they cost only $15 in quantities of one.

A final decoding circuit consists of a pair of 7485 comparator chips, for which the pin configuration and truth tables are given in Figs. 17-9 and 17-10. These 4-bit magnitude comparators perform comparison of straight binary and straight BCD (8-4-2-1) codes. Three fully decoded decisions about two 4-bit words (A, B) are made and are externally available at three outputs. These devices are fully expandable to any number of bits without external gates. Words of greater length may be compared by connecting comparators in cascade. The A > B, A < B, and A = B outputs of a stage handling less-significant bits are connected to the corresponding A > B, A< B, and A = B inputs of the next stage handling more-significant bits. The stage handling the least-significant bits must have a high-level voltage applied to the A = B input and additionally for the 74L85,

**7485, 74S85**
J OR N DUAL-IN-LINE OR
W FLAT PACKAGE (TOP VIEW)

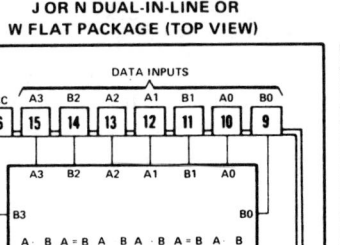

**74L85**
J OR N
DUAL-IN-LINE PACKAGE (TOP VIEW)

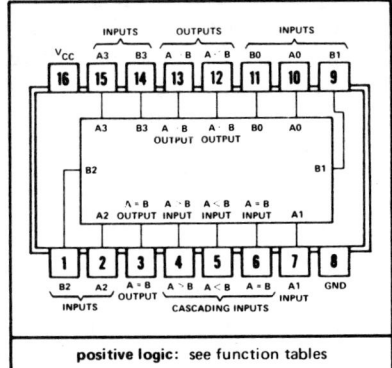

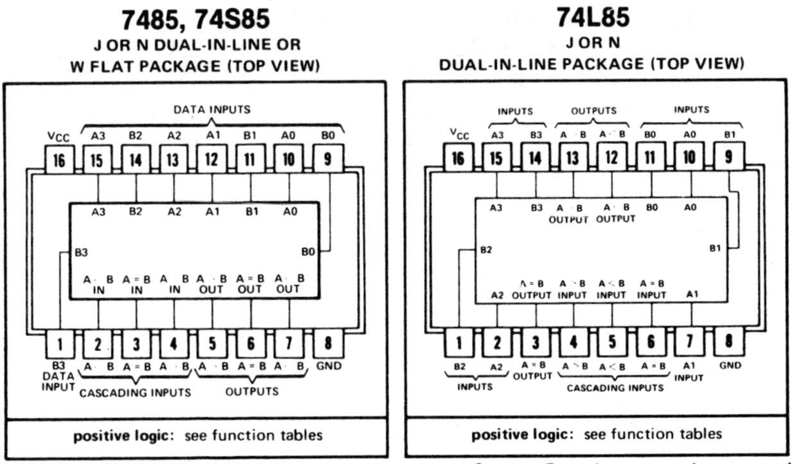

**Fig. 17-9. Pin configuration drawings.**

low-level voltages applied to the A > B and A < B inputs. The cascading paths of the 7485 and 74S85 are implemented with only a two-gate-level delay to reduce overall comparison times for long words.

**FUNCTION TABLES**

| COMPARING INPUTS | | | | CASCADING INPUTS | | | OUTPUTS | | |
|---|---|---|---|---|---|---|---|---|---|
| A3, B3 | A2, B2 | A1, B1 | A0, B0 | A > B | A < B | A = B | A > B | A < B | A = B |
| A3 > B3 | X | X | X | X | X | X | H | L | L |
| A3 < B3 | X | X | X | X | X | X | L | H | L |
| A3 = B3 | A2 > B2 | X | X | X | X | X | H | L | L |
| A3 = B3 | A2 < B2 | X | X | X | X | X | L | H | L |
| A3 = B3 | A2 = B2 | A1 > B1 | X | X | X | X | H | L | L |
| A3 = B3 | A2 = B2 | A1 < B1 | X | X | X | X | L | H | L |
| A3 = B3 | A2 = B2 | A1 = B1 | A0 > B0 | X | X | X | H | L | L |
| A3 = B3 | A2 = B2 | A1 = B1 | A0 < B0 | X | X | X | L | H | L |
| A3 = B3 | A2 = B2 | A1 = B1 | A0 = B0 | H | L | L | H | L | L |
| A3 = B3 | A2 = B2 | A1 = B1 | A0 = B0 | L | H | L | L | H | L |
| A3 = B3 | A2 = B2 | A1 = B1 | A0 = B0 | L | L | H | L | L | H |

**7485, 74S85**

| A3 = B3 | A2 = B2 | A1 = B1 | A0 = B0 | X | X | H | L | L | H |
| A3 = B3 | A2 = B2 | A1 = B1 | A0 = B0 | H | H | L | L | L | L |
| A3 = B3 | A2 = B2 | A1 = B1 | A0 = B0 | L | L | L | H | H | L |

**74L85**

| A3 = B3 | A2 = B2 | A1 = B1 | A0 = B0 | L | H | H | L | H | H |
| A3 = B3 | A2 = B2 | A1 = B1 | A0 = B0 | H | L | H | H | L | H |
| A3 = B3 | A2 = B2 | A1 = B1 | A0 = B0 | H | H | H | H | H | H |
| A3 = B3 | A2 = B2 | A1 = B1 | A0 = B0 | H | H | L | H | H | L |
| A3 = B3 | A2 = B2 | A1 = B1 | A0 = B0 | L | L | L | L | L | L |

H = high level, L = low level, X = irrelevant

Courtesy Texas Instruments Incorporated

**Fig. 17-10. Functional tables.**

We are providing information for both the high-power (7485, 74S85) and low-power (74L85) chips since they have different pin configurations and truth tables. The 74L85 has a typical power dissipation of 20 mW while the 7485 and 74S85 have dissipations of 275 mW and 365 mW, respectively. The typical delay (4-bit words) is 23 ns for the 7485, 90 ns for the 74L85, and 11 ns for the 74S85. You may wish to minimize fan-in through the use of the 74L85 chip.

As can be seen in Fig. 17-11, the only condition that you use is $A = B$. If the address bus byte A is equal to the byte B that you set at the B inputs to the 7485, you will obtain a logic 1 at the $A = B$ output from the 7485 chip (at the top right of Fig. 17-11). You invert this signal and then gate it with the $\overline{OUT}$ control signal to obtain the desired device select pulse. This circuit produces an absolutely decoded single device select pulse. However, you can change the device code simply by altering the 8-bit B input to the comparators. In this case, the B input corresponds to 11000110, or 306 in octal code.

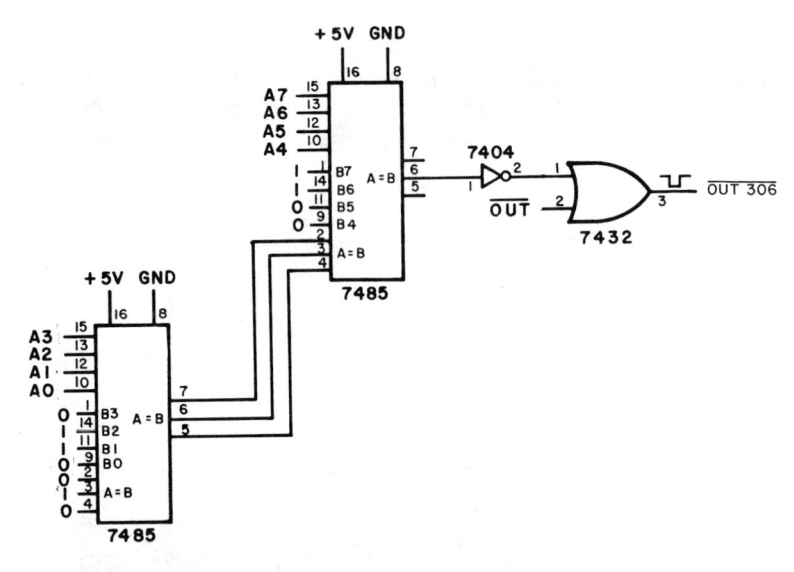

Fig. 17-11. Decoding circuit using a pair of 7485 4-bit comparator chips.

## I/O INSTRUCTIONS

There are only two 8080A input/output instructions:

323 &lt;B2&gt; OUT Place the 8-bit device code on the address bus, the accumulator contents on the bidirectional data bus, and generate an $\overline{OUT}$ control signal. The contents of the accumulator remain unchanged.

333 &lt;B2&gt; IN Place the 8-bit device code on the address bus, permit data on the bidirectional data bus to be input into the accumulator, and generate an $\overline{IN}$ control signal.

The second byte of each instruction is the 8-bit device code. You use the control signal and the information on the address bus to generate the required device select pulse. A more succinct way of stating the above two instructions is:

323 &lt;B2&gt; OUT Output the accumulator contents to the output device selected by the device code in the second byte.
333 &lt;B2&gt; IN Input into the accumulator the contents of the input device selected by the device code in the second byte.

Although device select pulses are frequently used to transfer information between the accumulator and an I/O device, they also are used to strobe the operation of I/O devices under conditions where data transfer to or from the accumulator does not occur.

## THE FETCH, INPUT AND OUTPUT MACHINE CYCLES

Having described several circuits that you can use to generate device select pulses, we will shortly provide you with several programming examples that illustrate the behavior of the IN and OUT instructions, both of which are two-byte instructions. If you execute either the IN or OUT instruction in the single-step mode and monitor the contents of the 8-bit data bus, you will observe something unusual: *a third byte appears that does not correspond to a byte present at that point in your program.* What is this extra byte? It is the 8-bit byte being transferred to or from the 8080A's accumulator register during the IN or OUT instruction cycle. It is during the execution of this third machine cycle that:

- Either an $\overline{\text{IN}}$ or $\overline{\text{OUT}}$ pulse is generated on the control bus.
- The device code appears on the 16-bit address bus as two identical 8-bit bytes.
- The external bidirectional data bus and the internal data bus within the microprocessor chip are connected to permit direct data communication between the accumulator and the I/O device, whether input or output.

When you single step through an 8080A microcomputer program, you single step through *machine cycles* rather than instruction bytes. Without going into great detail, we can define machine cycle as follows:

*machine cycle*—A subdivision of an instruction cycle during which time a related group of actions occur within the microprocessor chip. All instructions are combinations of one or more machine cycles.

As an example of a machine cycle, there is the *FETCH machine cycle,* during which the instruction code is fetched from the memory location addressed by the program counter. Simple arithmetic and logical operations involving the 8080A's internal registers are also performed during the FETCH cycle.

The output instruction, OUT, consists of two FETCH machine cycles in sequence, i.e., the instruction code and, then, the device code. These are followed by an *OUTPUT machine cycle*—the third step that you observe when you execute an OUT instruction. During this time, the contents of the accumulator are made available on the bi-directional data bus. The output device code appears as two identical 8-bit device code bytes on the address bus and an $\overline{OUT}$ pulse is generated. The IN instruction consists of two FETCH machine cycles in sequence, i.e., the instruction code and, then, the device code. These are followed by an *INPUT machine cycle*—the third step that you observe when you execute an IN instruction. During this time, the input buffer/latch within the 8080A chip is enabled to permit input data on the bidirectional data bus to be transferred to the accumulator. Two identical 8-bit device code bytes appear on the address bus, and an $\overline{IN}$ pulse is generated.

## FIRST PROGRAM

Let us first consider the program given in Experiment No. 5 in Unit 11:

| LO memory address | Instruction byte | Mnemonic | Description |
|---|---|---|---|
| 000 | 074 | INR A | Increment contents of accumulator by 1 |
| 001 | 323 | OUT | Output accumulator contents to device given in following byte |
| 002 | 002 | 002 | Device code for port 2 |
| 003 | 303 | JMP | Unconditional jump to the memory address given by the following two bytes |
| 004 | 000 | — | LO address byte |
| 005 | 003 | — | HI address byte |

If you would execute this program using the single-step circuit, you would observe the following bytes, in succession, on the bi-directional data bus:

| Address bus byte | Data bus byte | Comments |
|---|---|---|
| 000 | 074 | FETCH machine cycle for INR A instruction code. |
| 001 | 323 | FETCH machine cycle for OUT instruction code. |
| 002 | 002 | FETCH machine cycle for device code for port 2. |
| 002* | accumulator contents | OUTPUT machine cycle, during which the accumulator contents is made available on the bidirectional data bus and |

---

* This is the I/O device address that appears at bits A0 through A7 on the address bus.

|     |     | the device code appears on the address bus. An $\overline{OUT}$ pulse is also generated during this machine cycle. |
| --- | --- | --- |
| 003 | 303 | FETCH machine cycle for JMP instruction code. |
| 004 | 000 | FETCH machine cycle for LO address byte. |
| 005 | 003 | FETCH machine cycle for HI address byte. |

You observe such information on the data bus because (a) all instruction bytes move over the data bus from the memory to the instruction register within the 8080A chip, and (b) the contents of the accumulator is output on the data bus during the third machine cycle of the OUT instruction.

The program increments the contents of the accumulator during each loop. Also, it outputs that contents to Port 2 during each pass through the loop. You observe this at an 8-bit port that increments from $00000000_2$ to $11111111_2$ and then repeats the counting sequence.

## SECOND PROGRAM

In the second program, given below, you actually modify the device code in the OUT instruction. In this program, you encounter for the first time the use of a *register pair* instruction, LXI H; the use of register pair H to define the 16-bit address of a memory location M; and the use of a memory reference instruction, INR M. From this point forward, if you encounter an unfamiliar instruction, please refer to the instruction set summary provided in Unit 18 (Fig. 18-18).

| LO memory address | Instruction byte | Mnemonic | Description |
| --- | --- | --- | --- |
| 314 | 074 | INR A | Increment contents of accumulator by 1 |
| 315 | 323 | OUT | Output accumulator contents to device given by the device code stored at memory location HI = 003 and LO = 316 |
| 316 | \<B2\> | \<B2\> | Device code for output device |
| 317 | 041 | LXI H | Load immediate two bytes into register pair H |
| 320 | 316 | \<B2\> | L register byte |
| 321 | 003 | \<B3\> | H register byte |
| 322 | 064 | INR M | Increment the contents of the memory location pointed to by register pair H |
| 323 | 303 | JMP | Unconditional jump to the memory address given in the following two bytes |
| 324 | 314 | — | LO address byte |
| 325 | 003 | — | HI address byte |

This program permits you to output the accumulator contents to 256 different devices in sequence, starting with the device given at HI = 003 and LO = 316. On every loop of the program, the device code at LO = 316 is incremented by one. This is not a very useful

47

program, but it does demonstrate the fact that the device code is not inviolate within a program. With very few instructions, you can alter the device code and, thus, sequence through a series of devices. In practice, the device code of the output device of interest would probably be stored in a register, and a MOV instruction used to transfer the register contents to memory location M addressed by the register pair H.

Would you obtain a useful result if this program were in ROM, PROM, or EPROM? No, because then you would not be able to alter the contents of memory location LO = 316.

### INTRODUCTION TO THE EXPERIMENTS

The following experiments demonstrate how you can generate and use device select pulses. You will also wire a bus monitor and gain experience with the use of a single-step circuit.

| *Experiment No.* | *Comments* |
|---|---|
| 1 | Demonstrates a bus monitor circuit, based upon the TIL311 numeric indicator, that permits you to monitor all data that passes over the bidirectional data bus. |
| 2 | Demonstrates a bus monitor circuit, based upon the HP 5082-7300 numeric indicator, that permits you to monitor all data that passes over the bidirectional data bus. |
| 3 | Demonstrates the use of a single-step circuit for the microcomputer. You will single step through the first thirty-eight machine cycles of the KEX program. |
| 4 | You count $\overline{\text{IN}}$ and $\overline{\text{OUT}}$ strobe pulses with the aid of a 7490 counter while the microcomputer is operating in the single-step mode. You also determine the bit pattern for the keyboard. |
| 5 | You construct, operate, and test an interface circuit that will permit you to generate sixteen different device select pulses. |
| 6 | Demonstrates how the decoded addresses on the MMD-1 printed-circuit board can be used to generate device select pulses. |
| 7 | Demonstrates the use of a device select pulse to clear a 7490 counter. |
| 8 | Demonstrates the use of a pair of 7485 comparator chips to generate a single absolutely decoded device select pulse. |

9    Demonstrates the use of a 7430 8-input NAND gate to generate a single absolutely decoded device select pulse. Also demonstrates the use of two such pulses, a 7476 flip-flop, and a solid-state relay to turn a fan motor on and off.

### EXPERIMENT NO. 1

**Purpose**

The purpose of this experiment is to wire a *bus monitor,* a three-digit octal display that monitors the data that appears on the bi-directional data bus.

**Pin Configuration of Numeric Indicator (Fig. 17-12)**

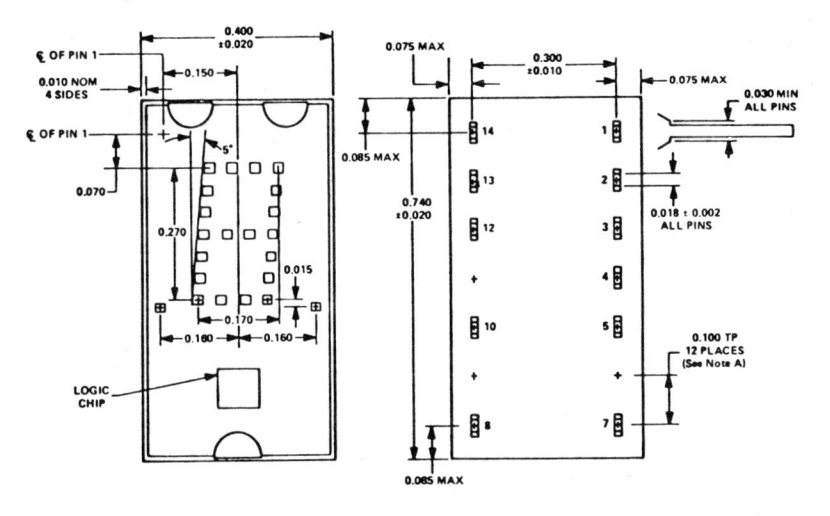

PIN 1  LED SUPPLY VOLTAGE
PIN 2  LATCH DATA INPUT B
PIN 3  LATCH DATA INPUT A
PIN 4  LEFT DECIMAL POINT CATHODE
PIN 5  LATCH STROBE INPUT
PIN 6  OMITTED
PIN 7  COMMON GROUND
PIN 8  BLANKING INPUT
PIN 9  OMITTED
PIN 10  RIGHT DECIMAL POINT CATHODE
PIN 11  OMITTED
PIN 12  LATCH DATA INPUT D
PIN 13  LATCH DATA INPUT C
PIN 14  LOGIC SUPPLY VOLTAGE, $V_{CC}$

Fig. 17-12.

**Step 1**

Wire the circuit shown in Fig. 17-13 *or* use a LR-27 Bus Monitor Outboard. When you wire the circuit, *as with any interface circuit in this book,* make sure the power to the microcomputer is turned off!

49

### Schematic Diagram of Circuit (Fig. 17-13)

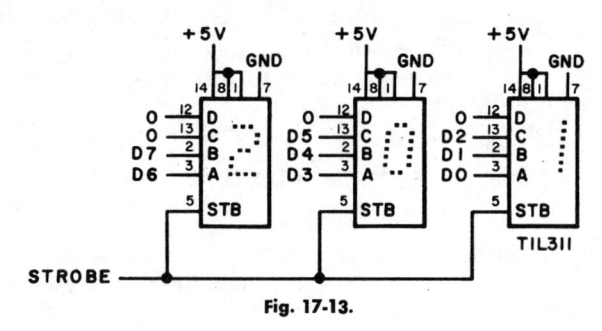

Fig. 17-13.

### Step 2

Connect the latch STROBE input to logic 0. This enables the displays so that each numeric indicator display follows the inputs. Apply power to the microcomputer. You should observe that most of the dots in the numeric indicators are lighted. What you are observing is the wait loop in the K̲eyboard EX̲ecutive (KEX) EPROM being executed at a clock rate of 750 kHz. The only thing that you can learn from the fact that most of the numeric indicator dots are lighted is that the microcomputer is executing a program.

### Step 3

Now connect the STROBE input to the $\overline{\text{OUT}}$ control signal line on the SK-10 bus socket. If you cannot find $\overline{\text{OUT}}$, please refer to Fig. 16-4.

### Step 4

Press the RESET key on the microcomputer. What three-octal-digit byte appears on the bus monitor? Write it in the space below.

What three-octal-digit byte appears at Port 2? Write it below.

Are they the same?

Yes.

## Step 5

What is the significance of the information that appears on the bus monitor when the STB input is connected to $\overline{\text{OUT}}$? If you are not certain how to answer this question, load some arbitrary octal values into read/write memory and observe what information appears on the bus monitor. Also examine your "program" and again observe the relationship between the byte on the bus monitor and the byte displayed at Port 2. What do you conclude?

We concluded that the information on the bus monitor (an output port for the microcomputer) and at Port 2 are identical. The two output ports give the contents of the memory location addressed by the 16-bit address given in Ports 0 and 1, or the byte to be loaded into the memory (the LO address register or the HI register). The bus monitor makes it easier to enter and check a program.

Save this circuit for *all* of the remaining experiments in this book. Experiment No. 2 is similar to this one, but employs a different numeric indicator, the HP 5082-7300.

## Additional Comments

A less expensive bus monitor circuit can be constructed from Port 1 on the microcomputer. Rather than three octal digits, the monitor consists of eight LEDs that continuously monitor the state of the bidirectional data bus, D0 through D7. To construct this bus monitor, place a switch in the ENABLE input line to the Port 1 7475 latch chips, as shown in Fig. 17-14. When the switch is closed, the 7475 chips, IC24 and IC25, operate normally as Port 1. When the switch is opened, they operate as an 8-bit bus monitor. This modification will probably be incorporated in future models of the MMD-1 microcomputer.

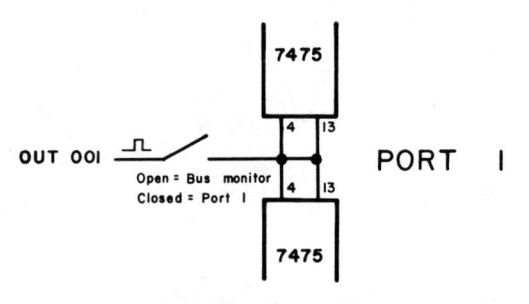

**Fig. 17-14. Bus monitor circuit modification.**

### EXPERIMENT NO. 2

**Purpose**

The purpose of this experiment is to wire a *bus monitor* using the HP 5082-7300 numeric indicator. This experiment is identical to Experiment No. 1.

**Pin Configuration of Numeric Indicator (Fig. 17-15)**

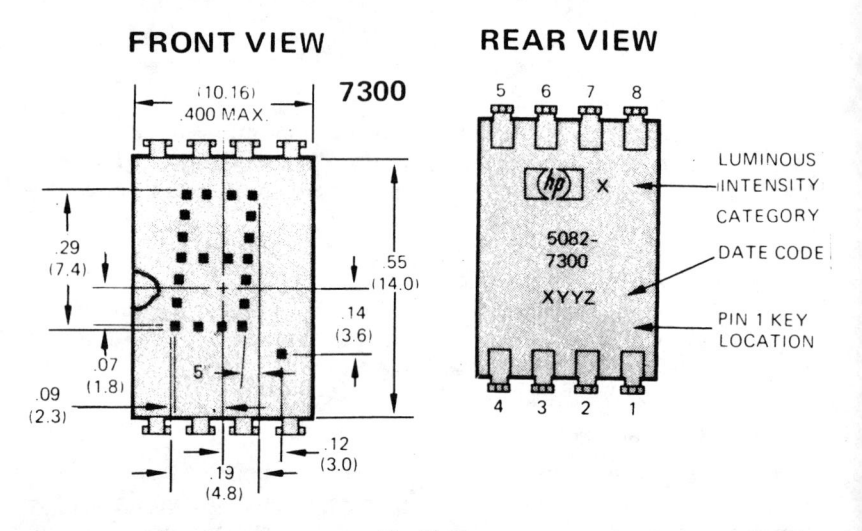

**Fig. 17-15.**

**Schematic Diagram of Circuit (Fig. 17-16)**

**Step 1**

Wire the circuit shown in Fig. 17-16. We recommend that you do so with the power to the microcomputer turned off.

Fig. 17-16.

STROBE (STB)

## Step 2

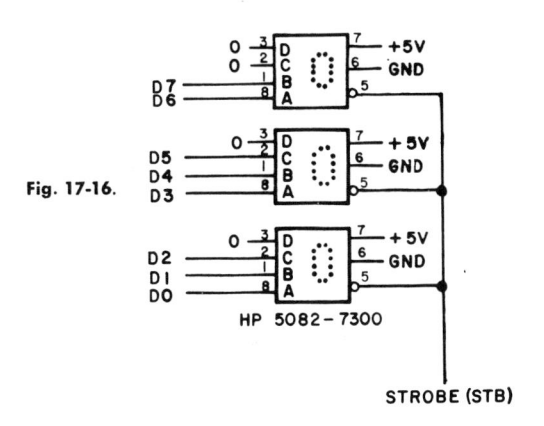

Connect the latch STROBE input to logic 0. This enables the displays so that each numeric indicator display follows the inputs.

Apply power to the microcomputer. You should observe that most of the dots in the numeric indicators are lighted. What you are observing is the wait loop in the <u>K</u>eyboard <u>EX</u>ecutive (KEX) EPROM being executed at a clock rate of 750 kHz. The only thing that you can learn from the fact that most of the numeric indicator dots are lighted is that the microcomputer is executing a program.

## Step 3

Now connect the STROBE input to the $\overline{\text{OUT}}$ control signal line on the breadboarding socket. If you cannot find $\overline{\text{OUT}}$, please refer to Fig. 16-4.

## Step 4

Press the RESET key on the microcomputer. What three-octal-digit byte appears on the bus monitor? Write it in the space below.

What three-octal-digit byte appears at Port 2? Write it below.

Are they the same?

Yes.

**Step 5**

What is the significance of the information that appears on the bus monitor when the STB input is connected to $\overline{\text{OUT}}$? If you are not certain how to answer this question, load some arbitrary octal values into read/write memory and observe what information appears on the bus monitor. Also examine your "program" and again observe the relationship between the byte on the bus monitor and the byte displayed at Port 2. What do you conclude?

We concluded that the information on the bus monitor (a microcomputer output port) and at Port 2 is identical. The two output ports give the contents of the memory location addressed by the 16-bit address given in Ports 0 and 1, *or* the byte to be loaded into memory (the LO address register or the HI address register on the microcomputer). The bus monitor makes it easier to enter and check a program.

Save either this circuit or the circuit given in Experiment No. 1 for all of the remaining experiments in this book.

### EXPERIMENT NO. 3

**Purpose**

The purpose of this experiment is to construct a single-step circuit for the microcomputer.

## Pin Configuration of Integrated-Circuit Chip (Fig. 17-17)

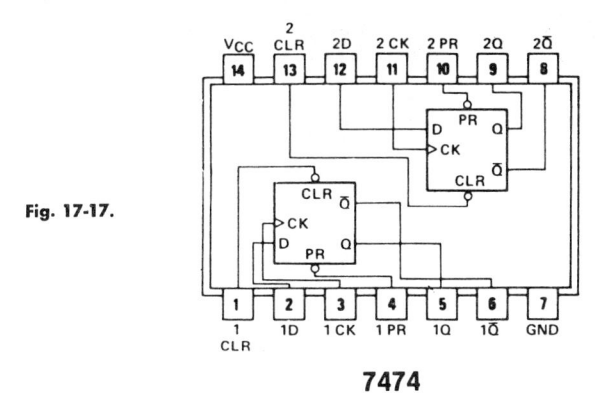

**Fig. 17-17.**

7474

## Schematic Diagram of Circuit (Fig. 17-18)

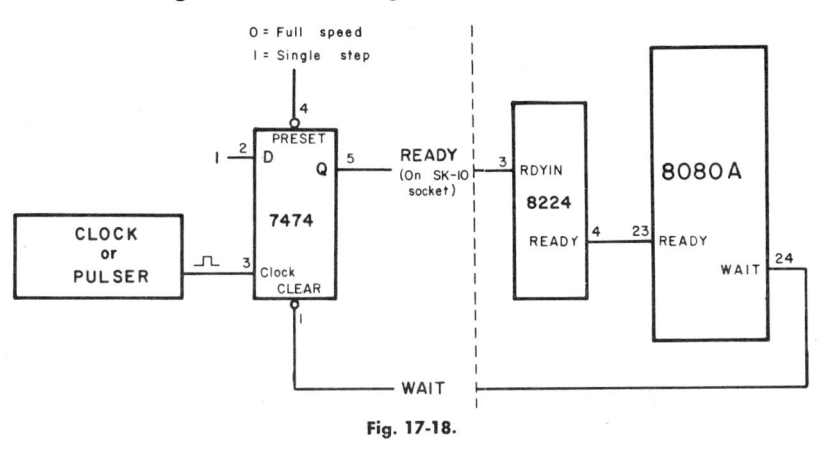

Fig. 17-18.

### Step 1

Wire the circuit shown in Fig. 17-18. The READY and WAIT signals are available on the breadboarding socket. Note in the diagram of Fig. 17-18 that the chips to the right of the dotted line are already wired on the printed-circuit board. The specific bus socket connections that you must make are shown in Fig. 17-19.

### Step 2

Connect the latch enable input (STB) of the bus monitor to logic 0 (GND). This permits you to observe all information that appears on the bidirectional data bus. Place the single-step circuit in the single-step mode (pin 4 of the 7474 chip connected to logic 1). If you

55

**7474**
(pin 5)

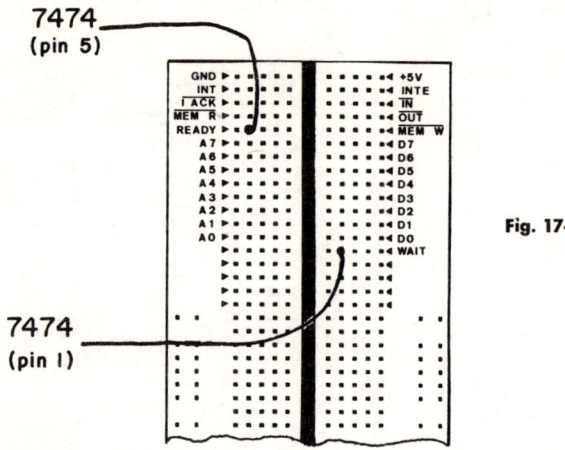

**7474**
(pin 1)

Fig. 17-19.

are using the single-step Outboard (LR-50), set the logic switch to the position that corresponds to single-step operation.

Press the RESET key. You should observe a 303 on the bus monitor. This is the *first* instruction byte in the KEX program.

## Step 3

Using the pulser, single step through the keyboard executive (KEX), which starts at memory location HI = 000 and LO = 000. Compare your observations with the sequence of bytes that we observed on the bus monitor, as given below. The purpose of this listing is to show you how the single-step operation works. You may not understand every instruction below.

| Memory Address | Instruction Byte | Mnemonic | Description |
|---|---|---|---|
| 000 000 | 303 | JMP | Unconditional jump to the memory address START given by the following two bytes |
| 000 001 | 070 | START | LO address byte of START |
| 000 002 | 000 | — | HI address byte of START |
| 000 070 | 061 | LXI SP | Load immediate two bytes into the stack pointer register |
| 000 071 | 000 | 000 | LO stack pointer byte |
| 000 072 | 004 | 004 | HI stack pointer byte |
| 000 073 | 041 | LXI H | Load immediate two bytes into register pair H |
| 000 074 | 000 | 000 | L register byte |
| 000 075 | 003 | 003 | H register byte |
| 000 076 | 116 | MOV C,M | Move contents of memory location M (which is pointed to by register pair H) to register C |

| Memory Address | Instruction Byte | Mnemonic | Description |
|---|---|---|---|
| 003 000 | XYZ | XYZ | **MEMORY READ machine cycle,** in which the contents of memory location HI = 003 and LO = 000 are moved to register C. You observe this memory byte on the bus monitor. Your **XYZ value** is the value contained in the first read/write memory location on your MMD-1 microcomputer. |
| 000 077 | 174 | MOV A,H | Move contents of register H to the accumulator |
| 000 100 | 323 | OUT | Output accumulator contents to the output port given in the following byte |
| 000 101 | 001 | 001 | Device code for output port 1 |
| 001 001 | 003 | 003 | **OUTPUT machine cycle,** during which the contents of the accumulator are output to port 1. The device code is output as **two identical 001 bytes** on the address bus of the 8080A chip. |
| 000 102 | 175 | MOV A,L | Move contents of register L to the accumulator |
| 000 103 | 323 | OUT | Output accumulator contents to the output port given in the following byte |
| 000 104 | 000 | 000 | Device code for output port 0 |
| 000 000 | 000 | 000 | **OUTPUT machine cycle,** during which the contents of the accumulator are output to port 0. The device code is output as **two identical 000 bytes** on the address bus. |
| 000 105 | 171 | MOV A,C | Move contents of register C to the accumulator |
| 000 106 | 323 | OUT | Output accumulator contents to the output port given in the following byte |
| 000 107 | 002 | 002 | Device code for output port 2 |
| 002 002 | XYZ | XYZ | **OUTPUT machine cycle,** during which the contents of the accumulator are output to port 2. This is the byte from read/write memory retrieved earlier. The device code is output as **two identical 002 bytes** on the address bus. |
| 000 110 | 315 | CALL | Call subroutine KBRD located at memory address given by the following two address bytes |
| 000 111 | 315 | KBRD | LO address byte of KBRD |
| 000 112 | 000 | — | HI address byte of KBRD |
| 003 377 | 000 | 000 | **STACK WRITE machine cycle,** during which the HI address byte in the program counter is moved to the stack in memory. |
| 003 376 | 113 | 113 | **STACK WRITE machine cycle,** during which the LO address byte in the program counter is moved to the stack in memory |

| Memory Address | Instruction Byte | Mnemonic | Description |
|---|---|---|---|
| 000 315 | 333 | IN | Input byte into the accumulator from the input port given by the following device code |
| 000 316 | 000 | 000 | Device code for the keyboard on the MMD-1 microcomputer |
| 000 000 | 160 | 160 | **INPUT machine cycle,** during which a byte is input from the keyboard. The byte, 160, is input if you do not press any key. The device code is output as **two identical 000 bytes** on the address bus. |
| 000 317 | 267 | ORA A | OR contents of accumulator with itself |
| 000 320 | 372 | JM | Jump if accumulator contents are minus (D7 bit is logic 1) back to memory location given by the following two address bytes |
| 000 321 | 315 | 315 | LO address byte |
| 000 322 | 000 | 000 | HI address byte |
| 000 323 | 315 | CALL | Call subroutine TIMOUT located at memory address given by the following two address bytes |
| 000 324 | 277 | TIMOUT | LO address byte of TIMOUT, a 10 msec time delay subroutine |
| 000 325 | 000 | — | HI address byte of TIMOUT |
| • | • | • | |
| • | • | • | |
| • | • | • | |

We will stop here as we enter the time-delay subroutine. Clearly, the value of the single-step and bus monitor circuits is that they permit you to observe data being transferred between memory or I/O devices and the interior of the 8080A microprocessor chip, i.e., you are able to observe machine cycles other than the FETCH cycle. This is particularly important when you check and test new programs.

**Step 4**

Return the microcomputer to its full operating speed, 750 kHz. Now load 377 into the HI register (Port 1), the LO register (Port 0), and the data register (Port 2). All twenty-four lamp monitors should now be lighted. Place the single-step circuit (or the LR-50 Outboard) in the STEP position, press the RESET button on the microcomputer keyboard, and single step through the KEX program. When does the HI register change from 377 to 003? You will have to single step the program at least to the first OUT instruction. Why?

At the machine cycle that immediately follows the execution of the instruction byte at HI = 000 and LO = 101. The only time that this data is changed is during the execution of an OUT 001 instruction. These LEDs are *not* connected to the address bus. They are only *used to represent* the HI address byte.

**Step 5**

Continue to single step through KEX. When does the LO register change from 377 to 000?




At the machine cycle that immediately follows the execution of the instruction byte at HI = 000 and LO = 104. The only time that this data is changed is during the execution of an OUT 000 instruction.

**Step 6**

Continue to single step through KEX. When does the data register change from 377 to whatever is stored at memory location HI = 003 and LO = 000?




At the machine cycle that immediately follows the execution of the instruction byte at LO = 107. The only time that this data is changed is during the execution of an OUT 002 instruction.

Save your bus monitor and single-step circuits and continue to the next experiment.

## EXPERIMENT NO. 4

**Purpose**

The purpose of this experiment is to count input and output strobe pulses, $\overline{IN}$ and $\overline{OUT}$. You will first count $\overline{OUT}$ strobe pulses. Then you will change the clock input to the 7490 chip to the $\overline{IN}$ signal and count $\overline{IN}$ strobe pulses.

## Pin Configuration of Integrated-Circuit Chip (Fig. 17-20)

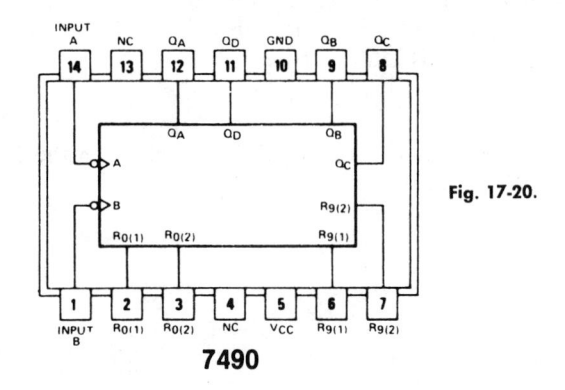

Fig. 17-20.

## Schematic Diagram of Circuit (Fig. 17-21)

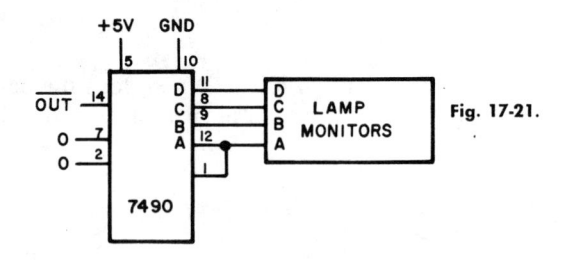

Fig. 17-21.

## Program

| LO Address Byte | Instruction Byte | Mnemonic | Description |
|---|---|---|---|
| 000 | 333 | IN | Input byte into the accumulator from the keyboard |
| 001 | 000 | 000 | Device code for keyboard |
| 002 | 323 | OUT | Output contents of accumulator to output port in following byte |
| 003 | 000 | 000 | Device code for port 0 |
| 004 | 303 | JMP | Unconditional jump to memory location given by the following two address bytes |
| 005 | 000 | — | LO address byte |
| 006 | 003 | — | HI address byte |

## Step 1

This program is an interesting one, since it demonstrates a number of important concepts associated with input and output instructions and the operation of the microcomputer.

## Step 2

Before you wire the 7490 circuit, load the above program into memory and execute it at 750 kHz. What bit pattern do you observe at Port 0?

We observed $01110000_2$ at Port 0.

## Step 3

Now press the following keys in sequence: 0, 1, 2, 3, 4, 5, 6, and 7. For each of these keys, write the bit pattern that you observe at Port 0 in the space below. What correlations do you observe between the key numbers and the bit pattern?

We observed the following:

| Key | Bit Pattern at Port 0 |
|-----|----------------------|
| 0   | 11110000             |
| 1   | 11110001             |
| 2   | 11110010             |
| 3   | 11110011             |
| 4   | 11110100             |
| 5   | 11110101             |
| 6   | 11110110             |
| 7   | 11110111             |

Note that whenever you press a key, Bit D7 becomes logic 1. For keys 0 through 7, the least significant three bits correspond to the octal equivalent of the key.

### Step 4

Press the remaining keys with the exception of RESET. Write the bit pattern that you observe in the space below.

We observed the following:

| Key | Bit Pattern at Port 0 |
|-----|----------------------|
| S   | 11111000             |
| C   | 11111010             |
| G   | 11111011             |
| H   | 11111100             |
| L   | 11111101             |
| A   | 11111110             |
| B   | 11111111             |

Again, whenever we pressed a key, Bit D7 became logic 1. This is the bit that is used by KEX to determine whether or not a key is pressed. Refer to Experiment No. 2 in this unit and the instruction at LO = 320. This is where KEX detects a key closure.

### Step 5

Wire the counter circuit and connect the $\overline{OUT}$ output to the counter. With the microcomputer executing the program at 750 kHz, switch the logic switch (or wire) connected to the 7474's Pin 4 input to the logic 1 state. The microcomputer is now in the single-step mode.

### Step 6

If you do not have a bus monitor, all that you will be able to do in this experiment is to count the $\overline{OUT}$ control signal pulses. Single step through the program and observe that you obtain a single count for every nine times that the single-step pulser is pressed in and out. If the use of the pulser becomes tedious, substitute a Clock Outboard that is operating at a frequency of approximately 0.3 to 1 Hz.

## Step 7

Remove the wire connecting $\overline{OUT}$ to Pin 14 of the 7490 counter. Connect $\overline{IN}$ to Pin 14. $\overline{IN}$ is adjacent to $\overline{OUT}$ on the breadboarding socket. Continue to execute the program in the single-step mode. You should observe a single count on the 7490 counter for every nine clock pulses applied to the single-step circuit.

## Step 8

If you have a bus monitor, connect the latch enable (STB) input to logic 0. Now single step through the program. You should observe the sequence of bytes given below. Even if you do not have a bus monitor, please study the following program.

| Memory Address | Instruction Byte | Mnemonic | Description |
|---|---|---|---|
| 003 000 | 333 | IN | Input byte into the accumulator from the keyboard |
| 003 001 | 000 | 000 | Device code for the keyboard on the MMD-1 microcomputer |
| 000 000 | 160 | 160 | **INPUT machine cycle,** during which a byte is input from the keyboard. The byte, 160, is input if you do not press any key. The device code is output as **two identical 000 bytes** on the address bus. |
| 003 002 | 323 | OUT | Output accumulator contents to the output port given in the following byte |
| 003 003 | 000 | 000 | Device code for port 0 |
| 000 000 | 160 | 160 | **OUTPUT machine cycle,** during which the contents of the accumulator are output to port 0. This is the byte that is input from the keyboard. The device code is output as **two identical 000 bytes** on the address bus. |
| 003 004 | 303 | JMP | Unconditional jump to the memory location given by the following two bytes. |
| 003 005 | 000 | 000 | LO address byte for the start of the program |
| 003 006 | 003 | 003 | HI address byte for the start of the program |
| • | • | • | |
| • | etc. | • | |
| • | • | • | |

This program repeats itself every nine machine cycles. Why does a seven-byte program take nine steps to execute?

Each step is a "machine cycle," and the IN and OUT instructions require an additional machine cycle for proper execution. This extra machine cycle is preset within the 8080A chip and is characteristic of other types of instructions as well, including memory-reference instructions, calls, returns, PUSH, and POP.

One source of difficulty in this experiment is that you must execute a program at 750 kHz initially *before* you enter the single-step mode of operation. If you forget to do so, you will *never* leave the KEX program.

**Step 9**

Press Key 7 and keep it pressed. If you have a bus monitor, what byte appears during the INPUT and OUTPUT machine cycles?

We observed the byte 367 during both the INPUT and OUTPUT machine cycles. This byte was also output to Port 0. By being able to monitor information on the bidirectional data bus, we were able to observe data moving into the accumulator from the keyboard, and data moving out of the accumulator into the Port 0 latch.

Save the 7490 counter, single-step and bus monitor circuits and continue to the following experiment.

### EXPERIMENT NO. 5

**Purpose**

The purpose of this experiment is to construct a decoder circuit, based upon the 74154 decoder chip, that can generate sixteen different output device select pulses.

**Pin Configurations of Integrated-Circuit Chips (Fig. 17-22)**

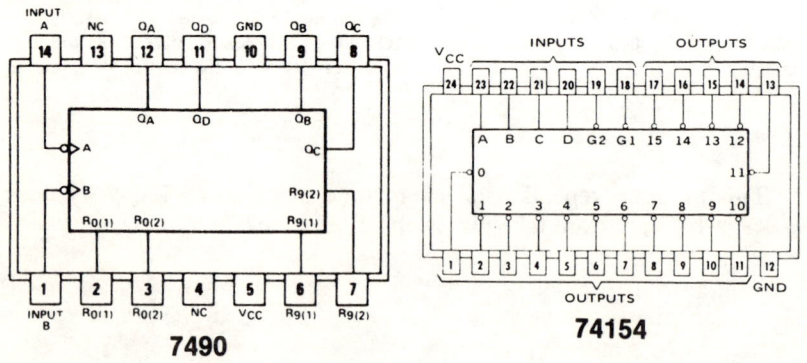

**7490**

**74154**

**Fig. 17-22.**

## Schematic Diagram of Circuit (Fig. 17-23)

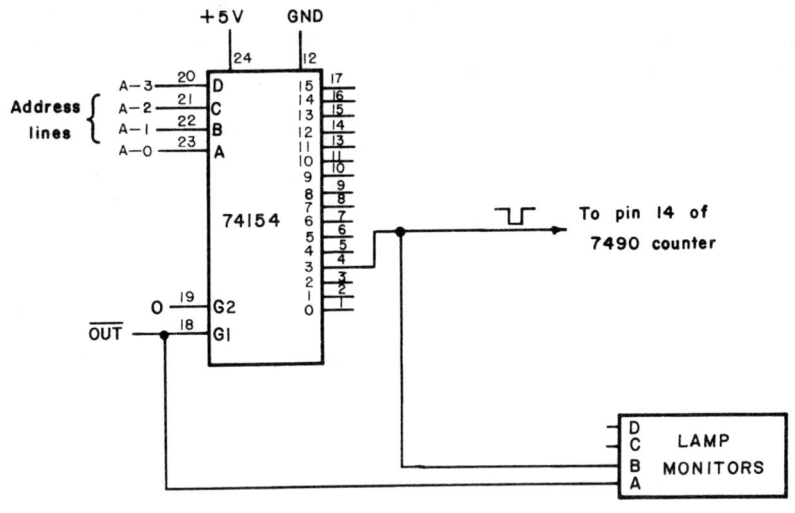

**Fig. 17-23.**

## Program

| LO Address Byte | Instruction Byte | Mnemonic | Description |
|---|---|---|---|
| 000 | 333 | IN | Generate device select pulse for the input device given in the following byte |
| 001 | 000 | 000 | Device code for input device 000 |
| 002 | 323 | OUT | Generate output device select pulse for the device given by the following byte |
| 003 | \<B2\> | \<B2\> | Device code for output device |
| 004 | 303 | JMP | Unconditional jump back to the beginning of this program, the address of which is given by the following two address bytes |
| 005 | 000 | — | LO address byte |
| 006 | 003 | — | HI address byte |

## Step 1

In the circuit shown in Fig. 17-23, you can generate sixteen "consecutive" output device select pulses. Wire the circuit using a 74154 decoder chip.

If you have a three-digit octal bus monitor, you may wish to observe the contents of the bidirectional data bus as you execute the program in the single-step mode. Wire the latch enable (STB) input on the bus monitor to logic 0.

**Step 2**

In the program, use device code 003 at LO = 003. Load the program into read/write memory. Be sure that the lamp monitor and Pin 14 of the 7490 counter are connected to Pin 4 of the 74154 decoder.

**Step 3**

Execute the program in the single-step mode. Count how many $\overline{\text{OUT}}$ control signal pulses occur every nine machine cycles, and write your answer in the space below.


You should observe one $\overline{\text{OUT}}$ pulse every nine machine cycles.

**Step 4**

It is common to denote a device select pulse by the notation, $\overline{\text{DS xxx}}$, if it is a logic 0 (or negative) pulse and by DS xxx if it is a logic 1 (or positive) pulse. The letters "xxx" denote the three-digit octal device code. The bar on the top of a functional pulse code is the standard notation for a logic zero active level.

In the schematic diagram, there is a wire connection between the No. 3 output of the 74154 decoder and the 7490 counter. With the 8080A operating at full speed, test some of the other decoder outputs and determine if any other channel generates an output $\overline{\text{DS 003}}$ pulse. Which channel is it?


Output No. 3 at Pin 4 on the integrated-circuit chip should be the only one to cause the counter to count at a rapid rate. All others are nonfunctional for a device code of 003.

**Step 5**

Now change the instruction byte at LO = 003 to 017. At which output on the 74154 decoder do you observe the device select pulse? What would be the proper way to denote such a pulse, i.e., as DS xxx or $\overline{\text{DS xxx}}$? What is "xxx"?

We observed the device select pulse at Channel $15_{10}$ (Pin 17). The proper way to denote this pulse is $\overline{\text{DS 017}}$.

**Step 6**

During the machine cycle when a device select pulse is generated, what is the logic state of lamp monitor A?


Lamp monitor A is at logic 0 whenever an OUT instruction is executed. A machine cycle is a subdivision of an instruction cycle during which time a related group of actions occur within the microprocessor chip. *When you single step through a microcomputer program, you single step through machine cycles, not instructions.*

**Step 7**

At which machine cycle for the OUT instruction is a device select pulse generated—the first, second, or third machine cycle?


The third machine cycle. The first two machine cycles are fetch cycles, which input the operation code, 323, and the device code, <B2>, from the memory locations in which they are stored.

**Step 8**

By varying the instruction byte at LO = 003, you can vary the 74154 decoder output channel at which a device select pulse appears. For the device code bytes given below, at which 74154 output channel does the device select pulse appear? Remember, when you change the program, you must be executing the KEX monitor at 750 kHz.

| Device code byte at LO = 003 | 74154 decoder output channel |
|---|---|
| 000 | 0 |
| 001 | 1 |
| 002 | 2 |
| 003 | 3 |
| 010 | — |
| 017 | — |
| 020 | — |
| 025 | — |
| 377 | — |

We observed the following results for the last five device code bytes:

|  |  |
|---|---|
| 010 | 8 |
| 017 | 15 |
| 020 | 0 |
| 025 | 5 |
| 377 | 15 |

If you did not observe similar results, please repeat this step.

**Step 9**

The 74154 decoder has only sixteen outputs, and you would initially expect that it could decode only the first sixteen output device codes: 000, 001, 002, 003, 004, . . . 015, 016, and 017. Why do you observe device select pulses for device codes greater than 017?

Because the 74154 decoder circuit does not *absolutely* decode the 8-bit device code byte. Only the four least-significant device code bits are decoded. Though you can generate only sixteen different device select pulses, *each unique device select pulse can be generated in sixteen different ways using sixteen different device codes*. This is not good engineering practice in microcomputer interface design. You should always *absolutely* decode both input and output device codes.

**Step 10**

How would you absolutely decode sixteen out of the 256 possible device codes using a 74154 decoder and one or more additional chips?

You can use another 74154 chip to enable the first 74154 decoder at the G1 input pin. Alternatively, you can use a 4-input OR gate to enable the 74154 decoder at the G1 input pin. Address Bits A-4, A-5, A-6, and A-7 would serve as inputs to the OR gate. There are many decoder schemes which might be used.

Remove the 74154 decoder and lamp monitor circuit used in this experiment. The 7490 counter and single-step circuit will be used in a subsequent experiment.

## EXPERIMENT NO. 6

### Purpose

The purpose of this experiment is to demonstrate how the decoded addresses on the microcomputer printed-circuit board can be used to generate input and output device select pulses.

### Pin Configuration of Integrated-Circuit Chip (Fig. 17-24)

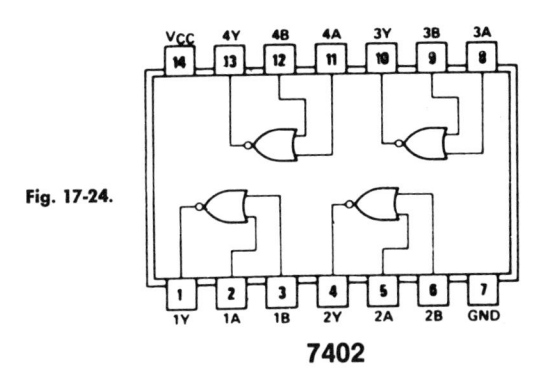

Fig. 17-24.

7402

### Schematic Diagram of Circuit (Fig. 17-25)

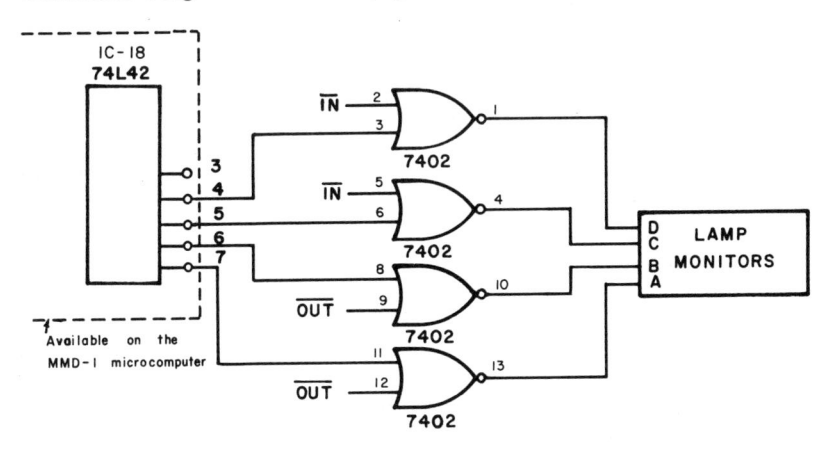

Fig. 17-25.

### Discussion

You will find five solderless breadboarding pins adjacent to the 74L42 chip in the I/O decoder section of the computer printed-circuit board. Look for integrated circuit IC-18 (or A18). The wiring diagram for the I/O decoder section is shown in Fig. 17-26.

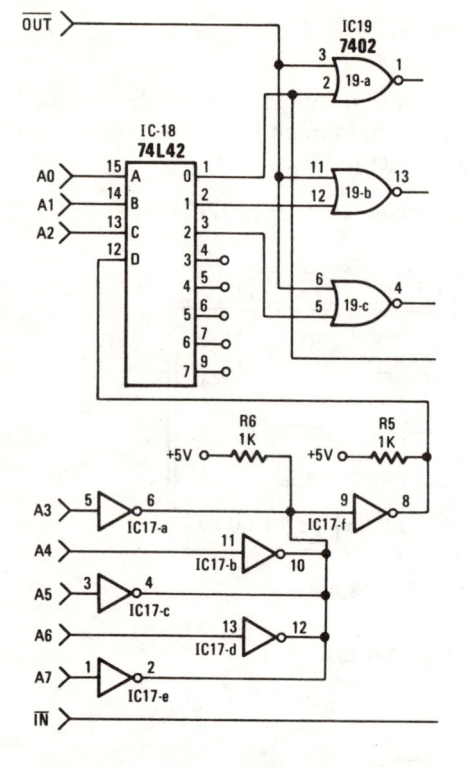

Fig. 17-26. Wiring diagram of the I/O decoder section.

The small circles associated with output channels 3 through 7 on the 74L42 chip represent either solder pads or breadboarding pins. Note that the computer employs 7402 2-input NOR gates, which generate positive device select pulses such as DS 000, DS 001, and DS 002. The circuit in the schematic diagram generates both input and output device select pulses which we can label "IN 004," "IN 005," "OUT 006," and "OUT 007," among others.

Note also that the 74L42 is an absolute decoder for the 8-bit device code. Bits A3, A4, A5, A6, and A7 on the address bus must all be at logic 0 in order for input D (Pin 12) on the 74L42 chip to be at logic 0. The three remaining address bus bits, A0, A1, and A2, are used in the decoding of eight device codes.

In contrast to the 74154 chip, there exists no G1 or G2 inputs to the 74L42 chip that can be used to enable and disable the chip. Thus, you will have to gate $\overline{\text{OUT}}$ and $\overline{\text{IN}}$ along with the decoded device code outputs if you wish to generate device select pulses.

A single 7402 chip allows you to generate four unique device select pulses. For most of the experiments in this book, four such pulses are all that you will require. If you wish, you can modify input and output device codes to correspond to those available through the use of this decoder. Such an action can simplify the interfacing task for many of the experiments and programs that are provided in this book.

## Program

| LO Address Byte | Instruction Byte | Mnemonic | Description |
|---|---|---|---|
| 000 | 074 | INR A | Increment contents of accumulator by 1 |
| 001 | 323 | OUT | Generate device select pulse for output device 004 |
| 002 | 004 | 004 | Device code for device 004 |
| 003 | 323 | OUT | Generate device select pulse $\overline{DS\ 005}$ |
| 004 | 005 | 005 | Device code for $\overline{DS\ 005}$ |
| 005 | 323 | OUT | Generate device select pulse $\overline{DS\ 006}$ |
| 006 | 006 | 006 | Device code for $\overline{DS\ 006}$ |
| 007 | 323 | OUT | Generate device select pulse $\overline{DS\ 007}$ |
| 010 | 007 | 007 | Device code for $\overline{DS\ 007}$ |
| 011 | 303 | JMP | Unconditional jump to memory location given by the following two address bytes |
| 012 | 000 | — | LO address byte |
| 013 | 003 | — | HI address byte |

### Step 1

Using a single 7402 chip, wire the circuit shown in the schematic diagram, in which both the $\overline{IN}$ and $\overline{OUT}$ control signals are employed as shown. Load the program into read/write memory.

### Step 2

Execute the program at 750 kHz, then move to single-step operation. What do you observe on the four lamp monitors?

You should observe that the lamp monitors associated with the OUT 006 and OUT 007 device select pulses are lighted, whereas, the other two are off at 750 kHz. These two lamp monitors are lighted once every sixteen machine cycles when the program is single stepped.

### Step 3

Why are the lamp monitors for the IN 004 and IN 005 device select pulses off?

Because there are no IN instructions in the program!

### Step 4

Change all of the OUT instructions to IN instructions in the program. Use the instruction code 333 for IN. Execute the program once again at 750 kHz. What do you observe?

Now, the lamp monitors for the IN 004 and IN 005 pulses are lighted, whereas the OUT 006 and OUT 007 lamp monitors are not lighted. The two input instructions in the program generate two pulses that are detected by the lamp monitors. The IN 006 and IN 007 device select pulses are not decoded by the circuit given in this experiment. Leave your experiment wired and continue to the following experiment.

### EXPERIMENT NO. 7

**Purpose**

The purpose of this experiment is to demonstrate the use of a device select pulse to clear a 7490 counter.

**Pin Configurations of Integrated-Circuit Chips (Fig. 17-27)**

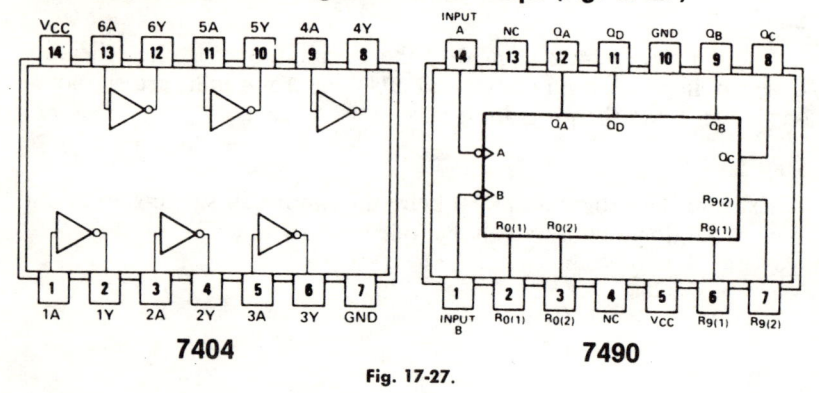

Fig. 17-27.

**Schematic Diagram of Circuit (Fig. 17-28)**

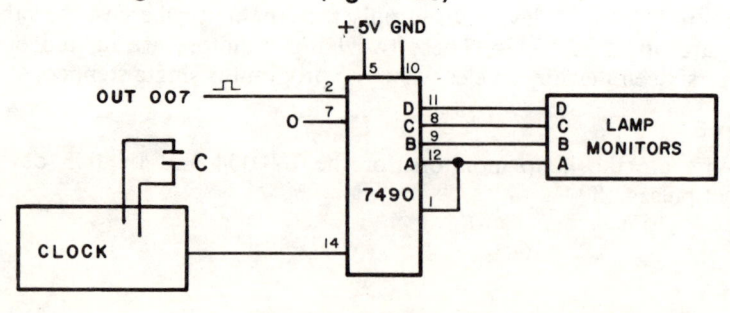

Fig. 17-28.

## Program

| LO Address Byte | Instruction Byte | Mnemonic | Description |
|---|---|---|---|
| 000 | 227 | SUB A | Clear the accumulator |
| 001 | 323 | OUT | Generate device select pulse for the device given by the following byte |
| 002 | 007 | 007 | Device code for device select pulse DS 007. |
| 003 | 303 | JMP | Unconditional jump to the memory location given by the following two address bytes |
| 004 | 000 | — | LO address byte |
| 005 | 003 | — | HI address byte |

### Step 1

To clear a 7490 decade counter, you will require a positive device select pulse. Thus, you will be able to use the OUT 007 pulse that you produced in Experiment No. 6. Wire the circuit shown in Fig. 17-28. Load the above program into read/write memory. Make certain that the OUT 007 connection is made between the 7402 gate output and the 7490 input at Pin 2.

### Step 2

Execute the program at 750 kHz. Then move to single step operation. The clock frequency to the 7490 counter should be approximately 10 Hz. We used a 0.05-$\mu$F timing capacitor with our Clock Outboard. Single step through the program with a pulser.

### Step 3

What behavior do you observe on the lamp monitors?

We observed a 10-Hz counting rate on the display LEDs until the third machine cycle of the OUT instruction, at which time a device select pulse was generated and the counter was cleared to zero. The counting resumed at the end of the third machine cycle.

### Step 4

Does the instruction 227, which clears the accumulator, have anything to do with the clearing of the 7490 counter?

No! In this program, it has no effect on the 7490 chip since we have not made any connection between the bidirectional data bus, D0 to D7, and the 7490 chip. Consequently, the 7490 does not know that the accumulator has been cleared.

You may remove all circuitry from your breadboarding socket except the single-step and counter circuits, which you will use in a subsequent experiment.

## EXPERIMENT NO. 8

### Purpose

The purpose of this experiment is to demonstrate the use of a pair of 7485 comparator chips to absolutely decode an 8-bit address.

### Pin Configurations of Integrated-Circuit Chips (Fig. 17-29)

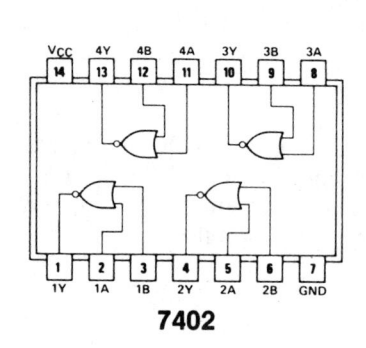

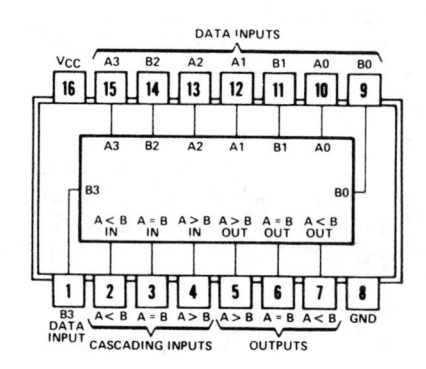

**7402**

**7485**

Fig. 17-29.

### Schematic Diagram of Circuit (Fig. 17-30)

### Program

| LO Address Byte | Instruction Byte | Mnemonic | Description |
|---|---|---|---|
| 000 | 227 | SUB A | Clear the accumulator |
| 001 | 323 | OUT | Generate a device select pulse for the device given by the following byte |
| 002 | 306 | 306 | Device code for output device 306 |
| 003 | 303 | JMP | Unconditional jump to the memory location given by the following two address bytes |
| 004 | 000 | — | LO address byte |
| 005 | 003 | — | HI address byte |

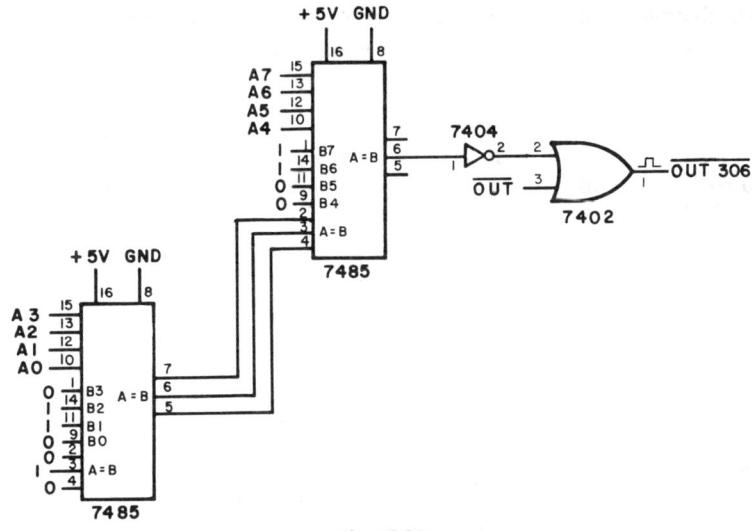

**Fig. 17-30.**

### Step 1

You will use the circuit shown in Experiment No. 4 in this unit to count the device select pulses that are produced by the pair of 7485 comparator chips. What change must you make to the 7490 counter circuit that you retained from the preceding experiment?

Reconnect the 7490 counter RESET input (Pin 2) to logic 0 (GND) and connect the CLOCK INPUT (Pin 14) to OUT 306 from the 7402 2-input NOR gate.

### Step 2

Wire the circuit shown in Fig. 17-30 and load the above program into read/write memory.

### Step 3

Execute the program at 750 kHz, and then move to the single-step mode of execution. Single step through the execution of the program, and explain in the space below what you observed on the counter's lamp monitor display.

We observed a single count each time the OUT instruction was executed, or one count every seven machine cycles.

### Step 4

Now change the device code at LO = 002 to 305. Execute the program at 750 kHz, and then single step through it once again. Do you now observe counting on the output lamp monitors connected to the 7490 chip?

We did not, because the address generated by the OUT instruction no longer matched the address *preset* at the comparator circuit.

### Step 5

Could the preset address be changed to correspond to a new software device code at LO = 002?

Yes. Now change the address byte at LO = 002 to 377. Does program execution cause any counting?

No. Now wire B0 through B7 on the comparator chips so that they are all at logic 1. Does this cause counting when you execute the program? Why?

Yes. Because, now, the hardware preset address and the software device code match.

Remove the 7485 decoder circuit from your breadboard, but save the counter and single-step circuits for the following experiment.

### EXPERIMENT NO. 9

**Purpose**

The purpose of this experiment is to demonstrate the use of a 7430 8-input NAND gate to absolutely decode the 8-bit address bus device code byte.

## Pin Configurations of Integrated-Circuit Chips (Fig. 17-31)

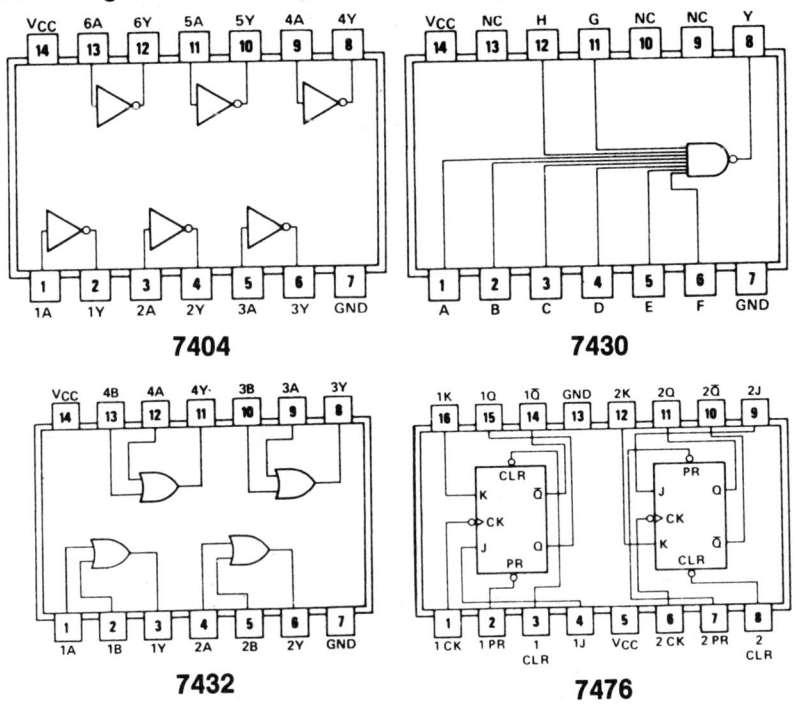

**7404**

**7430**

**7432**

**7476**

Fig. 17-31.

## Schematic Diagram of Circuit (Fig. 17-32)

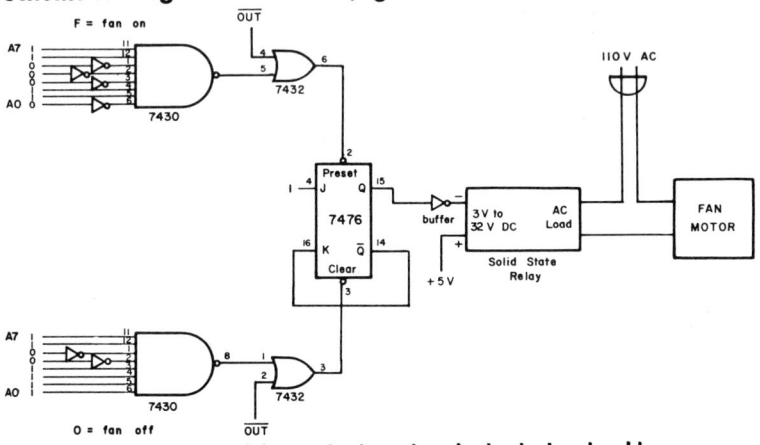

Fig. 17-32. Solid-state circuit used to absolutely decode address bus device code byte.

## Program

| LO Address Byte | Instruction Byte | Mnemonic | Description |
|---|---|---|---|
| 000 | 227 | SUB A | Clear the accumulator |
| 001 | 323 | OUT | Generate device select pulse to preset the 7476 flip-flop |
| 002 | 306 | 306 | Device code for the preset input to the 7476 flip-flop |
| 003 | 323 | OUT | Generate device select pulse to clear the 7476 flip-flop |
| 004 | 317 | 317 | Device code for the clear input to the 7476 flip-flop |
| 005 | 303 | JMP | Unconditional jump to the memory location given by the following two address bytes |
| 006 | 000 | — | LO address byte |
| 007 | 003 | — | HI address byte |

### Step 1

If you have a solid-state relay and a fan (or other appropriate ac power device, such as a lamp), we encourage you to wire the entire circuit shown in Fig. 17-32. Otherwise, wire the circuit up to, but not including, the buffer to the solid-state relay. Use a 7490 counter, as was done in Experiment No. 4, to count output pulses from the 7476 flip-flop.

### Step 2

Wire the circuit shown in the diagram of Fig. 17-32 and load the above program into memory.

### Step 3

You will single step through the execution of the program. Initiate execution at 750 kHz, then move to the single-step mode. If you are using a solid-state relay, you may wish to temporarily remove the wire connection between the 7476 flip-flop and the buffer until you are single stepping the program. Why?

The highest rate at which you can turn the relay on and off is double the 110-V ac line frequency, or 120 times per second. If you operate the microcomputer at 750 kHz with the above program, you will be attempting to turn the relay on and off at a rate that is greater than 10,000 times a second.

**Step 4**

As you execute the program in the single-step mode, explain what you observe.

We observed a single count each time we made a loop through the program. When we wired the solid-state relay circuit, we observed that the fan would turn on when we executed the instruction starting at LO = 001. When we executed the OUT instruction at LO = 003, the fan would turn off. We performed the solid-state relay experiment several times, and in each case observed the same result. This is an important experiment.

**Step 5**

What modifications to the program would you have to make in order to execute it at 750 kHz?

You would need to provide at least two time-delay loops to give the solid-state relay sufficient time to turn on and off. One delay loop would be located between the OUT 306 instruction and the OUT 317 instruction; the other loop would be located immediately after the OUT 317 instruction, but before the OUT 306 instruction.

## REVIEW QUESTIONS

The following questions will help you review device select pulses.
1. Device select pulses can be used to clear, strobe, trigger, etc., integrated-circuit chips. For the chips indicated below, identify the correct pin number—and whether a positive or negative device select pulse is required—at which the indicated operation must be performed.
    a. Clear the first flip-flop on a 7474 chip.
    b. Reset the second flip-flop on a 7474 chip.
    c. Reset a 7490 counter to Nine.
    d. Clear a 7493 binary counter.
    e. Enable a 74154 4-line-to-16-line decoder.
    f. Enable the first two latches on a 7475 chip.
    g. Clock the second flip-flop on a 7474 chip.
    h. Clear the first monostable on a 74123 chip.
    i. Strobe the first monostable on a 74123 chip.
    j. Strobe the first decoder on a 74155 chip.
    k. Trigger a 74122 monostable.
2. With the interface circuit shown in Experiment No. 4, it is possible to generate sixteen different device select pulses. For what output device codes will a device select pulse be generated at output 5 (Pin 6) on the 74154 decoder chip. Are you absolutely decoding the 8-bit device code byte?
3. How many machine cycles are there for the following 8080A instructions?
    a. JNZ <B2> <B3>
    b. CALL <B2> <B3>
    c. OUT <B2>
    d. IN <B2>
    e. MOV C,M
    f. LXI H <B2> <B3>
    g. MVI B <B2>
    h. JMP <B2> <B3>
    i. INR A
    j. DCR B

# 18

# The 8080A Instruction Set

### INTRODUCTION

This unit summarizes all of the important characteristics of each instruction in the 8080A instruction set: the number of machine cycles, the number of states, the type of memory addressing, and the flags that are influenced upon execution of the instruction. A description of each instruction is provided and, in some cases, examples of its use are given. Several programming experiments are provided at the end of the unit.

### OBJECTIVES

At the completion of this unit, you will be able to do the following:

- Indicate which flags are affected when a given instruction is executed.
- Subdivide the 8080A instruction set into five groups.
- Define program counter, register, accumulator, general-purpose register, stack, stack pointer, instruction register, instruction code, register pair, and nibble.
- List several sources of 8080 programming information.
- List different types of data transfer operations that occur within an 8080A-based microcomputer.
- Convert an 8-bit instruction code into both octal and hexadecimal code with the aid of a table provided in the unit.
- Distinguish between conditional and unconditional instructions.
- Describe the characteristics of the five condition flags in the 8080A microprocessor chip.
- Describe the operation of the stack and the instructions that influence its contents and the location of the stack.

- Describe the four different accumulator rotate instructions.
- Distinguish between LO and HI address bytes in instructions and programs.

## MICROCOMPUTER PROGRAMMING

Unless you have a background in computer science or possess a special knack for computer programming, you will probably find machine level and assembly level programming somewhat tedious and difficult initially. There does not appear to be any shortcut to learning programming. In due time, you will become sufficiently familiar with the 8080A's instruction set and with programming tricks to be able to write programs of modest size with little effort. You will be able to apply skills that you learn with one instruction set to other instruction sets, whether they are for microcomputers, minicomputers, or even mainframe computers.

The point that we would like to make is that you probably will need to learn some assembly language programming. Simple programs and subroutines can be written as easily and quickly in assembly language as they can in a higher level language. Such programs are also executed more quickly, require less memory, and are probably easier to understand. You will need to learn assembly language programming in order to understand other assembly programs that receive widespread distribution. Finally, a knowledge of assembly language programming provides the basis for understanding and comparing other instruction sets. If you have someone else do your programming, it will be expensive; if you do it yourself, it will also be expensive. However, if you can adapt other programs to your applications, your programming costs will be less.

## SOURCES OF 8080 PROGRAMMING INFORMATION

We would like to list some sources for 8080/8080A programming information that we have found to be useful:

1. *Intel 8080 Microcomputer Systems User's Manual,* Intel Corporation, 3605 Bowers Avenue, Santa Clara, California 95051.

    Chapter 4 provides a summary of the 8080/8080A instruction set. For each type of instruction, the number of machine cycles required to execute the instruction are listed. If the instruction has two possible execution times, both times are listed. Significant data addressing modes are listed, as are the flags that are affected by the execution of the instruction.

    Other chapters discuss the functions of a computer, the 8080 CPU, techniques of interfacing to the 8080, and the

8080 family of hardware components. If you are doing serious work with 8080 microcomputers, you should have this manual.
2. *Intel 8080 Assembly Language Programming Manual,* Intel Corporation, 3605 Bowers Avenue, Santa Clara, California 95051.

   An excellent manual that discusses such topics as the program counter, stack pointer, computer program representation in memory, memory addressing, condition bits, assembly language, and the entire 8080 instruction set.

   Also discussed is the use of *macros,* or macro instructions, which are extremely useful in assembly language programming. This manual is the one that you will need if you do programming with the 8080 Intel cross-assembler, or if you read programs that are cross-assembled using the Intel software package. Many of the programs in the Intel library can be understood with the aid of this manual.
3. *The μCOM-8 Software Manual,* NEC Microcomputers, Inc., 5 Militia Drive, Lexington, Massachusetts 02173.

   A superb manual that provides the following sample programming problems:
   - A simple sensing device
   - A gated counter
   - A programmed real-time motor controller
   - An N-way program branch
   - An interrupt subroutine program
   - A 10-cps teletypewriter I/O subroutine
   - A 16-digit BCD add or subtract subroutine
   - A data move-in-memory operation
   - Macro programming and conditional assembly

   Excellent descriptions are provided for individual 8080 instructions. Flow charts are provided for each programming problem.

   For the student who has some experience with 8080 assembly language programming, this manual will demonstrate a number of very useful programming techniques.
4. *Intel 8-Bit User's Program Library,* Intel Corporation, User's Library, Microcomputer Systems, 3065 Bowers Avenue, Santa Clara, California 95051. Membership is available on a 12-month basis to those contributing an acceptable program to the applicable library or by paying a membership fee.

   Programs submitted to the User's Library must be accompanied by the *Microcomputer User's Library Submittal Form,* a copy of which is shown in Figs. 18-1 and 18-2. Full-size copies may be ordered from the Software Marketing Group

# intel MICROCOMPUTER USER'S LIBRARY SUBMITTAL FORM

☐ 4004  ☐ 4040  ☐ 8008  ☐ 8080  ☐ 3000         (use additional sheets if necessary)

| | |
|---|---|
| Program Title | |
| Function | |
| Required Hardware | |
| Required Software | |
| Input Parameters | |
| Output Results | |

| | |
|---|---|
| Registers Modified: | Assembler/Compiler Used: |
| RAM Required: | Programmer: |
| ROM Required: | Company: |
| Maximum Subroutine Nesting Level: | Address: |

Courtesy Intel Corp.

**Fig. 18-1. Microcomputer User's Library Submittal Form (front).**

## INSTRUCTIONS FOR PROGRAM SUBMITTAL TO MCS USER'S LIBRARY

1. Complete Submittal Form as follows: (Please print or type)
   a. Processor (check appropriate box)
   b. Program title: Name or brief description of program function
   c. Function: Detailed description of operations performed by the program
   d. Required hardware:
      For example: TTY on port 0 and 1
      Interrupt circuitry
      I/O Interface
      Machine line and configuration for cross products
   e. Required software:
      For example: TTY routine
      Floating point package
      Support software required for cross products
   f. Input parameters: Description of register values, memory areas or values accepted from input ports
   g. Output results: Values to be expected in registers, memory areas or on output ports
   h. Program details (for resident products only)
      1. Registers modified
      2. RAM required (bytes)
      3. ROM required (bytes)
      4. Maximum subroutine nesting level
   i. Assembler/Compiler Used:
      For example: PL/M
      Intellec 8 Macro Assembler
      IBM 370 Fortran IV
   j. Programmer, company and address

2. A source listing of the program must be included. This should be the output listing of a compile or assembly. Extra information such as symbol table or code dumps is not necessary.

3. A test program which assures the validity of the contributed program must be included. This is for the user's verification after he has transcribed and assembled the program in question.

4. A source paper tape of the contributed program is required. This insures that a clear, original copy of the program is available to photo-copy for publication in a User's Library update publication.

Send completed documentation to:

   Intel Corporation
   User's Library
   Microcomputer Systems
   3065 Bowers Avenue
   Santa Clara, California 95051

Courtesy Intel Corp.

**Fig. 18-2. Microcomputer User's Library Submittal Form (back).**

at Intel. This form is used by the User's Library Manager in preparing the catalog and updates, and the description of the "Function" is used in preparation of the catalog index which is sent to prospective subscribers. This form is also used as the prefix to each program contained in the library and, therefore, should be carefully prepared. On the back of the Library Submittal Form are detailed instructions for program submittal which should be closely adhered to. These documentation standards are maintained to assure the usability of each library program by every interested member.

We refer you specifically to items 2, 3, and 4 in the instructions for program submittal to the User Library. *The program cannot be a duplication of a program that already is in the library.* The program should be error free and must be in standard Intel language (4004, 4040, 8008, 8080, or PL/M). Submit a typed source listing and a paper tape.

As of October, 1978, there were 300 programs in the library. It is the most extensive library of programs for any microcomputer. Source tapes and floppy disks are available for a small handling fee.

The User's Library saves development time in the development of 8080 programs. All of the programs can be modified or tailored to meet specific applications. During 1975, the Intel Corporation sponsored a 22-week User's Library Contest which rapidly expanded the User's Library. Some of the programs that you will find in the library include the following:

DATA ARRAY MOVE (8080). A contiguous array of data may be relocated in memory, regardless of the magnitude and direction of the move. The source and destination array locations may overlap. The maximum array size is $2^{16}$ bytes.

PAPER TAPE LABELER (8080). Accepts ASCII character from teletypewriter keyboard and punches corresponding alphanumeric character on tape.

TEXT STORAGE PROGRAM (8080). Allows text to be stored in memory using a letter of the alphabet as a pointer. After the message is stored, it can be retrieved by depressing a single key on the teletypewriter keyboard. Up to 32 messages may be stored and retrieved independently.

CLOCK SUBROUTINE (8080). Maintains a current time of day, decimal adjusted in BCD, of hours, minutes, and seconds. Must be invoked by external hardware once every

1.00000 second, usually by an external interrupt. Time is stored in three bytes of memory in the 24-hour system or, optionally, in the 12-hour system.

**TIMESHARING COMMUNICATIONS (8080).** To communicate with medium- to large-scale computer system as an external timeshare user.

**IBM SELECTRIC OUTPUT PROGRAM (8080).** Allows IBM Selectric Model 731 to be used as an output device.

**8080 IDLE ANALYZER FOR APPROXIMATING CPU UTILIZATION (8080).** Displays amount of time 8080 would have spent in an idle loop. When RUN time is compared with ideal time, the percent of CPU utilization can be calculated. Time display is in memory in ASCII.

**INTERRUPT SERVICE ROUTINE (8080).** Handles multiple-level interrupts, saving all resistors and flags and outputting the status of the current interrupt to an external status latch.

**8080 DISASSEMBLER (8080 PL/M).** This program inputs a hexadecimal tape and generates a symbolic assembly language program suitable for modifications and/or assembling.

**MEMORY DIAGNOSTIC PROGRAM (8080).** Writes test bytes in any range of memory and compares the written bit combination with what is read. Upon detection of a defective memory location, an error message is printed specifying the address, reference, and actual values.

**MATH (8080).** Routines for fixed and floating point arithmetic together with a demonstration program that performs algebraic evaluation (from left to right with no operator precedence) and allows unlimited parentheses nesting.

**ELEMENTARY FUNCTION PACKAGE (8080).** Calculates the following floating point values with five-decimal-digit precision: square root, logarithm, exponential function, sine, cosine, arc tangent, hyperbolic sine, and hyperbolic cosine. Adds, subtracts, multiplies, and divides with seven-decimal-digit floating point precision. [NOTE: We have used this program and like it very much. The entire program requires approximately 2½K of memory.]

**8080 FLOATING POINT PACKAGE WITH BCD CONVERSION ROUTINE (8080).** Performs floating addition, subtraction, multiplication, division, fixing, floating, negation, and conversion from floating point to BCD with exponent.

8080 LEAST SQUARES QUADRATIC FITTING ROUTINE (8080). Performs summations and matrix manipulation for fitting up to 256 floating point X-Y pairs to a function of the form:

$$aX^2 + bX + c = Y$$

N-BYTE BINARY MULTIPLICATION AND LEADING ZERO BLANKING (8080). The program performs binary multiplication on two numbers and returns a result that may be up to 255 bytes in length.

8080 CROSS COMPILER ON THE PDP-11 (8080). Accepts input in a format familiar to PDP-11 users and produces a fully coded listing, symbol table, and punched tape for use with the standard loader.

PAGE LISTING PROGRAM (8080). Provides facility for listing information in a pagenated, numbered format. This is accomplished through the system software with the console printer.

SOURCE PAPER TAPE TO MAGNETIC CASSETTE (8080). Will copy a source paper tape onto a magnetic cassette. End statement must be followed by a carriage return. Program will ignore leading blanks.

NATURAL LOGARITHM (8080). Computes the natural logarithm of a number between 1 and 65,535.

BCD MULTIPLICATION (8080). Multiplies up to a 6-digit BCD number by a 4-digit BCD number providing a 10-digit BCD result. All numbers are unsigned.

DOUBLE PRECISION MULTIPLY (8080 PL/M). To multiply two 16-bit numbers, returning the most significant 16 bits (in address form) through the appropriate registers to the calling program. The intrinsic PL/M multiply capability is employed for the byte-by-byte multiplications.

SUBROUTINE LOG. This subroutine takes the log to any integer base of any positive floating point number.

5. Scelbi Computer Consulting Inc., 1322 Rear Boston Post Road, Milford, Connecticut 06460.

The following software is available:

*Machine Language Programming for the 8008 (and similar microcomputers).*

*An 8080 Assembler Program.*

*An 8080 Editor Program.*

*8080 Monitor Routines.*

*SCELBAL. SCientific ELementary BAsic Language for 8008/8080 Systems.*

*SCELBI's First Book of Computer Games for the 8008/ 8080.*

*SCELBI's GALAXY GAME for the 8008/8080.*

Nat Wadsworth writes well. You can pick up many microcomputer programming techniques from the above software.

6. *Z80-CPU Technical Manual,* Zilog, Inc., 170 State Street, Los Altos, California 94022.

You do not obtain many programming hints from this manual, but it is very interesting to compare the Z80 chip with the 8080A in terms of the instruction set.

7. *BYTE,* Byte Publications, Inc., 70 Main Street, Peterborough, New Hampshire 03458. (Subscription rate is $15 for year, $27 for two years, or $39 for three years, from BYTE Subscriptions Dept., P.O. Box 590, Martinsville, New Jersey 08836.)

The quality of individual articles varies, but you will find useful programs and programming techniques discussed in this journal, which is one of the magazines that are aimed at the hobby microcomputer market.

8. *PACE Logic Designer's Guide to Programmed Equivalent of TTL Functions,* National Semiconductor Corporation, 2900 Semiconductor Drive, Santa Clara, California 95051.

Though it is for an entirely different microprocessor, the 8/16-bit PACE, this book does an excellent job of demonstrating the substitution of software for hardware. Hardware circuits are provided and described. Programs are then provided that duplicate the basic functions of the hardware. With some knowledge of the PACE instruction set, you should be able to convert the programs to Intel 8080 language. The advantages of 16-bit operations are certainly evident.

9. *73 Magazine,* Peterborough, New Hampshire 03458, $15 per year.

The magazine, which is directed toward radio amateurs, has a 40-page section entitled I/O that is devoted to practical uses for microcomputers. Since hams are very interested in communications, you should find increasing coverage of digital data communications in this magazine.

## 8080 INSTRUCTION SET SUMMARIES

Machine code and assembly language summaries of the 8080 instruction set are available from a number of different sources:

1. *Intel 8080 Assembly Language Reference Card,* Intel Corporation, 3065 Bowers Avenue, Santa Clara, California 95051.

    Provides a hexadecimal listing of the 8080 instruction set as well as a listing by instruction function. Hexadecimal-ASCII listing provided.

2. *8080 Hex/Octal Code Card,* Tychon, Inc., P. O. Box 242, Blacksburg, Virginia 24060.

    A sliding insert permits you to rapidly find either the 8-bit hex or octal instruction code for an assembly language instruction. Flag status after an instruction is executed is also indicated.

3. *8080 Instruction Set,* Martin Research, 3336 Commercial Ave., Northbrook, Illinois 60062.

    Subdivides the 8080 instruction set by function. Compact statement of flag status after different types of instructions are executed.

4. R. Baker, *Byte,* 84 (February 1976).

    Compact octal code listing of the 8080 instruction set.

5. P. R. Rony, D. G. Larsen, and J. A. Titus, *The 8080A Bugbook®\* Microcomputer Interfacing and Programming.*

    The 8080 instruction set is given as an instruction group listing, an alphabetic listing of mnemonics, and an octal/hexadecimal numerical listing. The octal/hexadecimal listing provides a handy conversion table for octal to hexadecimal, and vice versa.

## 8080 MICROPROCESSOR REGISTERS

The term *register* can be defined as follows:

*register*—A short-term digital electronic storage circuit, the capacity of which usually is one computer word.[15]

Single registers in the 8080 microprocessor chip store a single byte, i.e., eight contiguous bits.

---

\* Bugbook® is a registered trademark of E&L Instruments, Inc., Derby, CT 06418.

There are two different sets of registers in the 8080 chip. Those that we can address from a program and those that we cannot. The program addressable registers are shown in Fig. 18-3 and include the following:

- Six 8-bit general-purpose registers addressed singly or in pairs,
  - B register
  - C register
  - D register
  - E register
  - H register
  - L register
- The 8-bit *accumulator,* also known as *Register A.*
- The 16-bit *stack pointer register.*
- The 16-bit *program counter register.*

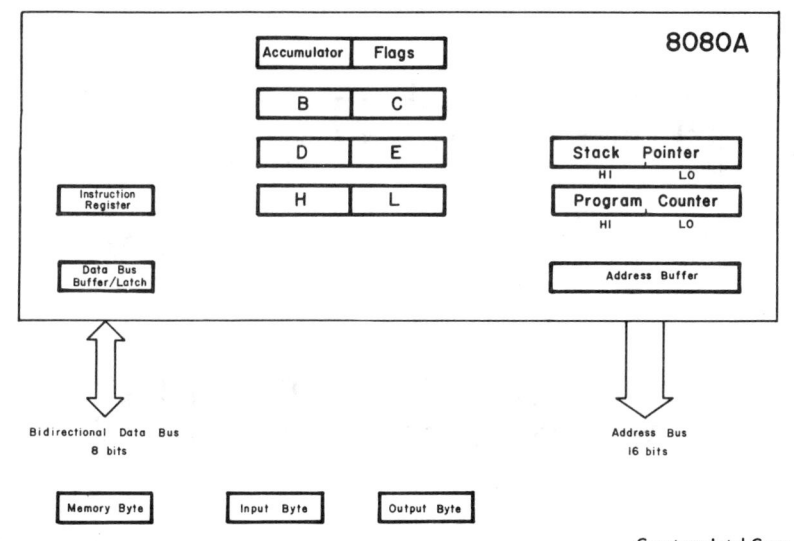

Fig. 18-3. The internal register architecture of an 8080A microprocessor chip.

Temporary registers over which you have no direct control are omitted from Fig. 18-3. The Data Bus Buffer/Latch and the Address Buffer provide the interface between the circuitry within the chip and the external buses. Fig. 18-4 shows the functional block diagram of the 8080A central processing unit (CPU). (Note the internal data bus, which communicates with the external bidirectional data bus through a data bus buffer/latch located within the 8080A chip.) Two other registers over which, in special cases, you will have some control include:

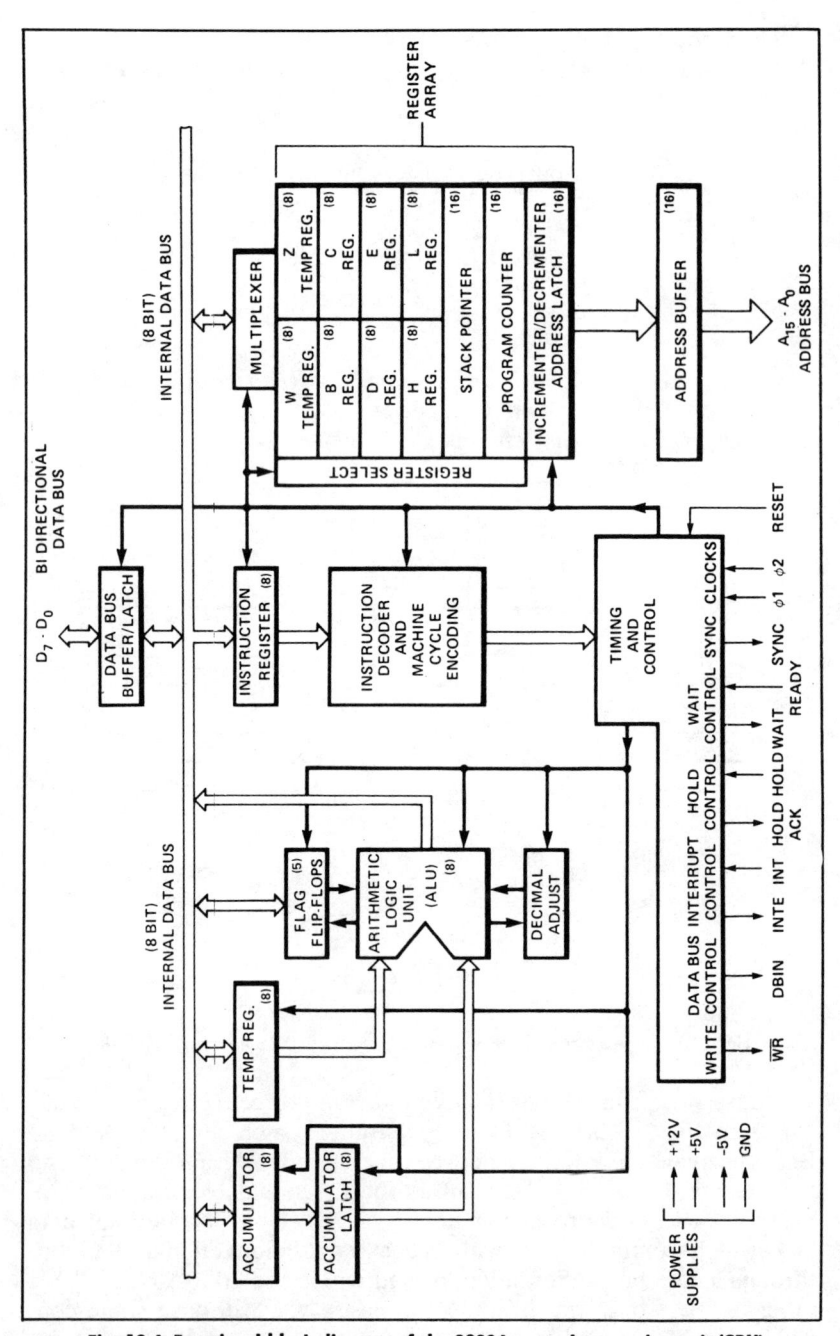

Fig. 18-4. Functional block diagram of the 8080A central processing unit (CPU).

- The 8-bit *instruction register*.
- A 5-bit *flag register* in the arithmetic/logic unit (ALU).

Additional registers that are required to allow the 8080A microprocessor chip to perform its internal operations include two 8-bit temporary registers used singly or as a pair, *W temporary register* and *Z temporary register;* an 8-bit *temporary accumulator* in the arithmetic/logic unit; and an 8-bit *temporary register* in the arithmetic/logic unit. You cannot address or control the contents of these temporary registers from a program and will not know when the 8080 uses them.

Some useful definitions include:

*program counter*—The 16-bit register in the 8080A microprocessor chip that contains the memory address of the next instruction byte that must be executed in a computer program.

*accumulator*—The register and associated digital electronic circuitry in the arithmetic/logic unit (ALU) of a computer in which arithmetic and logical operations are performed.

*general-purpose registers*—In the 8080A microprocessor chip, 8-bit registers that can participate in arithmetic and logical operations with the contents of the accumulator.

*stack pointer*—The 16-bit register in the 8080A microprocessor chip that stores the memory address of the stack, which is a region of memory that stores temporary information.

*instruction register*—The 8-bit register in the 8080A microprocessor chip that stores the instruction code of the instruction being executed.

*instruction code*—A unique 8-bit binary number that encodes an operation that the 8080A microprocessor chip can perform.

*instruction decoder*—A decoder within the 8080A microprocessor chip that decodes the instruction code into a series of actions that the microprocessor performs.

The Intel Corporation *Intellec 8/Mod 80 Microcomputer Development System Reference Manual* provides several well-written paragraphs that summarize the concepts of instruction code, instruction register, and instruction decoder. We quote these paragraphs in the following paragraphs. The illustration of Fig. 18-5 should also help you.

Every computer has a *word length* that is characteristic of that machine. In most eight-bit systems, it is most efficient to deal with eight-bit binary fields, and the memory associated with such a processor is therefore organized to store eight bits in each addressable memory location. Data and instructions are stored in memory as eight-bit binary numbers, or as numbers that are integral multiples of eight

**Fig. 18-5.**

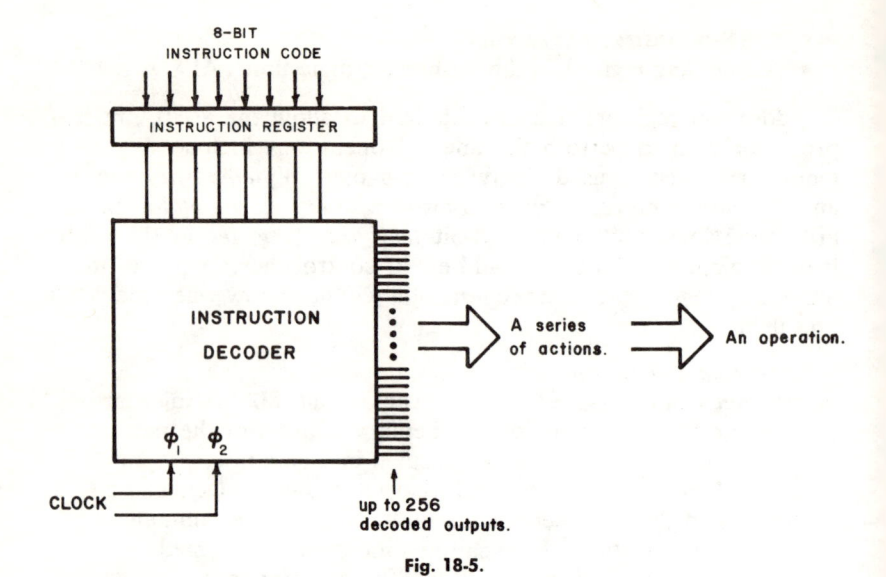

bits: 16 bits, 24 bits, and so on. This characteristic eight-bit field is sometimes referred to as a *byte*.

Each operation that the processor can perform is identified by a unique binary number known as an *instruction code*. An eight-bit word used as an instruction code can distinguish among 256 alternative actions, more than adequate for most processors.

The processor *fetches* an instruction in two distinct operations. In the first, it transmits the address in its program counter to the memory. In the second, the memory returns the addressed byte to the processor. The CPU stores this instruction byte in a register known as the *instruction register,* and uses it to direct activities during the remainder of the instruction cycle.

The mechanism by which the processor translates an instruction code into specific processing actions requires more elaboration than we can afford here. The concept, however, will be intuitively clear to an experienced logic designer. The eight bits stored in the instruction register can be decoded and used to activate selectively one of a number of output lines, in this case, up to 256 lines. Each line represents a set of activities associated with execution of a particular instruction code. The enabled line can be combined coincidentally with selected timing pulses to develop electrically sequential signals that can be used to initiate specific actions. This translation of code into action is performed by the *instruction decoder* and by the associated control circuitry.

The important point here is that the instruction code is translated into a sequence of specific actions. The two-phase clock is vital to this process. The actions may result in the moving of data from memory to the accumulator, or adding the contents of Register B to Register A, or complementing the accumulator, or any of the specific operations contained in the 8080A instruction set. Nevertheless, *each specific operation performed by an 8080A instruction is the result of one or more specific actions caused by the instruction decoder.*

### WHAT TYPES OF OPERATIONS DOES THE 8080A MICROPROCESSOR PERFORM?

The purpose of this section is not to subdivide the 8080A instruction set into categories, but rather to identify the basic types of operations that the chip actually performs.

1. Move a byte from one location to another.
   - From one general-purpose register to another.
   - From a general-purpose register to memory, and vice versa.
   - From the accumulator to memory, and vice versa.
   - From the accumulator to a general-purpose register, and vice versa.
   - From memory to the instruction register.
   - From memory to the program counter, and vice versa.
   - From memory to the stack pointer.
   - From the accumulator to an output latch.
   - From an input device to the accumulator.
   - From an external three-state buffer to the instruction register.
   - From the flag register to memory, and vice versa.
   - From a general-purpose register to the stack pointer.
   - From the program counter to the stack, and vice versa.
   - From the general-purpose registers to the stack, and vice versa.
   - From the accumulator to the stack, and vice versa.
   - From the flag register to the stack, and vice versa.
   - From an input device to a general-purpose register.
   - From a general-purpose register to an output device.
   - From a general-purpose register to the program counter.

2. Arithmetic and logical operations.
   - AND contents of register or memory with accumulator.
   - OR contents of register or memory with accumulator.
   - Exclusive-OR contents of register or memory with accumulator.

- Compare contents of register or memory with accumulator.
- Add contents of register or memory to accumulator (with or without carry).
- Subtract contents of register or memory from accumulator (with or without borrow).
- Rotate contents of accumulator.
- Increment contents of general-purpose register, register pair, accumulator, memory, or stack pointer.
- Decrement contents of general-purpose register, register pair, accumulator, memory, or stack pointer.
- Add contents of register pair to contents of register pair or stack pointer.
- Decimal adjust the contents of the accumulator.

3. Miscellaneous operations.
   - No operation.
   - Halt.
   - Enable the interrupt system.
   - Disable the interrupt system.
   - Complement the accumulator.
   - Set the carry.
   - Complement the carry.

Most of the time, all that the 8080A microprocessor chip does is to move a byte from one location to another or perform an arithmetic or logical operation. Rarely does it perform one of the miscellaneous operations. In other words, the chip does not just compute; it moves bytes around.

## 8080 MNEMONIC INSTRUCTIONS

We encourage you to learn as soon as possible the 8080 mnemonics, so that you can do assembly language programming, read other assembly language programs for the 8080, and improve your capability to understand the instruction sets for other microprocessor chips. The 8080 mnemonics are listed by groups in the *Intel 8080 Microcomputer Systems User's Manual,* which we recommend that you obtain. Here, we will first list the mnemonics in alphabetic order, and then proceed to describe them in detail. We gratefully acknowledge permission to use the following two reference sources for this material.

*Intel 8080 Microcomputer Systems User's Manual* by Intel Corporation.

*The μCOM-8 Software Manual* by NEC Microcomputers, Inc.

|  | Instruction Code |  |  |
|---|---|---|---|
| Mnemonic | Octal | Hexa-decimal | Description |
| ACI <B2> | 316 | CE | Add immediate byte to accumulator (with carry) |
| ADC M | 216 | 8E | Add memory contents to accumulator (with carry) |
| ADC r | 21S | † | Add register contents to accumulator (with carry) |
| ADD M | 206 | 86 | Add memory contents to accumulator |
| ADD r | 20S | † | Add register contents to accumulator |
| ADI <B2> | 306 | C6 | Add immediate byte to accumulator |
| ANA M | 246 | A6 | AND memory contents with accumulator |
| ANA r | 24S | † | AND register contents with accumulator |
| ANI <B2> | 346 | E6 | AND immediate byte with accumulator |
| CALL <B2> <B3> | 315 | CD | Call subroutine unconditionally |
| CC <B2> <B3> | 334 | DC | Call subroutine if carry flag is set |
| CM <B2> <B3> | 374 | FC | Call subroutine if sign flag is set |
| CMA | 057 | 2F | Complement contents of accumulator |
| CMC | 077 | 3F | Complement the carry |
| CMP M | 276 | BE | Compare memory contents with accumulator |
| CMP r | 27S | † | Compare register contents with accumulator |
| CNC <B2> <B3> | 324 | D4 | Call subroutine if carry flag is reset |
| CNZ <B2> <B3> | 304 | C4 | Call subroutine if zero flag is reset |
| CP <B2> <B3> | 364 | F4 | Call subroutine if sign flag is reset |
| CPE <B2> <B3> | 354 | EC | Call subroutine if parity flag is set |
| CPI <B2> | 376 | FE | Compare immediate byte with accumulator |
| CPO <B2> <B3> | 344 | E4 | Call subroutine if parity flag is reset |
| CZ <B2> <B3> | 314 | CC | Call subroutine if zero flag is set |
| DAA | 047 | 27 | Decimal adjust the accumulator contents |
| DAD B | 011 | 09 | Add register pair B to register pair H |
| DAD D | 031 | 19 | Add register pair D to register pair H |
| DAD H | 051 | 29 | Add register pair H to register pair H |
| DAD SP | 071 | 39 | Add register pair H to stack pointer |
| DCR M | 065 | 35 | Decrement memory contents |
| DCR r | 0D5 | † | Decrement register contents |
| DCX B | 013 | 0B | Decrement contents of register pair B |
| DCX D | 033 | 1B | Decrement contents of register pair D |
| DCX H | 053 | 2B | Decrement contents of register pair H |
| DCX SP | 073 | 3B | Decrement stack pointer |
| DI | 363 | F3 | Disable interrupt system |
| EI | 373 | FB | Enable interrupt system |
| HLT | 166 | 76 | Halt |
| IN <B2> | 333 | DB | Input data into accumulator |
| INR M | 064 | 34 | Increment memory contents |
| INR r | 0D4 | † | Increment register contents |
| INX B | 003 | 03 | Increment contents of register pair B |
| INX D | 023 | 13 | Increment contents of register pair D |

| Mnemonic | Instruction Code Octal | Hexa-decimal | Description |
|---|---|---|---|
| INX H | 043 | 23 | Increment contents of register pair H |
| INX SP | 063 | 33 | Increment stack pointer |
| JC <B2> <B3> | 332 | DA | Jump if carry flag is set |
| JM <B2> <B3> | 372 | FA | Jump if sign flag is set |
| JMP <B2> <B3> | 303 | C3 | Jump unconditionally |
| JNC <B2> <B3> | 322 | D2 | Jump if carry flag is reset |
| JNZ <B2> <B3> | 302 | C2 | Jump if zero flag is reset |
| JP <B2> <B3> | 362 | F2 | Jump if sign flag is reset |
| JPE <B2> <B3> | 352 | EA | Jump if parity flag is set |
| JPO <B2> <B3> | 342 | E2 | Jump if parity flag is reset |
| JZ <B2> <B3> | 312 | CA | Jump if zero flag is set |
| LDA <B2> <B3> | 072 | 3A | Load accumulator with contents of memory addressed by <B2> <B3> |
| LDAX B | 012 | 0A | Load accumulator with contents of memory addressed by register pair B |
| LDAX D | 032 | 1A | Load accumulator with contents of memory addressed by register pair D |
| LHLD <B2> <B3> | 052 | 2A | Load L and H with contents of M and M+1, respectively |
| LXI B <B2> <B3> | 001 | 01 | Load immediate bytes into register pair B |
| LXI D <B2> <B3> | 021 | 11 | Load immediate bytes into register pair D |
| LXI H <B2> <B3> | 041 | 21 | Load immediate bytes into register pair H |
| LXI SP <B2> <B3> | 061 | 31 | Load immediate bytes into stack pointer |
| MVI M <B2> | 066 | 36 | Move immediate byte into memory |
| MVI r <B2> | 0D6 | † | Move immediate byte into register |
| MOV M,r | 16S | † | Move register contents to memory |
| MOV r,M | 1D6 | † | Move memory contents to register |
| MOV rl,r2 | 1DS | † | Move register 2 contents to register 1 |
| NOP | 000 | 00 | No operation |
| ORA M | 266 | B6 | OR memory contents with accumulator |
| ORA r | 26S | † | OR register contents with accumulator |
| ORI <B2> | 366 | F6 | OR immediate byte with accumulator |
| OUT <B2> | 323 | D3 | Output accumulator contents |
| PCHL | 351 | E9 | Load program counter with contents of register pair H |
| POP B | 301 | C1 | Pop register pair B off stack |
| POP D | 321 | D1 | Pop register pair D off stack |
| POP H | 341 | E1 | Pop register pair H off stack |
| POP PSW | 361 | F1 | Pop program status word (accumulator and flags) off stack |
| PUSH B | 305 | C5 | Push register pair B on stack |
| PUSH D | 325 | D5 | Push register pair D on stack |
| PUSH H | 345 | E5 | Push register pair H on stack |
| PUSH PSW | 365 | F5 | Push program status word (accumulator and flags) on stack |
| RAL | 027 | 17 | Rotate accumulator contents left through carry |
| RAR | 037 | 1F | Rotate accumulator contents right through carry |

|  | Instruction Code | | |
|---|---|---|---|
| Mnemonic | Octal | Hexa-decimal | Description |
| RC | 330 | D8 | Return if carry flag is set |
| RET | 311 | C9 | Return unconditionally |
| RLC | 007 | 07 | Rotate accumulator contents left |
| RM | 370 | F8 | Return if sign flag is set |
| RNC | 320 | D0 | Return if carry flag is reset |
| RNZ | 300 | C0 | Return if zero flag is reset |
| RP | 360 | F0 | Return if sign flag is reset |
| RPE | 350 | E8 | Return if parity flag is set |
| RPO | 340 | E0 | Return if parity flag is reset |
| RRC | 017 | 0F | Rotate accumulator contents right |
| RST n | 3n7 | † | Call subroutine at location HI = 000 and LO = 0n0 |
| RZ | 310 | C8 | Return if zero flag is set |
| SBB M | 236 | 9E | Subtract memory contents from accumulator (with borrow) |
| SBB r | 23S | † | Subtract register contents from accumulator (with borrow) |
| SBI <B2> | 336 | DE | Subtract immediate byte from accumulator (with borrow) |
| SHLD <B2> <B3> | 042 | 22 | Store contents of register pair H into M and M+1, respectively, where M = <B2> <B3> |
| SPHL | 371 | F9 | Move register pair H contents to stack pointer |
| STA <B2> <B3> | 062 | 32 | Store accumulator contents into memory location address by <B2> <B3> |
| STAX B | 002 | 02 | Store accumulator contents into memory location addressed by register pair B |
| STAX D | 022 | 12 | Store accumulator contents into memory location addressed by register pair D |
| STC | 067 | 37 | Set carry flag |
| SUB M | 226 | 96 | Subtract memory contents from accumulator |
| SUB r | 22S | † | Subtract register contents from accumulator |
| SUI <B2> | 326 | D6 | Subtract immediate byte from accumulator |
| XCHG | 353 | EB | Exchange contents of register pair D with contents of register pair H |
| XRA M | 256 | AE | Exclusive-OR memory contents with accumulator |
| XRA r | 25S | † | Exclusive-OR register contents with accumulator |
| XRI <B2> | 356 | EE | Exclusive-OR immediate byte with accumulator |
| XTHL | 343 | E3 | Exchange top of stack with contents of register pair H |

NOTE: Instructions marked with a † are not easily translated into hexadecimal notation without register or other information. This is one reason why we have chosen to work with octal numbers, D = destination register code, S = source register code. Register codes are: A = 7, B = 0, C = 1, D = 2, E = 3, H = 4 and L = 5.

Not all possible 256 instruction codes are employed by the 8080A microprocessor chip. Missing codes include the following:

| Octal | Hexadecimal |
|-------|-------------|
| 010 | 08 |
| 020 | 10 |
| 030 | 18 |
| 040 | 20 |
| 050 | 28 |
| 060 | 30 |
| 070 | 38 |
| 313 | CB |
| 331 | D9 |
| 335 | DD |
| 355 | ED |
| 375 | FD |

## THE INSTRUCTION SET

We now shall proceed to describe the 8080 instruction set in detail. We shall use material from both the *Intel 8080 Microcomputer Systems User's Manual* and *The μCOM-8 Software Manual,* courtesy of the Intel Corporation and NEC Microcomputers, Inc., respectively. To help you clearly understand the significance of the terms, symbols, and abbreviations used in the description of each instruction, several pages from the Intel manual are included. We shall group the 8080 instruction set as Intel does:

>Data Transfer Group
>Arithmetic Group
>Logical Group
>Branch Group
>Stack, I/O, and Machine Control Group

A computer, no matter how sophisticated, can only do what it is "told" to do. One "tells" the computer what to do via a series of coded instructions referred to as a *Program*. The realm of the programmer is referred to as *Software,* in contrast to the *Hardware* that comprises the actual computer equipment. A computer's software refers to all of the programs that have been written for that computer.

When a computer is designed, the engineers provide the Central Processing Unit (CPU) with the ability to perform a particular set of operations. The CPU is designed so that a specific operation is performed when the CPU control logic decodes a particular instruction. Consequently, the operations that can be performed by a CPU define the computer's *Instruction Set*.

Each computer instruction allows the programmer to initiate the performance of a specific operation. All computers implement certain arithmetic operations in their instruction set, such as an instruction to add the contents of two registers. Often logical operations (e.g., OR the contents of two registers) and register operate instructions (e.g., increment a register) are included in the instruction set. A computer's instruction set will also have instructions that move data between registers, between a register and memory, and between a register and an I/O device. Most instruction sets also provide *Conditional Instructions*. A conditional instruction specifies an operation to be performed only if certain conditions have been met; for example, jump to a particular instruction if the result of the last operation was zero. Conditional instructions provide a program with a decision-making capability.

By logically organizing a sequence of instructions into a coherent program, the programmer can "tell" the computer to perform a very specific and useful function.

The computer, however, can only execute programs whose instructions are in a binary coded form (i.e., a series of 1s and 0s), that is called *Machine Code*. Because it would be extremely cumbersome to program in machine code, programming languages have been developed. There are programs available which convert the programming language instructions into machine code that can be interpreted by the processor.

One type of programming language is *Assembly Language*. A unique assembly language mnemonic is assigned to each of the computer's instructions. The programmer can write a program (called the *Source Program*) using these mnemonics and certain operands; the source program is then converted into machine instructions (called the *Object Code*). Each assembly language instruction is converted into one machine code instruction (1 or more bytes) by an *Assembler* program. Assembly languages are usually machine dependent (i.e., they are usually able to run on only one type of computer).

## THE 8080 INSTRUCTION SET

The 8080 instruction set includes five different types of instructions:

- Data Transfer Group—Move data between registers or between memory and registers.
- Arithmetic Group—Add, subtract, increment or decrement data in registers or in memory.
- Logical Group—AND, OR, EXCLUSIVE-OR, compare, rotate or complement data in registers or in memory.
- Branch Group—Conditional and unconditional jump instructions, subroutine call instructions and return instructions.
- Stack, I/O and Machine Control Group—Includes I/O instruc-

tions, as well as instructions for maintaining the stack and internal control flags.

### Instruction and Data Formats

Memory for the 8080 is organized into 8-bit quantities, called bytes. Each byte has a unique 16-bit binary address corresponding to its sequential position in memory. The 8080 can directly address up to 65,536 bytes of memory, which may consist of both read-only memory (ROM) elements and random-access memory (RAM) elements (read/write memory).

Data in the 8080 is stored in the form of 8-bit binary integers:

```
              DATA WORD
     | D7 | D6 | D5 | D4 | D3 | D2 | D1 | D0 |
      MSB                              LSB
```

When a register or data word contains a binary number, it is necessary to establish the order in which the bits of the number are written. In the Intel 8080, BIT 0 is referred to as the *Least Significant Bit* (*LSB*), and BIT 7 (of an 8-bit number) is referred to as the *Most Significant Bit* (*MSB*).

The 8080 program instructions may be one, two or three bytes in length. Multiple byte instructions must be stored in successive memory locations; the address of the first byte is always used as the address of the instructions. The exact instruction format will depend on the particular operation to be executed.

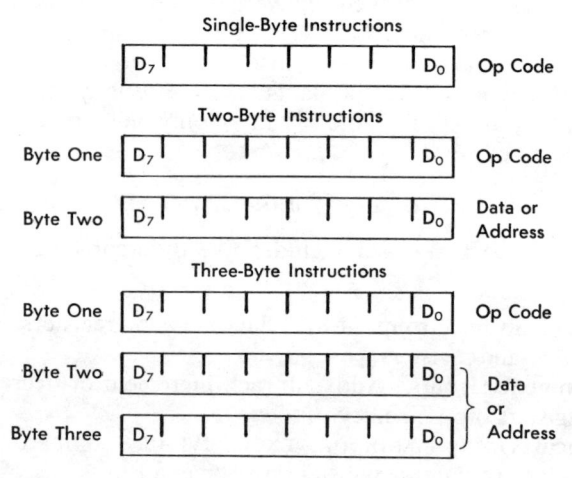

```
              Single-Byte Instructions
              | D7 |   |   |   |   |   | D0 |   Op Code

              Two-Byte Instructions
  Byte One    | D7 |   |   |   |   |   | D0 |   Op Code

  Byte Two    | D7 |   |   |   |   |   | D0 |   Data or
                                                Address

              Three-Byte Instructions
  Byte One    | D7 |   |   |   |   |   | D0 |   Op Code

  Byte Two    | D7 |   |   |   |   |   | D0 | ⎫
                                              ⎬ Data
  Byte Three  | D7 |   |   |   |   |   | D0 | ⎭ or
                                                Address
```

### Addressing Modes

Often the data that is to be operated on is stored in memory. When multibyte numeric data is used, the data, like instructions, is stored in successive memory locations, with the least significant byte first, followed by increasingly significant bytes. The 8080 has four different modes for addressing data stored in memory or in registers:

*Direct*—Bytes 2 and 3 of the instruction contain the exact memory address of the data item (the low-order bits of the address are in byte 2, the high-order bits in byte 3).

*Register*—The instruction specifies the register or register-pair in which the data is located.

*Register Indirect*—The instruction specifies a register-pair which contains the memory address where the data is located (the high-order bits of the address are in the first register of the pair, the low-order bits in the second).

*Immediate*—The instruction contains the data itself. This is either an 8-bit quantity or a 16-bit quantity (least significant byte first, most significant byte second).

Unless directed by an interrupt or branch instruction, the execution of instructions proceeds through consecutively increasing memory locations. A branch instruction can specify the address of the next instruction to be executed in one of two ways:

*Direct*—The branch instruction contains the address of the next instruction to be executed. (Except for the "RST" instruction, byte 2 contains the low-order address and byte 3 the high-order address.)

*Register Indirect*—The branch instruction indicates a register-pair which contains the address of the next instruction to be executed. (The high-order bits of the address are in the first register of the pair, the low-order bits in the second.)

The RST instruction is a special one-byte call instruction (usually used during interrupt sequences). RST includes a three-bit field; program control is transferred to the instruction whose address is eight times the contents of this three-bit field.

## Condition Flags

There are five condition flags associated with the execution of instructions on the 8080. They are Zero, Sign, Parity, Carry, and Auxiliary Carry and are each represented by a 1-bit register in the CPU. A flag is "set" by forcing the bit to 1; "reset" by forcing the bit to 0.

Unless indicated otherwise, when an instruction affects a flag, it affects it in the following manner:

*Zero*—If the result of an instruction has the value 0, this flag is set; otherwise it is reset.

*Sign*—If the most significant bit of the result of the operation has the value 1, this flag is set; otherwise it is reset.

*Parity*—If the modulo 2 sum of the bits of the result of the operation is 0 (i.e., if the result has even parity), this flag is set; otherwise it is reset (i.e., if the result has odd parity).

*Carry*—If the instruction resulted in a carry (from addition), or a borrow (from subtraction or a comparison) out of the high-order bit, this flag is set; otherwise it is reset.

*Auxiliary Carry*—If the instruction caused a carry out of bit 3 and into bit 4 of the resulting value, the auxiliary carry is set; otherwise it is reset. This flag is affected by single precision additions, subtractions, increments, decrements, comparisons, and logical operations, but is only used with additions preceding a DAA (Decimal Adjust Accumulator) instruction.

## Symbols and Abbreviations

The following symbols and abbreviations are used in the subsequent description of the 8080 instructions:

| Symbols | Meaning |
|---|---|
| accumulator | Register A |
| addr | 16-bit address quantity |
| data | 8-bit data quantity |
| data 16 | 16-bit data quantity |
| byte 2 | The second byte of the instruction |
| byte 3 | The third byte of the instruction |
| port | 8-bit address of an I/O device |
| r, r1, r2 | One of the registers A,B,C,D,E,H,L |
| DDD, SSS | The bit pattern designating one of the registers A,B,C,D,E,H,L (DDD=destination, SSS=source): |

| DDD or SSS | REGISTER NAME |
|---|---|
| 111 | A |
| 000 | B |
| 001 | C |
| 010 | D |
| 011 | E |
| 100 | H |
| 101 | L |

| | |
|---|---|
| rp | One of the register pairs: |
| | B represents the B,C pair with B as the high-order register and C as the low-order register; |
| | D represents the D,E pair with D as the high-order register and E as the low-order register; |
| | H represents the H,L pair with H as the high-order register and L as the low-order register; |
| | SP represents the 16-bit stack pointer register. |

| Symbols | Meaning |
|---|---|
| RP | The bit pattern designating one of the register pairs B,D,H,SP: |

| RP | REGISTER PAIR |
|---|---|
| 00 | B-C |
| 01 | D-E |
| 10 | H-L |
| 11 | SP |

| Symbols | Meaning |
|---|---|
| rh | The first (high-order) register of a designated register pair. |
| rl | The second (low-order) register of a designated register pair. |
| PC | 16-bit program counter register (PCH and PCL are used to refer to the high-order and lower-order 8 bits, respectively). |
| SP | 16-bit stack pointer register (SPH and SPL are used to refer to the high-order and low-order 8 bits, respectively). |
| $r_m$ | Bit m of the register r (bits are numbered 7 through 0 from left to right). |
| Z,S,P,CY,AC | The condition flags: <br> Zero, <br> Sign, <br> Parity, <br> Carry, <br> and Auxiliary Carry, respectively. |
| ( ) | The contents of the memory location or registers enclosed in the parentheses. |
| ← | "Is transferred to" |
| ∧ | Logical AND |
| ∀ | Exclusive-OR |
| ∨ | Inclusive-OR |
| + | Addition |
| − | Two's complement subtraction |
| * | Multiplication |
| ↔ | "Is exchanged with" |
| ‾ | The one's complement (e.g., $\overline{A}$). |
| n | The restart number 0 through 7 |
| NNN | The binary representation 000 through 111 for restart number 0 through 7, respectively. |

## Description Format

The following pages provide a detailed description of the instruction set of the 8080. Each instruction is described in the following manner:

1. The MAC 80 assembler format, consisting of the instruction mnemonic and operand fields, is printed in boldface on the left side of the first line.
2. The name of the instruction is enclosed in parentheses on the right side of the first line.
3. The next line(s) contain a symbolic description of the operation of the instruction.
4. This is followed by a narrative description of the operation of the instruction.
5. The following line(s) contain the binary fields and patterns that comprise the machine instruction.

6. The last four lines contain incidental information about the execution of the instruction. The number of machine cycles and states required to execute the instruction are listed first. If the instruction has two possible execution times, as in a Conditional Jump, both times will be listed, separated by a slash. Next, any significant data addressing modes are listed. The last line lists any of the five Flags that are affected by the execution of the instruction.

### DATA TRANSFER GROUP

This group of instructions transfers data to and from registers and memory. *Condition flags are not affected* by any instruction in this group.

### MOV r1, r2

The MOV r1, r2 instruction transfers data from the specified source register S (or r2) to the specified destination register D (or r1). The source or destination may be any of the single registers B, C, D, E, H,

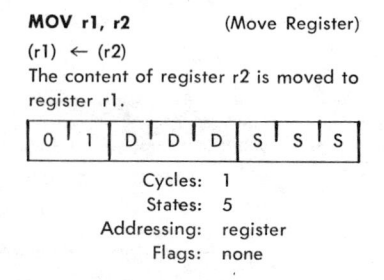

**MOV r1, r2**   (Move Register)
(r1) ← (r2)
The content of register r2 is moved to register r1.

| 0 | 1 | D | D | D | S | S | S |
|---|---|---|---|---|---|---|---|

Cycles: 1
States: 5
Addressing: register
Flags: none

or L, the accumulator A, and M (the contents of the memory address specified by the register pair H, L). In the three-octal-digit byte, the first digit is always a 1. The second and third octal digits vary depending upon the source and destination. The octal instruction, 166, is a Halt rather than a MOV M,m instruction. The contents of the source register are not changed during a MOV instruction, you are duplicating the register contents somewhere else.

### MOV r, M

The MOV r, M instruction transfers data from M (the contents of the memory address specified by the register pair H, L) to the specified destination register D, which may be any of the single registers B, C, D, E, H, or L or the accumulator A. You duplicate the contents of the memory address into a register; the contents of memory remain unchanged.

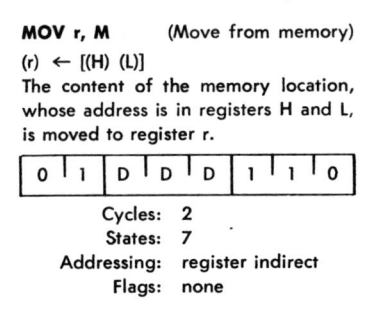

**MOV r, M**  (Move from memory)
(r) ← [(H) (L)]
The content of the memory location, whose address is in registers H and L, is moved to register r.

| 0 | 1 | D | D | D | 1 | 1 | 0 |
|---|---|---|---|---|---|---|---|

Cycles: 2
States: 7
Addressing: register indirect
Flags: none

## MOV M, r

The MOV M, r instruction transfers data from the specified source register S to M (the memory address specified by the register pair H, L). The source register may be any of the single registers B, C, D, E, H, or L or the accumulator A. The register contents are duplicated in memory; the contents of the register remain unchanged.

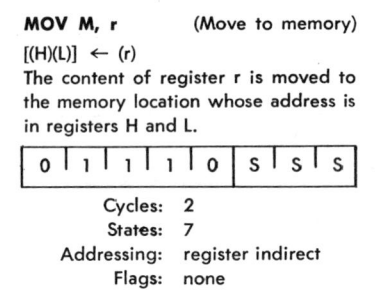

**MOV M, r**  (Move to memory)
[(H)(L)] ← (r)
The content of register r is moved to the memory location whose address is in registers H and L.

| 0 | 1 | 1 | 1 | 1 | 0 | S | S | S |
|---|---|---|---|---|---|---|---|---|

Cycles: 2
States: 7
Addressing: register indirect
Flags: none

## MVI r, data

The MVI r, data instruction transfers data from the second byte of the two-byte instruction to the specified destination register D (or r). The term *immediate* refers to the fact that the data byte is contained within the multibyte instruction. The specified destination register may be any of the single registers B, C, D, E, H, or L, the accumulator A, and M (the contents of the memory address specified by the register pair H, L). When the destination is M, you have the instruction MVI M, data, which is discussed below. The data can be any 8-bit binary number between 00000000 and 111111111.

**MVI r, data**         (Move Immediate)

(r) ← (byte 2)

The content of byte 2 of the instruction is moved to register r.

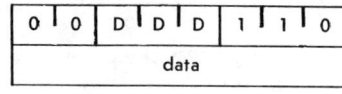

| 0 | 0 | D | D | D | 1 | 1 | 0 |
|---|---|---|---|---|---|---|---|
| data ||||||||

Cycles: 2
States: 7
Addressing: immediate
Flags: none

## MVI M, data

The MVI M, data instruction transfers data from the second byte of the instruction to M (the memory address specified by the register pair H, L). The data can be any 8-bit binary number between 00000000 and 11111111.

**MVI M, data**     (Move to memory immediate)

[(H)(L)] ← (byte 2)

The content of byte 2 of the instruction is moved to the memory location whose address is in registers H and L.

| 0 | 0 | 1 | 1 | 0 | 1 | 1 | 0 |
|---|---|---|---|---|---|---|---|
| data ||||||||

Cycles: 3
States: 10
Addressing: immediate/register indirect
Flags: none

## LXI rp, data 16

The LXI rp, data instruction causes a 16-bit data quantity contained in the second and third bytes of the instruction to be loaded into the register pair specified by RP. RP can be any of the double registers HL, DE, or BC or the stack pointer, which are represented by the mnemonics H, D, B, and SP, respectively. The second instruction byte is loaded into the LO registers L, E, C, or the LO eight bits of the stack pointer; the third instruction byte is loaded into the HI registers H, D, B, or the HI eight bits of the stack pointer. The 16-bit data word can vary from 0000000000000000 to 1111111111111111, in binary notation.

**LXI rp, data 16**  (Load register pair immediate)

(rh) ← (byte 3),
(rl) ← (byte 2)

Byte 3 of the instruction is moved into the high-order register (rh) of the register pair rp. Byte 2 of the instruction is moved into the low-order register (rl) of the register pair rp.

| 0 | 0 | R | P | 0 | 0 | 0 | 1 |
|---|---|---|---|---|---|---|---|
| low-order data ||||||||
| high-order data ||||||||

Cycles: 3
States: 10
Addressing: immediate
Flags: none

### Instruction Characteristics

The diagrams shown in Figs. 18-6 and 18-7 illustrate some of the characteristics of the MOV, MVI, and LXI instructions. Only

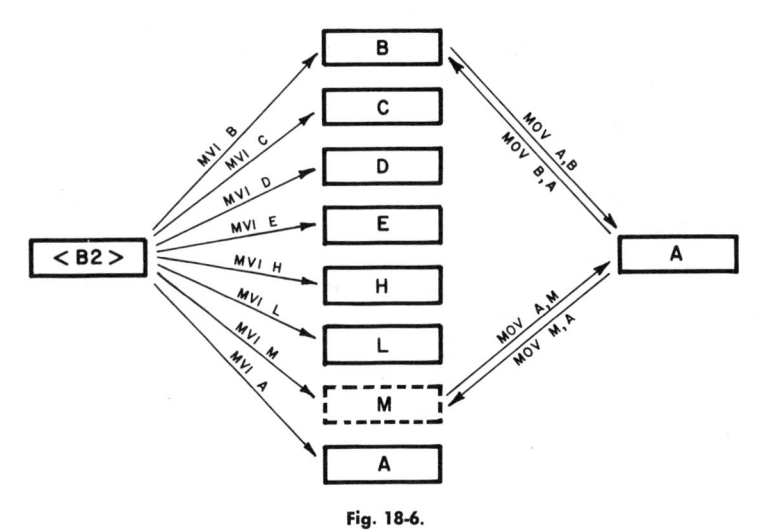

**Fig. 18-6.**

two sets of MOV rl, r2 instructions are shown. Note that LXI rp, data is equivalent to two MVI r, data instructions. Thus,

LXI B
<B2>
<B3>

**Fig. 18-7.**

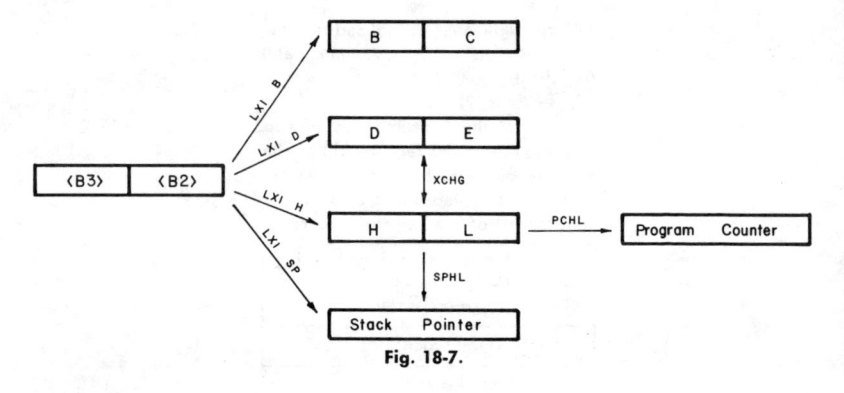

is equivalent to

```
MVI B
<B2>    (corresponds to <B3> in the LXI B instruction)
MVI C
<B2>    (corresponds to <B2> in the LXI B instruction)
```

The second byte in a two-byte instruction is always referred to as <B2>. A single LXI rp, data instruction requires 10 states for its execution, whereas two MVI r, data instructions require a total of 14 states execution time. Thus, by using the LXI rp, data instruction, you save 4 states of execution time. In many cases, such a saving is unimportant.

## STA addr

The STA addr instruction permits you to store the contents of the accumulator directly into a memory location without the use of the register pair H, L. The address of the memory location is specified in the second and third bytes of the instruction. The LO address byte is byte 2 and the HI address byte is byte 3.

**STA addr** (Store Accumulator direct)
[(byte 3)(byte 2)] ← (A)
The content of the accumulator is moved to the memory location whose address is specified in byte 2 and byte 3 of the instruction.

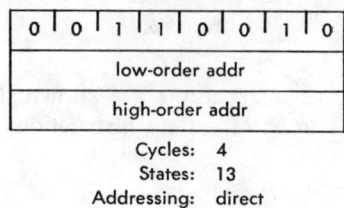

| 0 | 0 | 1 | 1 | 0 | 0 | 1 | 0 |
|---|---|---|---|---|---|---|---|
| low-order addr ||||||||
| high-order addr ||||||||

Cycles: 4
States: 13
Addressing: direct
Flags: none

## LDA addr

The LDA addr instruction permits you to load the accumulator with the contents of the memory location specified by bytes <B2> and <B3> in the instruction. You need not use the H, L register pair. The LO address byte is <B2> and the HI address byte is <B3>.

**LDA addr**  (Load Accumulator direct)
(A) ← [(byte 3)(byte 2)]
The content of the memory location, whose address is specified in byte 2 and byte 3 of the instruction, is moved to register A.

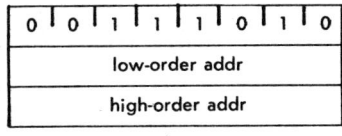

| 0 | 0 | 1 | 1 | 1 | 1 | 0 | 1 | 0 |
|---|---|---|---|---|---|---|---|---|
| low-order addr ||||||||||
| high-order addr |||||||||

Cycles: 4
States: 13
Addressing: direct
Flags: none

## LHLD addr

This instruction is useful when memory locations contain address information. Thus, LHLD addr causes the L register to be loaded with the memory byte addressed by bytes <B2> and <B3> in the

**LHLD addr**  (Load H and L direct)
(L) ← [(byte 3)(byte 2)]
(H) ← [(byte 3)(byte 2) + 1]
The content of the memory location, whose address is specified in byte 2 and byte 3 of the instruction, is moved to register L. The content of the memory location at the succeeding address is moved to register H.

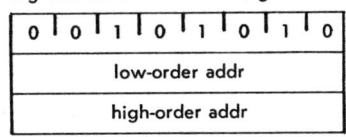

| 0 | 0 | 1 | 0 | 1 | 0 | 1 | 0 |
|---|---|---|---|---|---|---|---|
| low-order addr ||||||||
| high-order addr ||||||||

Cycles: 5
States: 16
Addressing: direct
Flags: none

instruction, i.e., addr. The H register is loaded with the memory byte located at addr + 1. Thus, you perform a 16-bit transfer from memory to the register pair H, L. Once you learn XCHG, you will observe that the section of code,

```
LHLD
<B2>
<B3>
XCHG
```

is functionally equivalent to

```
LXI H
<B2>
<B3>
MOV E,M
INX H
MOV D,M
```

The first section of code requires 20 states for execution; the second section of code requires 29 states.

## XCHG

The XCHG instruction causes the contents of the register pairs D, E and H, L to be exchanged. To be specific, the contents of registers D and H are exchanged, and the contents of registers E and L are exchanged. This instruction permits you to use register pair H, L as a memory address while another address is held in register pair D, E. You can modify the contents of register pair D, E without changing register pair H, L. For example, register pair H, L may specify a memory location that you use to modify register pair D, E. Two XCHG instructions in sequence,

```
XCHG
XCHG
```

are equivalent to a "no operation."

**XCHG**  (Exchange H and L with D and E)

$(H) \leftrightarrow (D)$
$(L) \leftrightarrow (E)$

The contents of registers H and L are exchanged with the contents of registers D and E.

| 1 | 1 | 1 | 0 | 1 | 0 | 1 | 1 |
|---|---|---|---|---|---|---|---|

Cycles: 1
States: 4
Addressing: register
Flags: none

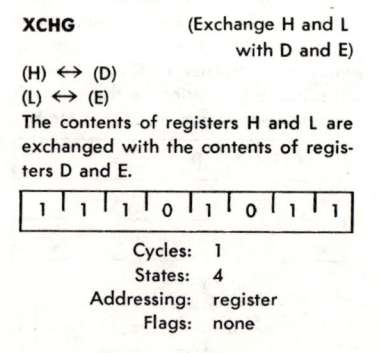

## SHLD addr

The SHLD addr instruction causes the contents of the L register to be stored at the memory location given by bytes <B2> and <B3>

in the instruction, i.e., addr. The contents of the H register are stored in the memory location, addr + 1. In other words, you perform a 16-bit transfer of register pair H, L to two successive memory locations, addr and addr + 1. This instruction is useful in creating a group of memory locations that contain address information rather than data. As with most 8080A instruction, byte <B2> is the LO address byte and byte <B3> is the HI address byte of addr.

The section of code,

```
XCHG
SHLD
<B2>
<B3>
```

is equivalent to the section of code,

```
LXI H
<B2>
<B3>
MOV M,E
INX H
MOV M,D
```

**SHLD addr**   (Store H and L direct)
[(byte 3)(byte 2)] ← (L)
[(byte 3)(byte 2) + 1] ← (H)
The content of register L is moved to the memory location whose address is specified in byte 2 and byte 3. The content of register H is moved to the succeeding memory location.

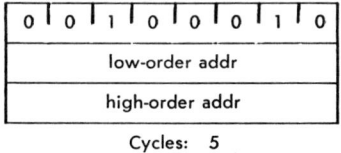

| 0 | 0 | 1 | 0 | 0 | 0 | 1 | 0 |
|---|---|---|---|---|---|---|---|
| low-order addr |||||||||
| high-order addr |||||||||

Cycles: 5
States: 16
Addressing: direct
Flags: none

## LDAX rp

The LDAX rp instruction permits you to load the accumulator with the contents of the memory location addressed by a register pair other than register pair H, L. Thus, with LDAX B, you use register pair B, C to supply the 16-bit memory address; with LDAX D, you use register pair D, E to supply the address. The section of code,

```
LXI D
<B2>
<B3>
LDAX D
```

will load the accumulator identical to,

```
LXI H
<B2>
<B3>
MOV A,M
```

**LDAX rp** (Load accumulator indirect)
(A) ← (rp)
The content of the memory location, whose address is in the register pair rp, is moved to register A. Note: only register pairs rp=B (registers B and C) or rp=D (registers D and E) may be specified.

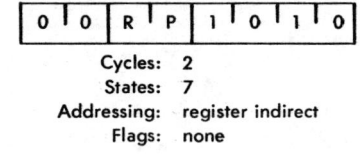

| 0 | 0 | R | P | 1 | 0 | 1 | 0 |
|---|---|---|---|---|---|---|---|

Cycles: 2
States: 7
Addressing: register indirect
Flags: none

## STAX rp

The STAX rp instruction permits you to store the contents of the accumulator in the memory location addressed by either register pair B, D or register pair D, E. The section of code,

```
LXI B
<B2>
<B3>
STAX B
```

will store the accumulator contents similar to,

```
LXI H
<B2>
<B3>
MOV M,A
```

The significance of the STAX rp and LDAX rp instructions is that you can have three independent 16-bit memory addresses stored in the general-purpose registers inside the 8080A microprocessor chip. Enough instructions are available to permit you to use all three addresses.

**STAX rp** (Store accumulator indirect)
(rp) ← (A)

The content of register A is moved to the memory location whose address is in the register pair rp. Note: only register pairs rp=B (registers B and C) or rp=D (registers D and E) may be specified.

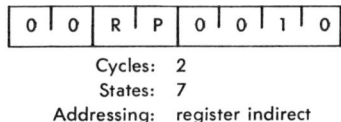

| 0 | 0 | R | P | 0 | 0 | 1 | 0 |
|---|---|---|---|---|---|---|---|

Cycles: 2
States: 7
Addressing: register indirect
Flags: none

## Flags

The condition flags are not affected by any of the instructions in the following list:

```
MOV, r1,r2
MOV r,M
MOV M,r
MVI r, data
MVI M, data
LXI rp, data 16
STA addr
LDA addr
XCHG
LHLD addr
SHLD addr
LDAX rp
STAX rp
```

These instructions comprise the data transfer group in the 8080A microprocessor.

## ARITHMETIC GROUP

This group of instructions performs arithmetic operations on data in registers and memory. *Unless indicated otherwise, all instructions in this group affect the Zero, Sign, Parity, Carry, and Auxiliary Carry flags according to standard rules.* All subtraction operations are performed via two's complement arithmetic and set the carry flag to one to indicate a borrow and clear it to indicate no borrow.

## ADD r

The ADD r instruction causes the contents of the source register S to be added to the contents of the accumulator. The source register can be any of the general-purpose registers B, C, D, E, H, L, the accumulator A, or M (the contents of memory as addressed by register pair H, L). The ADD M instruction is described below. The instruction affects all four of the testable flag bits: Carry, Parity, Zero, and Sign. The Auxiliary Carry flag is also affected.

**Add r** (Add Register)
(A) ← (A) + (r)
The content of register r is added to the content of the accumlator. The result is placed in the accumulator.

| 1 | 0 | 0 | 0 | 0 | S | S | S |

Cycles: 1
States: 4
Addressing: register
Flags: Z, S, P, CY, AC

## ADD M

The ADD M instruction causes the contents of the memory location M, which is addressed by register pair H, L, to be added to the contents of the accumulator. The memory contents remain unchanged after the addition. The instruction affects all five flags.

**ADD M** (Add Memory)
(A) ← (A) + [(H) (L)]
The content of the memory location whose address is contained in the H and L registers is added to the content of the accumulator. The result is placed in the accumulator.

| 1 | 1 | 1 | 0 | 0 | 0 | 1 | 1 | 1 | 0 |

Cycles: 2
States: 7
Addressing: register indirect
Flags: Z, S, P, CY, AC

## ADI data

The ADI data instruction causes the data present in the second byte of the instruction to be added to the contents of the accumulator. The instruction affects all five flags.

**ADI data** (Add immediate)
(A) ← (A) + (byte 2)
The content of the second byte of the instruction is added to the content of the accumulator. The result is placed in the accumulator.

| 1 | 1 | 1 | 0 | 0 | 0 | 1 | 1 | 1 | 0 |
| data |

Cycles: 2
States: 7
Addressing: immediate
Flags: Z, S, P, CY, AC

## ADC r and ADC M

To quote the *μCOM-8 Software Manual*: "In order to perform add and subtract operations, some special arithmetic instructions are required. Multiple digit arithmetic requires that two items be monitored and saved somewhere. These two items are the sum of the digits as they are added, and the presence or absence of a carry bit. When a carry bit is produced, it must be added to the sum of the next digits. Similarly, with subtract operations, the existence of a borrow must be detected so it can be deducted from the difference of the next digits. The Add with Carry and Subtract with Borrow instructions provide simple monitoring and saving of carry bits, making multi-digit addition and subtraction quite straightforward. ADC r, ADC M, and ACI data are the Add with Carry instructions. ADC r causes the contents of the source S to be added to the sum of the accumulator contents and the carry bit."

The ADC r and ADC M instructions are similar to the ADD r and ADD M instructions; the only difference is that the carry bit is added to the least-significant bit in the 8-bit accumulator byte. All flags are affected by these instructions. Memory location M is addressed by the contents of register pair H, L.

**ADC r** (Add Register with carry)
(A) ← (A) + (r) + (CY)
The content of register r and the content of the carry bit are added to the content of the accumulator. The result is placed in the accumulator.

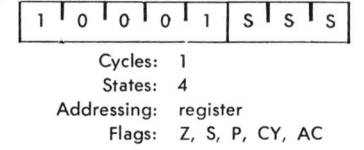

| 1 | 0 | 0 | 0 | 1 | S | S | S |
|---|---|---|---|---|---|---|---|

Cycles: 1
States: 4
Addressing: register
Flags: Z, S, P, CY, AC

**ADC M** (Add memory with carry)
(A) ← (A) + [(H) (L)] + (CY)
The content of the memory location whose address is contained in the H and L registers and the content of the CY flag are added to the accumulator. The result is placed in the accumulator.

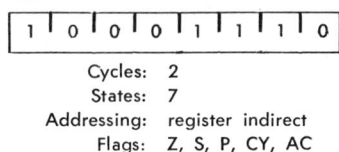

| 1 | 0 | 0 | 0 | 1 | 1 | 1 | 0 |
|---|---|---|---|---|---|---|---|

Cycles: 2
States: 7
Addressing: register indirect
Flags: Z, S, P, CY, AC

## ACI data

The ACI data instruction causes the 8-bit data quantity present in the second byte of the instruction to be added to the sum of the accumulator contents and the carry bit. The instruction affects all five flags.

**ACI data** (Add immediate with carry)
(A) ← (A) + (byte 2) + (CY)
The content of the second byte of the instruction and the content of the CY flag are added to the contents of the accumulator. The result is placed in the accumulator.

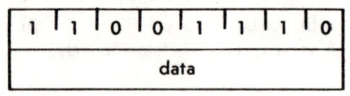

| 1 | 1 | 0 | 0 | 1 | 1 | 1 | 0 |
|---|---|---|---|---|---|---|---|
| data |||||||||

Cycles: 2
States: 7
Addressing: immediate
Flags: Z, S, P, CY, AC

## SUB r and SUB M

The SUB r instruction causes the contents of the source register S to be subtracted from the accumulator. The source register can be any of the general-purpose registers B, C, D, E, H, and L, the accumulator A, or M (the contents of memory as addressed by register pair H, L). All five flags are affected by the execution of this instruction. If you wish to clear the accumulator, the single instruction,

  SUB A

which has an instruction code of 277, will do it.

**SUB r**  (Subtract Register)
(A) ← (A) − (r)
The content of register r is subtracted from the content of the accumulator. The result is placed in the accumulator.

| 1 | 0 | 0 | 1 | 0 | S | S | S |
|---|---|---|---|---|---|---|---|

Cycles: 1
States: 4
Addressing: register
Flags: Z, S, P, CY, AC

**SUB M**  (Subtract memory)
(A) ← (A) − [(H)(L)]
The content of the memory location whose address is contained in the H and L registers is subtracted from the content of the accumulator. The result is placed in the accumulator.

| 1 | 0 | 0 | 1 | 0 | 1 | 1 | 0 |
|---|---|---|---|---|---|---|---|

Cycles: 2
States: 7
Addressing: register indirect
Flags: Z, S, P, CY, AC

## SUI data

The SUI data instruction causes the 8-bit data quantity specified in the second instruction byte to be subtracted from the accumulator. All five flags are affected.

**SUI data** (Subtract immediate)
(A) ← (A) − (byte 2)
The content of the second byte of the instruction is subtracted from the content of the accumulator. The result is placed in the accumulator.

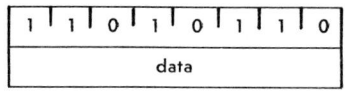

| 1 | 1 | 0 | 1 | 0 | 1 | 1 | 0 |
|---|---|---|---|---|---|---|---|
| data ||||||||

Cycles: 2
States: 7
Addressing: immediate
Flags: Z, S, P, CY, AC

## SBB r and SBB M

The SBB r instruction causes the contents of the source S to be subtracted from the difference of the accumulator contents and the borrow bit. The source register can be any of the general-purpose registers B, C, D, E, H, and L; the accumulator A; or M, the contents of memory addressed by register pair H, L. All five flags are affected by the SBB r and SBB M instructions.

**SBB r** (Subtract Register with borrow)
(A) ← (A) − (r) − (CY)
The content of register r and the content of the CY flag are both subtracted from the accumulator. The result is placed in the accumulator.

| 1 | 0 | 0 | 1 | 1 | S | S | S |
|---|---|---|---|---|---|---|---|

Cycles: 1
States: 4
Addressing: register
Flags: Z, S, P, CY, AC

**SBB M** (Subtract memory with borrow)
(A) ← (A) − [(H) (L)] − (CY)
The content of the memory location whose address is contained in the H and L registers and the content of the CY flag are both subtracted from the accumulator. The result is placed in the accumulator.

| 1 | 0 | 0 | 1 | 1 | 1 | 1 | 0 |
|---|---|---|---|---|---|---|---|

Cycles: 2
States: 7
Addressing: register indirect
Flags: Z, S, P, CY, AC

## SBI data

The SBI data instruction causes the 8-bit data quantity specified in the second instruction byte to be subtracted from the difference of the accumulator contents and the borrow bit. All five flags are affected.

**SBI data**     (Subtract immediate with borrow)
(A) ← (A) − (byte 2) − (CY)
The contents of the second byte of the instruction and the contents of the CY flag are both subtracted from the accumulator. The result is placed in the accumulator.

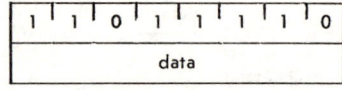

| 1 | 1 | 0 | 1 | 1 | 1 | 1 | 0 |
|---|---|---|---|---|---|---|---|
| data ||||||||

      Cycles: 2
      States: 7
  Addressing: immediate
       Flags: Z, S, P, CY, AC

## Addition and Subtraction Operations

Some examples of the various addition and subtraction operations would be appropriate. Consider the following program:

```
ADD B
ADD C
```

If the initial register contents are A = 00111110, B = 11100000, and C = 00101111, and if the carry bit were initially zero, then the above section of code would yield the following result in the accumulator:

```
Carry bit
   0        00111110     Accumulator contents
          + 11100000     Register B contents
   1        00011110     Sum stored in accumulator
          + 00101111     Register C contents
   0        01001101     Sum stored in accumulator
```

Note carefully the behavior of the carry bit in this situation. If there is no carry out of the most-significant bit (MSB) in the accumulator, the carry bit is cleared; if there is a carry out of the most-significant bit in the accumulator during the addition, the carry bit is set. When you added B to the accumulator, you had a carry. When you added the contents of C to the sum, there was no carry. The carry from previous operations is not preserved, or "carried forward."

Now let us contrast the above results with the behavior of the following section of code:

```
ADC B
ADC C
```

Assume the same initial values for registers A, B, C, and the carry bit. You would obtain the following results:

```
Carry bit
  0       00111110      Accumulator contents
        + 11100000      Register B contents
  1       00011110      Sum stored in accumulator
```

So far, there is no difference. However, when we add the contents of register C to the above sum, we do observe a difference:

```
          00011110      Sum stored in accumulator
        +        1      Carry bit
        + 00101111      Register C contents
  0       01001110      Sum stored in accumulator
```

Now consider the following section of code,

```
SUB B
SUB C
```

for the same initial values of registers A, B, C, and the carry bit. Note that if you perform a borrow out of the MSB of the accumulator, the carry bit is set; if no borrow occurs, the carry bit is cleared. You, therefore, should observe the following:

```
Carry bit
  0       00111110      Accumulator contents
        - 11100000      Register B contents
  1       01011110      Difference stored in accumulator
        - 00101111      Register C contents
  0       00101111      Difference' stored in accumulator
```

Now, let us perform subtraction operations using the SBB r instructions,

```
SBB B
SBB C
```

We have the following results:

```
Carry bit
  0       00111110      Accumulator contents
        - 11100000      Register B contents
  1       01011110      Difference stored in accumulator
```

When we perform the SBB C operation, we subtract the contents of register C from the difference between the borrow bit and the contents of the accumulator:

```
          01011110      Difference stored in accumulator
        -        1
        - 00101111      Register C contents
  0       00101110      Difference' stored in accumulator
```

The ADC r and SBB r instructions are used whenever you perform double or triple precision arithmetic operations. A *double precision* arithmetic operation is one which is performed on two 16-bit quantities to yield a 16-bit result. A *triple precision* is one which is performed on two 24-bit quantities to yield a 24-bit result. The above examples of addition and subtraction operations are courtesy of NEC Microcomputers, Inc. in their *μCOM-8 Software Manual.*

## DAA

**DAA** (Decimal Adjust Accumulator) The 8-bit number in the accumulator is adjusted to form two 4-bit Binary Coded Decimal digits by the following process:
1. If the value of the least-significant 4 bits of the accumulator is greater than 9 **or** if the AC flag is set, 6 is added to the accumulator.
2. If the value of the most-significant 4 bits of the accumulator is now greater than 9, **or** if the CY flag is set, 6 is added to the most-significant 4 bits of the accumulator.

NOTE: All flags are affected.

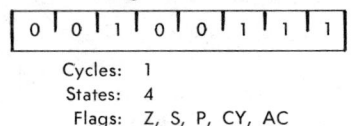

| 0 | 0 | 1 | 0 | 0 | 1 | 1 | 1 |
|---|---|---|---|---|---|---|---|

Cycles: 1
States: 4
Flags: Z, S, P, CY, AC

To quote the *μCOM-8 Software Manual:*

In order to perform operations in binary coded decimal (BCD), one special instruction is needed. When the 8080A CPU performs an arithmetic operation, it produces the result in binary. When working in BCD, this does not produce the correct result. To remedy this, a DAA instruction is used. DAA stands for Decimal Adjust Accumulator, which is exactly what DAA does. The DAA instruction treats the 8-bit Accumulator as two 4-bit Accumulators. Through the use of a nontestable flag known as the Auxiliary Carry, the DAA operation adjusts the result of a binary addition operation to packed BCD.

For addition, the DAA instruction causes the following operation. If the Auxiliary Carry is set to one or the least significant nibble (LSN) is greater than 9, six is added to the least significant nibble. Then, if the Carry flag is set to one or the most significant nibble is greater than 9, six is added to the most significant nibble (MSN).

The term *nibble* is defined as,

*nibble*—A group of four contiguous bits that usually represent a BCD digit.

The least significant nibble (LSN), most significant nibble (MSN), Accumulator, Auxiliary Carry flag (ACy), and Carry flag (Cy) can be represented as shown in Fig. 18-8. Assume that the Accumulator contains the BCD representation for 75 (MSN = 0111 and LSN = 0101) and that the B register contains the BCD representation for 38 (MSN = 0011 and LSN = 1000) and the Carry flag is logic zero.

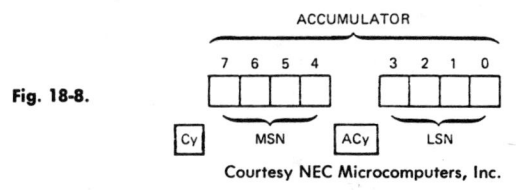

**Fig. 18-8.**

ACCUMULATOR

| Cy | MSN (7 6 5 4) | ACy | LSN (3 2 1 0) |

Courtesy NEC Microcomputers, Inc.

The instruction, ADC B, produces the following result in the Accumulator:

| Carry bit | Auxiliary Carry bit | | |
|---|---|---|---|
| 0 | — | 0 1 1 1 0 1 0 1 | Accumulator contents |
| | | + 0 0 1 1 1 0 0 0 | Register B contents |
| 0 | 0 | 1 0 1 0 1 1 0 1 | Sum stored in the accumulator |

With the Auxiliary Carry, if the instruction causes a carry out of bit 3 and into bit 4 of the resulting value, the Auxiliary Carry flag is set; otherwise it is reset. In the above example, there is no carry out of bit 3 and into bit 4, so the Auxiliary Carry bit is zero after the operation.

The DAA command finds ACy reset to 0 and LSN = 1101. Because the LSN is greater than nine, six is added to it and the result is 0011 and ACy set to 1. Because the MSN is greater than nine, both six and the ACy is added to it to yield a result of 0001. The final decimal adjusted result after the DAA operation is,

| 1 | 1 | 0 0 0 1 0 0 1 1 | Decimal adjusted sum |
| 1 | — | 1   3 | Decimal number |

which is equivalent to the decimal number 113. The DAA operation can be written as follows:

| Carry bit | Auxiliary Carry bit | | | |
|---|---|---|---|---|
| 0 | 0 | 1 0 1 0 | 1 1 0 1 | Sum |
| | | + 0 1 1 0 | + 0 1 1 0 | DAA Operation |
| 1 | [1] | 0 0 0 1 | 0 0 1 1 | Result of DAA Operation |
| 1 | 1 | 0 0 0 1 | 0 0 1 1 | BCD |
| 1 | — | 1 | 3 | Decimal number |

Thus, 75 + 38 = 113.

In actual operation, the DAA adjustment is done in parallel, rather than in the serial manner illustrated. However, this serial explanation, courtesy of the *μCOM-8 Software Manual* of NEC Microcomputers,

Inc., is easier to understand and illustrates the adjustment better. *The DAA instruction should immediately follow an addition operation, as certain 8080A instructions alter the state of the auxiliary carry flag. Such an alteration could result in incorrect results.*

There is an important difference between the Intel 8080A microprocessor chip and the equivalent chip, the µCOM-8 chip of NEC Microcomputers, Inc. The µCOM-8 chip has an extra nontestable flag called Subtract. We quote from the NEC Manual: "For addition, the Sub flag is set to zero. . . . For subtraction, Sub is set to one causing the following DAA operation. If ACy is set to one (a borrow occurred), six is subtracted from the LSN. Then if the Cy is set to one (a borrow occurred), six is subtracted from the MSN. The use of a DAA instruction immediately after an operation on two bytes in packed BCD format adjusts the result to two BCD digits and a carry or borrow in packed BCD format. Note that the DAA operations performs directly after subtraction, *eliminating the need for 100s complement arithmetic for subtraction.*"

If you are doing considerable amounts of BCD manipulation, you might be interested in the µCOM-8 chip in preference to the 8080A. However, such only would be the case if you require the full speed of the microcomputer. With additional instructions, the 8080A can easily accomplish the same task of producing a packed BCD format after a subtraction.

### INR r and INR M

The INR r instruction causes a one to be added to the destination register D. The destination register can be any of the general-purpose registers B, C, D, E, H, and L; the accumulator A; or M, the contents of memory as addressed by register pair H, L. All flags are affected except the carry flag.

**INR r** (Increment Register)
(r) ← (r) + 1
The content of register r is incremented by one. Note: All condition flags **except CY** are affected.

| 0 | 0 | D | D | D | 1 | 0 | 0 |
|---|---|---|---|---|---|---|---|

Cycles: 1
States: 5
Addressing: register
Flags: Z, S, P, AC

**INR M** (Increment memory)
[(H) (L)] ← [(H) (L)] + 1
The content of the memory location whose address is contained in the H and L registers is incremented by one. Note: All condition flags **except CY** are affected.

| 0 | 0 | 1 | 1 | 0 | 1 | 0 | 0 |
|---|---|---|---|---|---|---|---|

Cycles: 3
States: 10
Addressing: register indirect
Flags: Z, S, P, AC

### DCR r and DCR M

The DCR r instruction causes a one to be subtracted from the destination register D. The destination register can be any of the general-purpose registers B, C, D, E, H, and L; the accumulator A; or M, the contents of memory as addressed by register pair H, L. Only four of the five flags are affected; the carry flag remains unchanged.

**DCR r**     (Decrement Register)
(r) ← (r) − 1
The content of register r is decremented by one. Note: All condition flags **except CY** are affected.

| 0 | 0 | D | D | D | 1 | 0 | 1 |
|---|---|---|---|---|---|---|---|

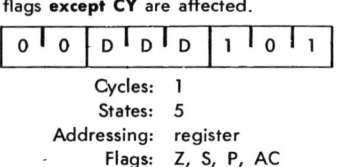

Cycles: 1
States: 5
Addressing: register
Flags: Z, S, P, AC

**DCR M**     (Decrement memory)
[(H) (L)] ← [(H) (L)] − 1
The content of the memory locations whose address is contained in the H and L registers is decremented by one. Note: All condition flags **except CY** are affected.

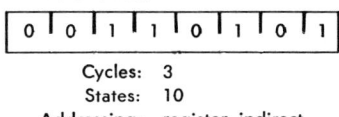

| 0 | 0 | 1 | 1 | 1 | 0 | 1 | 0 | 1 |
|---|---|---|---|---|---|---|---|---|

Cycles: 3
States: 10
Addressing: register indirect
Flags: Z, S, P, AC

### INX rp and DCX rp

The INX rp causes the register pair specified by RP to be incremented by one; the DCX rp causes the register pair specified by RP to be decremented by one. RP can be register pair B, D, or H (corresponding to BC, DE, or HL) or the 16-bit stack pointer specified by SP. INX and DCX do not affect any flag bits. They are usually not used in arithmetic operations, their main use being to increment or decrement 16-bit memory addresses.

**INX rp**     (Increment register pair)
(rh) (rl) ← (rh) (rl) + 1
The content of the register pair rp is incremented by one. Note: **No condition flags are affected.**

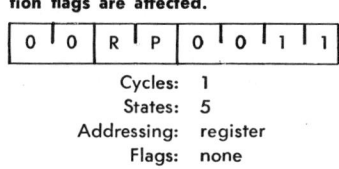

| 0 | 0 | R | P | 0 | 0 | 1 | 1 |
|---|---|---|---|---|---|---|---|

Cycles: 1
States: 5
Addressing: register
Flags: none

**DCX rp**     (Decrement register pair)
(rh) (rl) ← (rh) (rl) − 1
The content of the register pair rp is decremented by one. Note: **No condition flags are affected.**

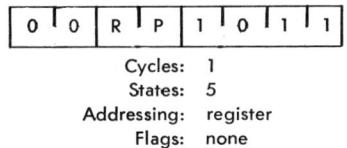

| 0 | 0 | R | P | 1 | 0 | 1 | 1 |
|---|---|---|---|---|---|---|---|

Cycles: 1
States: 5
Addressing: register
Flags: none

### DAD rp

According to the NEC Manual: "While the INX and DCX instructions allow incrementing and decrementing register pairs, the DAD, Double Add, instruction allows adding register pairs together. DAD rp causes the register pair specified by RP to be added to the contents of the HL register pair, with the result remaining in the HL pair. The Carry Flag is the only status flag affected by the DAD instruction. The instructions INX, DCX, and DAD allow the calculation of table lookup." They are also used for indexed addressing and data file manipulation.

**DAD rp** (Add register pair to H and L)
(H) (L) ← (H) (L) + (rh) (rl)
The content of the register pair rp is added to the content of the register pair H and L. The result is placed in the register pair H and L. Note: **Only the CY flag is affected.** It is set if there is a carry out of the double precision add; otherwise it is reset.

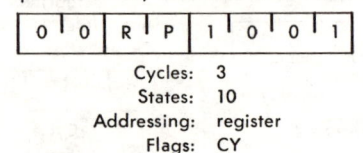

| 0 | 0 | R | P | 1 | 0 | 0 | 1 |
|---|---|---|---|---|---|---|---|

Cycles: 3
States: 10
Addressing: register
Flags: CY

## CMP r and CMP M

To quote the μCOM-8 Software Manual: "CMP r and CMP M are used to compare two data quantities *without altering them*. CMP r compares the contents of the accumulator with one of the single registers B, C, D, E, H, and L; the accumulator A; or M, the memory location addressed by the H, L register pair. The instruction does not affect any of the data registers, but affects the four flag bits, Carry, Zero, Sign, and Parity. The compare instructions actually perform an internal subtraction of the source S from the accumulator. The flags are set on the basis of what would have been the result of the subtraction. Thus, Zero is set if the quantities were equal, Sign is set if the result was negative (the most-significant bit is logic 1), Parity is set if the result has even parity, and Carry is set if there is a borrow out of bit 7 (source data greater than Accumulator data)."

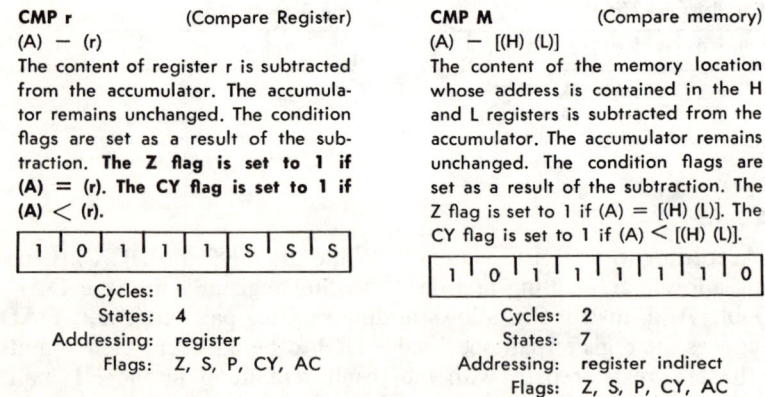

**CMP r** (Compare Register)
(A) − (r)
The content of register r is subtracted from the accumulator. The accumulator remains unchanged. The condition flags are set as a result of the subtraction. **The Z flag is set to 1 if (A) = (r). The CY flag is set to 1 if (A) < (r).**

| 1 | 0 | 1 | 1 | 1 | S | S | S |
|---|---|---|---|---|---|---|---|

Cycles: 1
States: 4
Addressing: register
Flags: Z, S, P, CY, AC

**CMP M** (Compare memory)
(A) − [(H) (L)]
The content of the memory location whose address is contained in the H and L registers is subtracted from the accumulator. The accumulator remains unchanged. The condition flags are set as a result of the subtraction. The Z flag is set to 1 if (A) = [(H) (L)]. The CY flag is set to 1 if (A) < [(H) (L)].

| 1 | 0 | 1 | 1 | 1 | 1 | 1 | 0 |
|---|---|---|---|---|---|---|---|

Cycles: 2
States: 7
Addressing: register indirect
Flags: Z, S, P, CY, AC

Thus, in every case:

Carry is set if a borrow occurs; else reset;
Sign is set equal to the MSB of the result;

Zero is set if the result is zero; else reset;
Parity is set if the parity of the result is even; else reset.

The Compare instructions are best used for unsigned arithmetic comparison (numbers in the range of 0 to $255_{10}$), also called logical or character comparisons. For this case, the results for the Zero and Carry flags may be interpreted as follows:

### Result of Compare Operation (X = Don't Care)

| Zero Flag | Carry Flag | Relationship Between Accumulator and Register |
|---|---|---|
| 1 | X | Accumulator $=$ register |
| X | 1 | Accumulator $<$ register |
| 1 | 1 | Accumulator $\leq$ register |
| 0 | 0 | Accumulator $>$ register |
| X | 0 | Accumulator $\geq$ register |

Thus, the relations $\{=,<,\geq\}$ may be tested using a single jump instruction, while $\{\leq,>\}$ require two. Note that if the operands are reversed, $>$ replaces $\leq$ and $<$ replaces $\geq$.

### CPI data

The CPI data instruction is an immediate operation which compares the contents of the accumulator with the 8-bit quantity in the second byte of the instruction. The instruction affects all five flags, but only four of the flags produce useful results. The flags are set or cleared on the basis of what would have been the result of the subtraction. *The contents of the accumulator remain unchanged.* See the preceding discussion on the CMP r instruction for additional details.

It can be argued that the CMP r and CPI data instructions are logical rather than arithmetic operations. In view of the fact that an

> **CPI data** (Compare immediate)
> (A) − (byte 2)
> The content of the second byte of the instruction is subtracted from the accumulator. The condition flags are set by the result of the subtraction. The Z flag is set to 1 if (A) = (byte 2). The CY flag is set to 1 if (A) < (byte 2).
>
> | 1 | 1 | 1 | 1 | 1 | 1 | 1 | 0 |
> |---|---|---|---|---|---|---|---|
> | | | | data | | | | |
>
> Cycles: 2
> States: 7
> Addressing: immediate
> Flags: Z, S, P, CY, AC

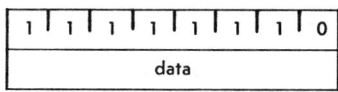

arithmetic operation—subtraction—is performed, we would include it in the group of arithmetic operations. The objective of the compare instructions is to produce decisions that are reflected in the logic states of the flag bits.

## LOGICAL GROUP

This group of instructions performs logical, i.e., Boolean, operations on data in registers and memory and on condition flags. Unless indicated otherwise, all instructions in this group affect the Zero, Sign, Parity, Auxiliary Carry, and Carry flags according to the standard rules.

### ANA r and ANA M

The ANA r instruction performs a *parallel bit-by-bit logical* AND of the contents of the accumulator and the contents of the source register S. The source register can be any of the general-purpose registers B, C, D, E, H, and L; the accumulator A; or M, the contents of the memory location addressed by the register pair H, L. For example, the ANA B operation performs a bit-by-bit logic AND operation with the contents of register B and the contents of the accumulator. The special case of

ANA A

clears the Carry flag and causes the Zero flag to be set if the result is zero, cleared if the result is not zero. All of the flags are affected by the ANA r instruction. Since $A \cdot A = A$, the data in the accumulator is not changed. This is a "trick" to clear the carry or simply test for zero in the accumulator.

**ANA r** (AND Register)
(A) ← (A) ∧ (r)
The content of register r is logically ANDed with the content of the accumulator. The result is placed in the accumulator. **The CY and AC flags are cleared.**

| 1 | 0 | 1 | 0 | 0 | S | S | S |
|---|---|---|---|---|---|---|---|

Cycles: 1
States: 4
Addressing: register
Flags: Z, S, P, CY, AC

**ANA M** (AND memory)
(A) ← (A) ∧ [(H) (L)]
The contents of the memory location whose address is contained in the H and L registers is logically ANDed with the content of the accumulator. The result is placed in the accumulator. **The CY and AC flags are cleared.**

| 1 | 0 | 1 | 0 | 0 | 1 | 1 | 0 |
|---|---|---|---|---|---|---|---|

Cycles: 2
States: 7
Addressing: register indirect
Flags: Z, S, P, CY, AC

## ANI data

The ANI data instruction performs a bit-by-bit logical AND of the contents of the accumulator with the contents of the second byte of the instruction. All flags are affected by the instruction.

> **ANI data** (AND immediate)
> (A) ← (A) ∧ (byte 2)
> The content of the second byte of the instruction is logically ANDed with the contents of the accumulator. The result is placed in the accumulator.
> **The CY and AC flags are cleared.**
>
> | 1 | 1 | 1 | 0 | 0 | 1 | 1 | 0 |
> |---|---|---|---|---|---|---|---|
> | data |||||||||
>
> Cycles: 2
> States: 7
> Addressing: immediate
> Flags: Z, S, P, CY, AC

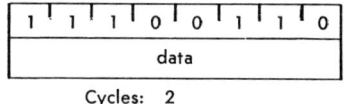

## ORA r and ORA M

The ORA r instruction performs a *parallel bit-by-bit logical* OR of the contents of the accumulator and the contents of the source register S. The source register can be any of the general-purpose registers B, C, D, E, H, and L; the accumulator A; or M, the contents of the memory location addressed by the register pair H, L. The command,

ORA A

which has the octal instruction code 267, is a convenient way to clear the Carry flag without affecting anything else. Both ORA r and a related two-byte instruction, ORI data, clear the Carry flag and cause the Zero flag to be set if the result is zero, cleared if the result is not zero.

> **ORA r** (OR Register)
> (A) ← (A) V (r)
> The content of register r is inclusive-OR'd with the content of the accumulator. The result is placed in the accumulator. **The CY and AC flags are cleared.**
>
> | 1 | 0 | 1 | 1 | 0 | S | S | S |
> |---|---|---|---|---|---|---|---|
>
> Cycles: 1
> States: 4
> Addressing: register
> Flags: Z, S, P, CY, AC

> **ORA M** (OR memory)
> (A) ← (A) V [(H) (L)]
> The content of the memory location whose address is contained in the H and L registers is inclusive-OR'd with the content of the accumulator. The result is placed in the accumulator. **The CY and AC flags are cleared.**
>
> | 1 | 0 | 1 | 1 | 0 | 1 | 1 | 0 |
> |---|---|---|---|---|---|---|---|
>
> Cycles: 2
> States: 7
> Addressing: register indirect
> Flags: Z, S, P, CY, AC

## ORI data

The ORI data instruction performs a bit-by-bit logical OR of the contents of the accumulator with the contents of the second byte of the instruction. All flags are affected by the instruction.

> **ORI data** (OR immediate)
> (A) ← (A) V (byte 2)
> The content of the second byte of the instruction is inclusive-OR'd with the content of the accumulator. The result is placed in the accumulator. **The CY and AC flags are cleared.**
>
> | 1 | 1 | 1 | 1 | 0 | 1 | 1 | 0 |
> |---|---|---|---|---|---|---|---|
> 
> | data |
> |------|
>
> Cycles: 2
> States: 7
> Addressing: immediate
> Flags: Z, S, P, CY, AC

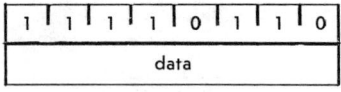

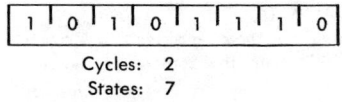

## XRA r and XRA M

The XRA r instruction performs a *parallel bit-by-bit Exclusive-*OR of the contents of the accumulator and the contents of the source register S. The source register can be any of the general-purpose registers B, C, D, E, H, and L; the accumulator A; or M, the memory location addressed by the register pair H, L. All flags are affected by the instruction.

> **XRA r** (Exclusive-OR Register)
> (A) ← (A) ∀ (r)
> The content of register r is Exclusive-OR'd with the content of the accumulator. The result is placed in the accumulator. **The CY and AC flags are cleared.**
>
> | 1 | 0 | 1 | 0 | 1 | S | S | S |
> |---|---|---|---|---|---|---|---|
>
> Cycles: 1
> States: 4
> Addressing: register
> Flags: Z, S, P, CY, AC

> **XRA M** (Exclusive-OR memory)
> (A) ← (A) ∀ [(H) (L)]
> The content of the memory location whose address is contained in the H and L registers is Exclusive-OR'd with the content of the accumulator. The result is placed in the accumulator. **The CY and AC flags are cleared.**
>
> | 1 | 0 | 1 | 0 | 1 | 1 | 1 | 0 |
> |---|---|---|---|---|---|---|---|
>
> Cycles: 2
> States: 7
> Addressing: register indirect
> Flags: Z, S, P, CY, AC

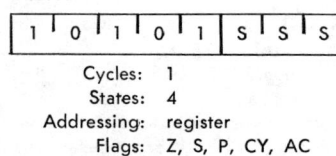

## XRI data

The XRI data instruction performs a bit-by-bit logical Exclusive-OR of the contents of the accumulator with the contents of the second byte of the instruction. All flags are affected by the instruction.

**XRI data** (Exclusive-OR immediate)
(A) ← (A) ∀ (byte 2)
The content of the second byte of the instruction is Exclusive-OR'd with the content of the accumulator. The result is placed in the accumulator. **The CY and AC flags are cleared.**

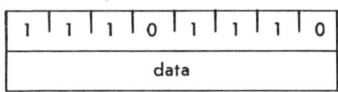

| 1 | 1 | 1 | 0 | 1 | 1 | 1 | 0 |
|---|---|---|---|---|---|---|---|
| data |||||||||

Cycles: 2
States: 7
Addressing: immediate
Flags: Z, S, P, CY, AC

To quote the NEC Microcomputers, Inc., $\mu COM\text{-}8$ *Software Manual:* "The above logic instructions will be used to implement a programming technique known as *masking*. Masking is a technique by which bits of an operand are selectively modified for use in a later operation. There are three general types of masking:

- Clear all bits not operated upon.
- Set all bits not operated upon (seldom used).
- Leave unaltered all bits not operated upon.

"The first two approaches are called *exclusive masking* and the third approach is called *inclusive masking*. For example, assume that the accumulator contains the following value,

Bit  7 6 5 4 3 2 1 0
       1 1 0 1$^1{}_0$ 1 1 0   Accumulator contents

To test bit 3 for a zero or one and simultaneously clear the other bits, the accumulator is masked with 00001000. By using the instruction

ANI
010

the accumulator will contain zeros with the Zero flag set if bit 3 had been a zero, and it will contain 010, in octal code, with the Zero flag cleared if bit 3 had been one.

"In order to set bit 3 to one and leave the other bits alone, the same bit pattern is used and the instruction

ORI
010

is used. The result in this case is 11011110 in the accumulator.

"In order to set bit 3 to zero and leave the other bits alone, the accumulator is ANDed with 11110111, the complement of the mask of the first example. With the instruction

ANI
367

the accumulator result is 11010110. These are the most commonly used bit manipulation operations, since masking is accomplished in one step. Many others are possible, but they often require more than one instruction for implementation."

## ROTATE INSTRUCTIONS

All of the 8080A rotate instructions are summarized in the diagrams of Fig. 18-9.

### RAL and RAR

The RAL instruction, or Rotate Accumulator Left, causes the accumulator to rotate all bits one position to the left through the Carry bit, i.e., a 9-bit rotate. Bit 7 transfers to the Carry, the Carry bit transfers to bit 0, bit 0 transfers to bit 1, bit 1 transfers to bit 2, and so on, as shown in Fig. 18-9.

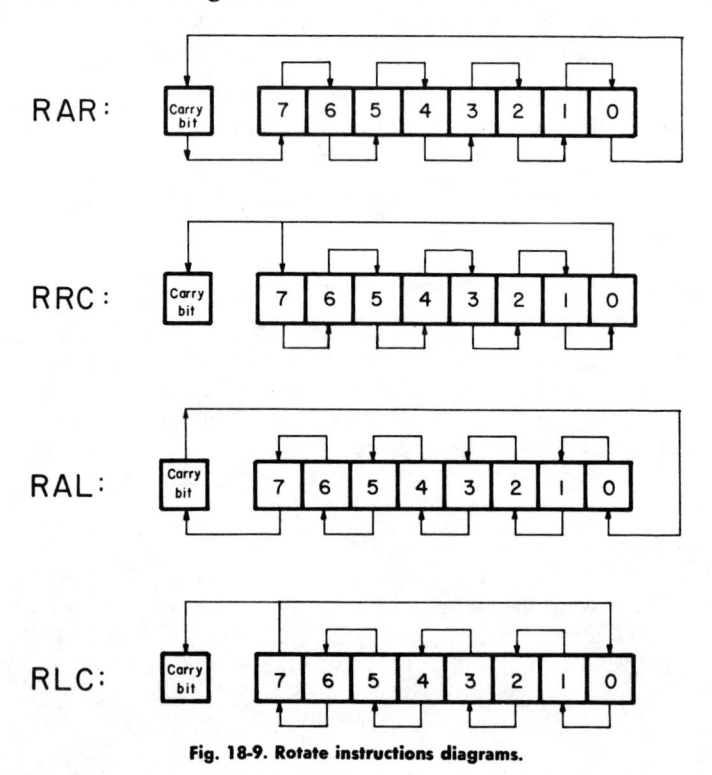

**Fig. 18-9. Rotate instructions diagrams.**

**RAL** (Rotate left through carry)
$(A_{n+1}) \leftarrow (A_n) ; (CY) \leftarrow (A_7)$
$(A_0) \leftarrow (CY)$
The content of the accumulator is rotated left one position through the CY flag. The low-order bit is set equal to the CY flag and the CY flag is set to the value shifted out of the high-order bit. **Only the CY flag is affected.**

| 0 | 0 | 0 | 1 | 0 | 1 | 1 | 1 |
|---|---|---|---|---|---|---|---|

Cycles: 1
States: 4
Flags: CY

**RAR** (Rotate right through carry)
$(A_n) \leftarrow (A_{n+1}) ; (CY) \leftarrow (A_0)$
$(A_7) \leftarrow (CY)$
The content of the accumulator is rotated right one position through the CY flag. The high-order bit is set to the CY flag and the CY flag is set to the value shifted out of the low-order bit. **Only the CY flag is affected.**

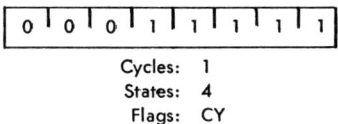

| 0 | 0 | 0 | 1 | 1 | 1 | 1 | 1 |
|---|---|---|---|---|---|---|---|

Cycles: 1
States: 4
Flags: CY

The RAR instruction, or Rotate Accumulator Right, causes the accumulator to rotate all bits one position to the right through the Carry bit, i.e., a 9-bit rotate. Bit 0 transfers to the Carry, the Carry bit transfers to bit 7, bit 7 transfers to bit 6, and so on, as shown in Fig. 18-9.

### RLC and RRC

The RLC instruction, or Rotate Left Circular, rotates the accumulator one bit to the left and into the Carry, as shown in the diagram of Fig. 18-9.

**RLC** (Rotate left)
$(A_{n+1}) \leftarrow (A_n) ; (A_0) \leftarrow (A_7)$
$(CY) \leftarrow (A_7)$
The content of the accumulator is rotated left one position. The low-order bit and the CY flag are both set to the value shifted out of the high-order bit position. **Only the CY flag is affected.**

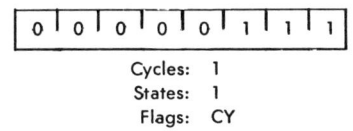

| 0 | 0 | 0 | 0 | 0 | 1 | 1 | 1 |
|---|---|---|---|---|---|---|---|

Cycles: 1
States: 1
Flags: CY

**RRC** (Rotate right)
$(A_n) \leftarrow (A_{n-1}); (A_7) \leftarrow (A_0)$
$(CY) \leftarrow (A_0)$
The content of the accumulator is rotated right one position. The high-order bit and the CY flag are both set to the value shifted out of the lower-order bit position. **Only the CY flag is affected.**

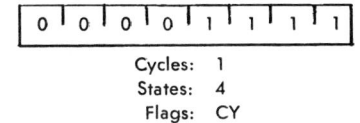

| 0 | 0 | 0 | 0 | 1 | 1 | 1 | 1 |
|---|---|---|---|---|---|---|---|

Cycles: 1
States: 4
Flags: CY

The RRC instruction, or Rotate Right Circular, rotates the accumulator one bit to the right and into the Carry, as also shown in Fig. 18-9.

In both of these instructions, the original information appearing in the Carry is lost.

### CMA

The CMA instruction complements the contents of the accumulator without affecting any of the flag bits. For example, if the accumulator contained 11010001, the

CMA

instruction would convert it to 00101110. Each individual bit is complemented.

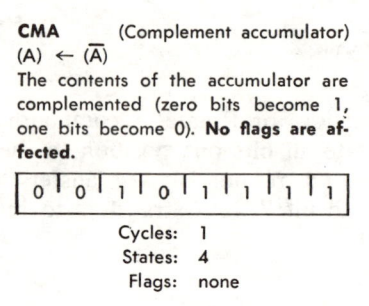

**CMA** (Complement accumulator)
$(A) \leftarrow (\overline{A})$
The contents of the accumulator are complemented (zero bits become 1, one bits become 0). **No flags are affected.**

| 0 | 0 | 1 | 0 | 1 | 1 | 1 | 1 |
|---|---|---|---|---|---|---|---|

Cycles: 1
States: 4
Flags: none

### STC and CMC

The STC instruction sets the Carry to logic 1; the CMC instruction complements the Carry. Only the Carry flag is affected.

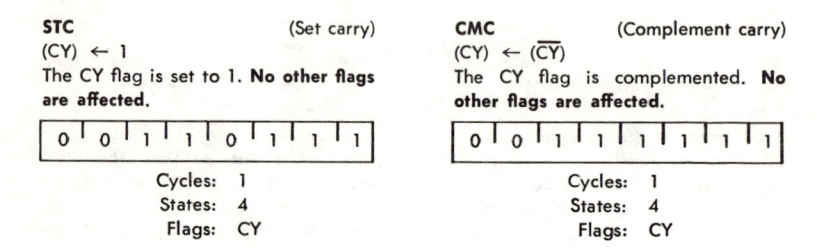

**STC** (Set carry)
$(CY) \leftarrow 1$
The CY flag is set to 1. **No other flags are affected.**

| 0 | 0 | 1 | 1 | 0 | 1 | 1 | 1 |
|---|---|---|---|---|---|---|---|

Cycles: 1
States: 4
Flags: CY

**CMC** (Complement carry)
$(CY) \leftarrow (\overline{CY})$
The CY flag is complemented. **No other flags are affected.**

| 0 | 0 | 1 | 1 | 1 | 1 | 1 | 1 |
|---|---|---|---|---|---|---|---|

Cycles: 1
States: 4
Flags: CY

### BRANCH GROUP

This group of instructions alters normal sequential program execution. *Condition flags are not affected by any instruction in this group.* The two types of branch instructions are unconditional and conditional. Unconditional transfers simply perform the specified operation on register PC, the program counter. Conditional transfers examine the status of one of the four flags—Zero, Sign, Parity, or Carry—to determine if the specified branch operation is to be executed. The conditions that may be specified are as follows:

| CONDITION | CCC |
|---|---|
| NZ—not zero (Z = 0) | 000 |
| Z—zero (Z = 1) | 001 |
| NC—no carry (CY = 0) | 010 |
| C—carry (CY = 1) | 011 |
| PO—parity odd (P = 0) | 100 |
| PE—parity even (P = 1) | 101 |
| P—plus (S = 0) | 110 |
| M—minus (S = 1) | 111 |

NOTE: CCC is the 3-bit code for the condition of the flags.

## JMP addr

The *program counter* is the 16-bit register in the 8080A microprocessor chip that contains the memory address of the next instruction byte that must be executed in a program. The JMP addr instruction is simply a byte transfer instruction, in which the second and third instruction bytes are transferred directly to the program counter. No arithmetic or logical operations are involved, and no flag bits are affected. The JMP instruction is a three-byte instruction that contains the 16-bit memory address to which program control is transferred. You can jump forwards or backwards to any of the 65,536 possible memory locations. The microprocessor chip does not remember the point from which it jumped, in distinct contrast to the behavior of the CALL and RET instructions discussed next.

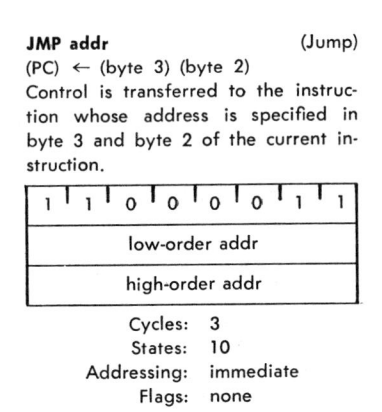

**JMP addr** (Jump)
(PC) ← (byte 3) (byte 2)
Control is transferred to the instruction whose address is specified in byte 3 and byte 2 of the current instruction.

| 1 | 1 | 0 | 0 | 0 | 0 | 1 | 1 |
|---|---|---|---|---|---|---|---|
| low-order addr |||||||||
| high-order addr |||||||||

Cycles: 3
States: 10
Addressing: immediate
Flags: none

The behavior of the JMP instruction can be understood with the aid of the diagram shown in Fig. 18-10. The first JMP instruction, ①, is a backwards jump that creates a loop. JMP ② and JMP ③ transfer program control to the subprogram. The exit from the subprogram is designated by the JMP ④ instruction.

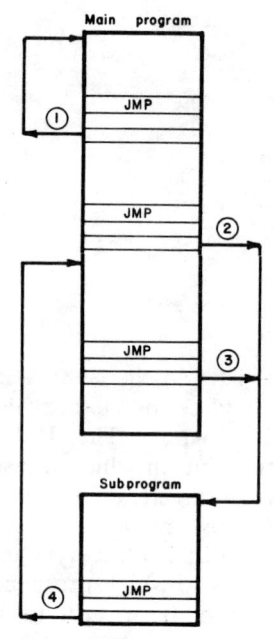

Fig. 18-10. Diagram of the JMP instruction.

## CALL addr and RET

**CALL addr** (Call)
[(SP) − 1] ← (PCH)
[(SP) − 2] ← (PCL)
(SP) ← (SP) − 2
(PC) ← (byte 3) (byte 2)

The high-order eight bits of the next instruction address are moved to the memory location whose address is one less than the content of register SP. The low-order eight bits of the next instruction address are moved to the memory location whose address is two less than the content of register SP. The content of register SP is decremented by 2. Control is transferred to the instruction whose address is specified in byte 3 and byte 2 of the current instruction.

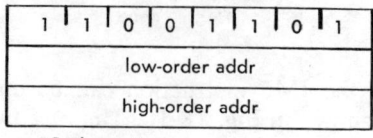

| 1 | 1 | 0 | 0 | 1 | 1 | 0 | 1 |
|---|---|---|---|---|---|---|---|
| low-order addr ||||||||
| high-order addr ||||||||

Cycles: 5
States: 17
Addressing: immediate/register indirect
Flags: none

**RET**                                (Return)
(PCL) ← (SP);
(PCH) ← [(SP) + 1];
(SP) ← (SP) + 2

The content of the memory location whose address is specified in register SP is moved to the low-order eight bits of register PC. The content of the memory location whose address is one more than the content of register SP is moved to the high-order eight bits of register PC. The content of register SP is incremented by 2.

| 1 | 1 | 0 | 0 | 1 | 1 | 0 | 0 | 1 |
|---|---|---|---|---|---|---|---|---|

        Cycles: 3
        States: 10
Addressing: register indirect
         Flags: none

Many times you may want to branch out of a main program but return to it later. To do so, you must not only know your new destination, but you must somehow also remember your original location. To accomplish this, you have two types of instructions, call subroutine and return from subroutine. Here we shall discuss the unconditional instructions—CALL addr and RET. (See Fig. 18-11.) To quote the NEC Microcomputers, Inc., manual:

> The call instruction transfers control to a subroutine. The instruction CALL addr saves the incremented program counter on the pushdown *stack* and places the address in the program counter. The pushdown stack is a block of read/write memory addressed by a special 16-bit register known as the Stack Pointer which can be loaded by the user (LXI SP, data 16). The stack operates as a last-in-first-out memory (LIFO), with the Stack Pointer register addressing the most recent entry into the stack. The Return instruction causes the entry at the top of the stack to be placed into the Program Counter. Thus, a CALL instruction transfers program control from the main program into the subroutine and a RET instruction transfers control back to the main program.

The location of the *stack* is usually at the higher memory addresses in the available memory of an 8080A-based microcomputer. In the diagram of Fig. 18-12, the stack is some distance from the main program or subroutines.

### JNZ, JZ, JNC, JC, JPO, JPE, JP, and JM addr

In a conditional jump instruction, if the condition is satisfied, the second and third bytes of the instruction are transferred to the program counter and a jump occurs. If the condition is not satisfied, no

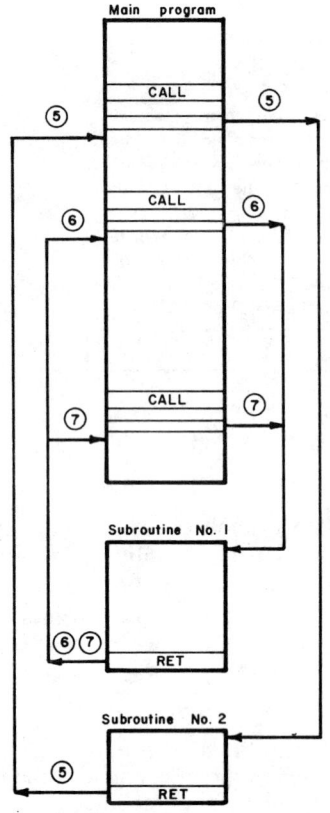

**Fig. 18-11.**

change occurs to the program counter; program control passes to the instruction immediately following the jump.

**Jcondition addr** (Conditional jump)
If (CCC),
 (PC) ← (byte 3) (byte 2)
If the specified condition is true, control is transferred to the instruction whose address is specified in byte 3 and byte 2 of the current instruction; otherwise, control continues sequentially.

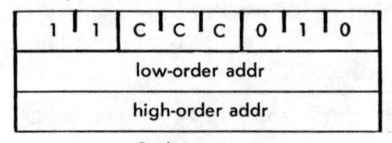

| 1 | 1 | C | C | C | 0 | 1 | 0 |
|---|---|---|---|---|---|---|---|
| low-order addr ||||||||
| high-order addr ||||||||

Cycles: 3
States: 10
Addressing: immediate
Flags: none

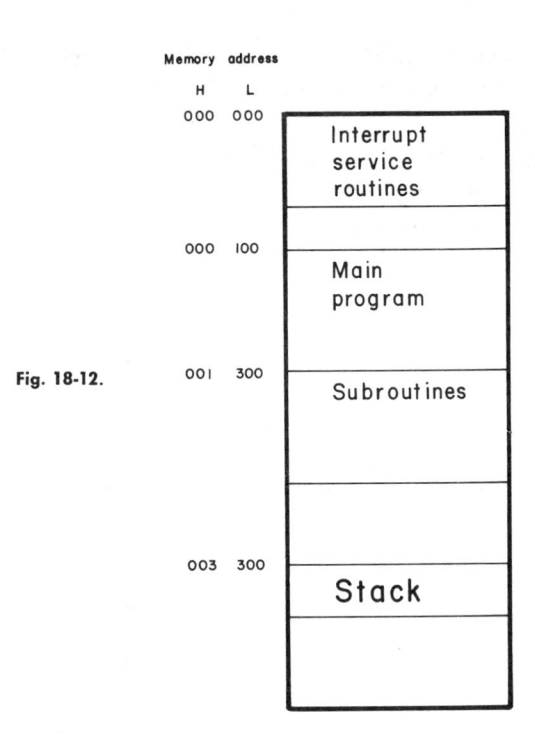

```
                Memory address
                 H      L
                000    000  ┌─────────────────┐
                            │   Interrupt     │
                            │   service       │
                            │   routines      │
                000    100  ├─────────────────┤
                            │                 │
                            │   Main          │
                            │   program       │
                            │                 │
  Fig. 18-12.   001    300  ├─────────────────┤
                            │                 │
                            │   Subroutines   │
                            │                 │
                            │                 │
                            ├─────────────────┤
                            │                 │
                003    300  ├─────────────────┤
                            │                 │
                            │     Stack       │
                            │                 │
                            └─────────────────┘
```

The various conditions can be summarized as follows:

**NZ**—The 8-bit result of the immediately preceding arithmetic or logical operation is Not equal to Zero, i.e., the Zero flag is cleared.

**Z**—The 8-bit result of the immediately preceding arithmetic or logical operation is equal to Zero, i.e., the Zero flag is set.

**NC**—The 8-bit result of the immediately preceding arithmetic or logical operations produces No Carry out of the most significant bit; or, the Carry flag is cleared.

**C**—The 8-bit result of the immediately preceding arithmetic or logical operation produces a Carry out of the most significant bit; or, the Carry flag is set.

**PO**—The 8-bit result of the immediately preceding arithmetic or logical operation has a Parity that is Odd, i.e., the Parity flag is cleared.

**PE**—The 8-bit result of the immediately preceding arithmetic or logical operation has a Parity that is Even, i.e., the Parity flag is set.

**P**—The 8-bit result of the immediately preceding arithmetic or logical operation produces an MSB that has a Plus sign, i.e., the Sign flag is cleared.

**M**—The 8-bit result of the immediately preceding arithmetic or logical operation produces an MSB that has a Minus sign, i.e., the Sign flag is set.

The value of CCC that corresponds to each of the conditions was given at the beginning of this Branch Group section. The behavior of two of the conditional instructions, JNZ and JZ, can be understood with the aid of the diagrams given in Fig. 18-13.

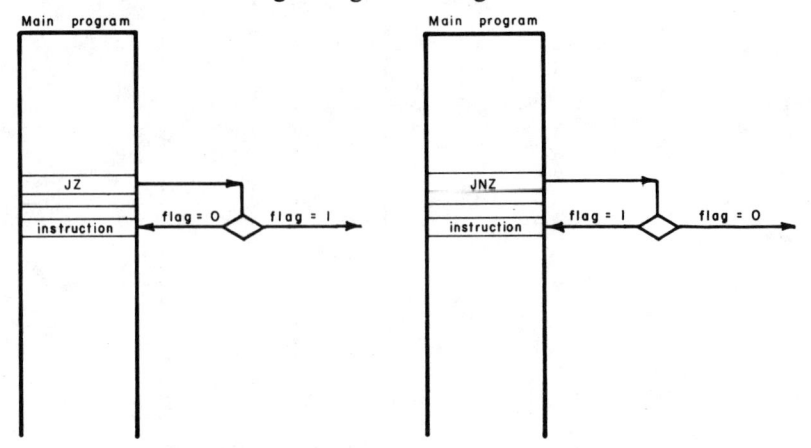

**Fig. 18-13. Diagram showing behavior of the conditional instructions.**

In the JNZ instruction, the jump occurs only if the 8-bit result of an arithmetic or logical operation is Not Zero. The decision symbol, which is used in flowcharting, indicates that what happens next depends upon the state of the Zero flag. For JNZ, a jump occurs only if the Zero flag is cleared, i.e., at logic 0. For JZ, a jump occurs if the 8-bit result is equal to zero; in such a case, the Zero flag is at logic 1.

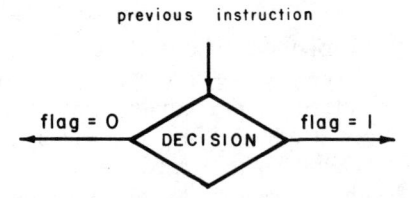

It is possible to become confused concerning the conditions NZ and Z. Note that NZ and Z refer to the 8-bit result of an operation, not to the logic state of the Zero flag. NZ means that the 8-bit result of an operation is not zero; Z means that the 8-bit result of an operation is zero (though the Zero flag is at logic 1). In this discussion, we have tried to demonstrate that a condition can be viewed in terms of the 8-bit result of an arithmetic/logic operation (NZ, Z, NC, C, PO, PE, P, or M) *or* in terms of the logic state of the individual flags that test the result of an arithmetic/logic operation. We prefer the use of the 8-bit result of an ALU operation, including the letter symbols NZ, Z, NC, etc. We hope that we have not confused you.

### CNZ, CZ, CNC, CC, CPO, CPE, CP, and CM addr

In a conditional call instruction, if the condition is satisfied, the subroutine at the memory location given in the second and third instruction bytes is called. The contents of the program counter are placed on the stack, so that a return instruction can return program control to the instruction immediately following the conditional call instruction.

If the condition is not satisfied, program execution passes to the instruction immediately following the conditional call instruction.

**Ccondition addr**  (Condition call)
**If** (CCC),
  [SP] − 1] ← (PCH)
  [(SP) − 2] ← (PCL)
  (SP) ← (SP) − 2
  (PC) ← (byte 3) (byte 2)
If the specified condition is true, the actions specified in the CALL instruction (see above) are performed; otherwise, control continues sequentially.

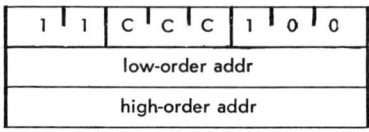

| 1 | 1 | C | C | C | 1 | 0 | 0 |
|---|---|---|---|---|---|---|---|
| low-order addr ||||||||
| high-order addr ||||||||

Cycles: 3/5
States: 11/17
Addressing: immediate/register indirect
Flags: none

### RNZ, RZ, RNC, RC, RPO, RPE, RP, and RM

In a conditional return instruction, if the condition is satisfied, a return occurs from the subroutine; the program counter contents on the stack are transferred to the program counter and program execution resumes at the instruction immediately after the subroutine call instruction.

If the condition is not satisfied, program execution passes to the instruction immediately following the conditional return instruction.

**Rcondition** (Conditional return)
**If** (CCC),
  (PCL) ← (SP)
  (PCH) ← [(SP) + 1]
  (SP) ← (SP) + 2

If the specified condition is true, the actions specified in the RET instruction (see above) are performed; otherwise, control continues sequentially.

| 1 | 1 | C | C | C | 0 | 0 | 0 |
|---|---|---|---|---|---|---|---|

Cycles: 1/3
States: 5/11
Addressing: register indirect
Flags: none

The conditional instructions, CZ, CNZ, RZ, and RNZ are depicted schematically in the diagrams given in Figs. 18-14 and 18-15. Remember, Z means that the Zero flag must be at logic 1 for a call or return to occur; otherwise, program control passes to the next instruction. NZ means that the Zero flag must be at logic 0 for a call or return to occur; otherwise, program control passes to the next instruction.

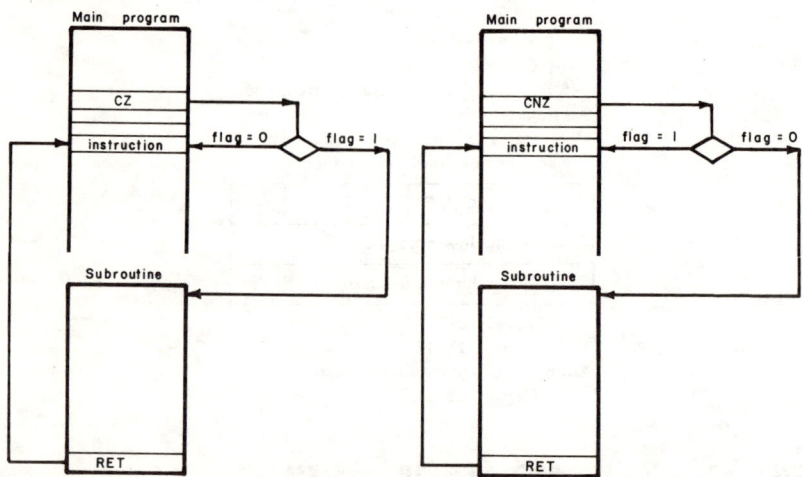

**Fig. 18-14. Diagram depicting conditional instructions RZ and RNZ.**

## RST n

To quote the *µCOM-8 Software Manual:*

The EI (enable interrupt) and the DI (disable interrupt) instructions provide control over the acceptance of an interrupt request. With this control established, the next problem to be resolved is how does the external device indicate to the processor where the desired interrupt routine is located. The 8080A accomplishes this identification by al-

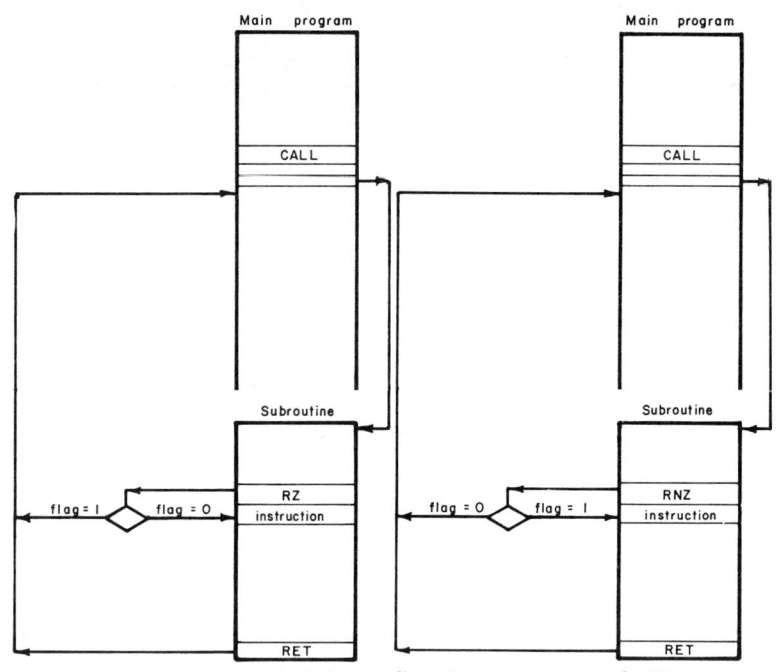

**Fig. 18-15. Diagrams depicting conditional instructions CZ and CNZ.**

lowing the device to supply one instruction when the interrupt is acknowledged. Although any 8080A instruction can be specified, only two are of practical value: a Call instruction, CALL, and a Restart instruction, RST. . . . An RST instruction is actually a specialized type of CALL. The instruction RST n is a call to one of eight locations in memory specified by an integer expression in the range, 0 through 7 in octal code, indicated by N. The locations specified by the integers 0 through 7 are listed below.

| Value of N | Location called |
|---|---|
| 0 | HI = 000 and LO = 000 |
| 1 | HI = 000 and LO = 010 |
| 2 | HI = 000 and LO = 020 |
| 3 | HI = 000 and LO = 030 |
| 4 | HI = 000 and LO = 040 |
| 5 | HI = 000 and LO = 050 |
| 6 | HI = 000 and LO = 060 |
| 7 | HI = 000 and LO = 070 |

An RST instruction causes the incremented program counter to be pushed onto the stack exactly as a CALL instruction does. It then loads the program counter with HI = 000 and LO = ONO, where N is 0 through 7. Thus, RST 4 causes the program counter to be pushed onto the stack and HI = 000 and LO = 040 to be entered into the program counter.

Program execution then continues from the restart location. If the device service routine requires more than eight bytes to service (as most do), the instruction placed at the Restart point must jump to the interrupt service subroutine. Since RST is actually a specialized subroutine call, the interrupt service subroutine *must end with a return instruction,* to return control to the interrupted program by popping the return address.

Since the 8080A has only eight RST instructions, any additional levels of interrupt must be implemented using CALL instructions. This means a CALL addr instruction must be supplied by the interrupting device, which is somewhat more difficult to implement in hardware because CALL is a 3-byte instruction. However, once implemented, a direct call to a routine is slightly faster than a Restart and subsequent jump operation. Although this is not a major factor, this difference in response speed should be considered when determining how to implement interrupt service routines. The primary benefit realized by using the CALL approach is that n-way interrupt vectoring is achieved in hardware, eliminating the need for software in low order memory (for RST processing). This frees those memory locations for use by user programs and removes a constraint from the system memory design.

**RST n** (Restart)
[(SP) − 1] ← (PCH)
[(SP) − 2] ← (PCL)
(SP) ← (SP) − 2
(PC) ← 8*(NNN)

The high-order eight bits of the next instruction address are moved to the memory location whose address is one less than the content of register SP. The low-order eight bits of the next instruction address are moved to the memory location whose address is two less than the content of register SP. The content of register SP is decremented by two. Control is transferred to the instruction whose address is eight times the content of NNN.

| 1 | 1 | N | N | N | 1 | 1 | 1 |
|---|---|---|---|---|---|---|---|

Cycles: 3
States: 11
Addressing: register indirect
Flags: none

| 15 | 14 | 13 | 12 | 11 | 10 | 9 | 8 | 7 | 6 | 5 | 4 | 3 | 2 | 1 | 0 |
|----|----|----|----|----|----|---|---|---|---|---|---|---|---|---|---|
| 0 | 0 | 0 | 0 | 0 | 0 | 0 | 0 | 0 | 0 | N | N | N | 0 | 0 | 0 |

Program Counter After Restart

## PCHL

The PCHL instruction causes the program counter to be loaded with the contents of the HL register pair. Program execution then continues at the point designated by the content of HL. In effect, this is a jump instruction, but since the HL register pair can be operated upon arithmetically, it allows the implementation of a variety of calculated jumps. The instruction sequence

```
LXI H
<B2>
<B3>
PCHL
```

is identical in function to

```
JMP
<B2>
<B3>
```

PCHL     (Jump H and L indirect—move H and L to PC)

$(PCH) \leftarrow (H)$
$(PCL) \leftarrow (L)$

The content of register H is moved to the high-order eight bits of register PC. The content of register L is moved to the low-order eight bits of register PC.

| 1 | 1 | 1 | 0 | 1 | 0 | 0 | 1 |
|---|---|---|---|---|---|---|---|

Cycles: 1
States: 5
Addressing: register
Flags: none

## STACK, I/O AND MACHINE CONTROL GROUP

This group of instructions performs I/O, manipulates the Stack, and alters internal control flags. Unless otherwise specified, *condition flags are not affected by any instructions in this group.*

### PUSH rp and POP rp

To quote the *μCOM-8 Software Manual:*

Two special instructions enable programmers to save and restore the registers using the stack, PUSH and POP. PUSH rp causes the register pair specified by RP to be placed at the top of the stack. The stack is a special portion of read/write memory designated by the user and treated as a last-in-first-out (LIFO) memory through the use of a 16-bit Stack Pointer. A PUSH operation causes the Stack Pointer to decrement by one and store the most-significant register (the HI register) in memory at this new location specified by the Stack Pointer. The Stack Pointer is then decremented again and the least-significant register (the LO register) is then stored in memory at that address. For a POP operation, the data at the memory location addressed by the Stack Pointer is moved into the least-significant register (the LO register, which can be C, E, or L); the Stack Pointer is incremented and the data at the new memory location is loaded into the most-significant register (the HI register, which can be B, D, or H). The Stack Pointer is then incremented again.

For both PUSH and POP operations, the register pair, RP, may be one of the three double registers BC, DE, or HL (identified as B, D, and H, respectively) or the contents of the Flag register and the Accumulator, indicated by PSW (which stands for processor status word).

**PUSH rp**  (Push)
[(SP) − 1] ← (rh)
[(SP) − 2] ← (rl)
(SP) ← (SP) − 2
The content of the high-order register of register pair rp is moved to the memory location whose address is one less than the content of register SP. The content of the low-order register of register pair rp is moved to the memory location whose address is two less than the content of register SP. The content of register SP is decremented by 2. **Note: Register pair rp = SP may not be specified.**

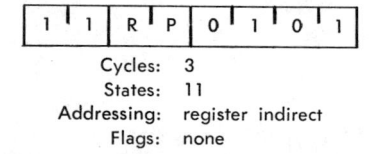

| 1 | 1 | R | P | 0 | 1 | 0 | 1 |
|---|---|---|---|---|---|---|---|

Cycles: 3
States: 11
Addressing: register indirect
Flags: none

**POP rp**  (Pop)
(rl) ← (SP)
(rh) ← [(SP) + 1)]
(SP) ← (SP) + 2
The content of the memory location, whose address is specified by the content of register SP, is moved to the low-order register of register pair rp. The content of the memory location, whose address is one more than the content of register SP, is moved to the high-order register of register pair rp. The content of register SP is incremented by 2. **Note: Register pair rp = SP may not be specified.**

| 1 | 1 | R | P | 0 | 0 | 0 | 1 |
|---|---|---|---|---|---|---|---|

Cycles: 3
States: 10
Addressing: register indirect
Flags: none

The PUSH and POP instructions are represented schematically in Fig. 18-16. In this diagram, SP is the original stack pointer location before the PUSH or POP instruction.

**PUSH psw and POP psw**

The letters psw stand for *processor status word,* which is the contents of the accumulator and the five status flags. We refer you to the description of the PUSH rp and POP rp instructions given earlier. The flag register, F, is regarded as the least-significant register and the accumulator, A, is regarded as the most-significant register.

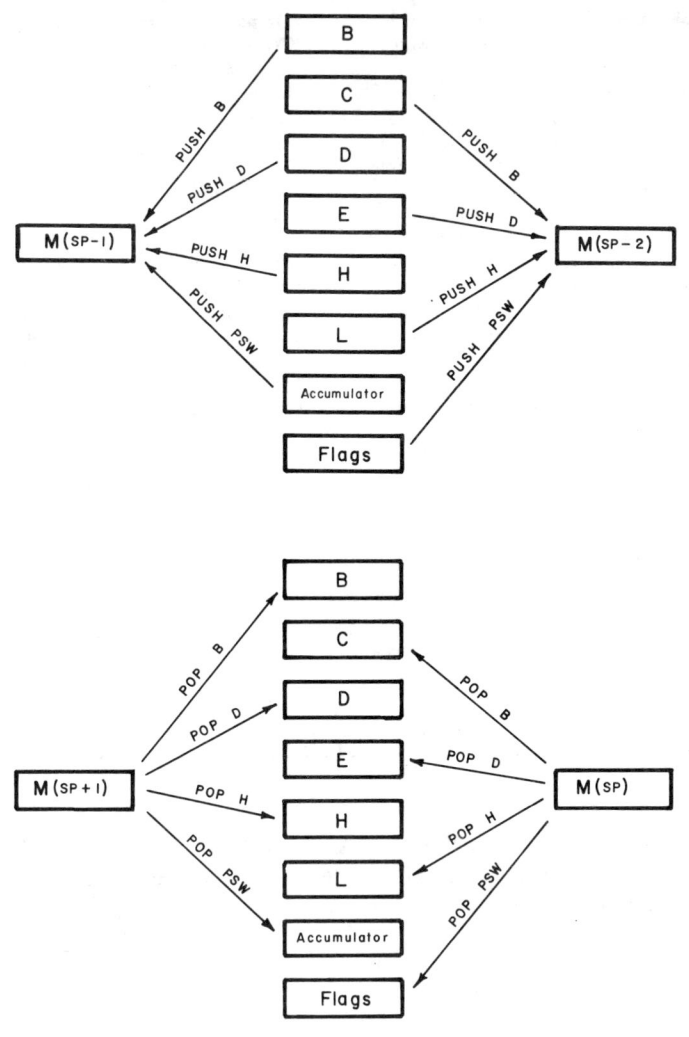

Fig. 18-16. Diagrams representing the PUSH and POP instructions.

The processor status word is important because it saves the actual machine status as determined by the five flag bits. When it is restored, machine operation can resume in the correct state, regardless of how a subroutine affects the flags.

In the μCOM-8 integrated-circuit chip, which is essentially identical in function to the 8080A microprocessor chip, there is an extra status flag, SUB. In the flag register, SUB occupies the $D_5$ bit position. In addition, the $D_3$ bit position is at logic 1 rather than at logic

**PUSH psw** (Push processor status word)
[(SP) − 1] ← (A)
[(SP) − 2]$_0$ ← (CY), [(SP) − 2]$_1$ ← 1
[(SP) − 2]$_2$ ← (P), [(SP) − 2]$_3$ ← 0
[(SP) − 2]$_4$ ← (AC), [(SP) − 2]$_5$ ← 0
[(SP) − 2]$_6$ ← (Z), [(SP) − 2]$_7$ ← (S)
(SP) ← (SP) − 2

The content of register A is moved to the memory location whose address is one less than register SP. The contents of the condition flags are assembled into a processor status word and the word is moved to the memory location whose address is two less than the content of register SP. The content of register SP is decremented by two.

| 1 | 1 | 1 | 1 | 0 | 1 | 0 | 1 |
|---|---|---|---|---|---|---|---|

Cycles: 3
States: 11
Addressing: register indirect
Flags: none

**POP psw** (Pop processor status word)
(CY) ← (SP)$_0$
(P) ← (SP)$_2$
(AC) ← (SP)$_4$
(Z) ← (SP)$_6$
(S) ← (SP)$_7$
(A) ← [(SP) + 1]
(SP) ← (SP) + 2

The content of the memory location whose address is specified by the content of register SP is used to restore the condition flags. The content of the memory location whose address is one more than the content of register SP is moved to register A. The content of register SP is incremented by 2.

| 1 | 1 | 1 | 1 | 0 | 0 | 0 | 1 |
|---|---|---|---|---|---|---|---|

Cycles: 3
States: 10
Addressing: register indirect
Flags: Z, S, P, CY, AC

### FLAG WORD

| $D_7$ | $D_6$ | $D_5$ | $D_4$ | $D_3$ | $D_2$ | $D_1$ | $D_0$ |
|---|---|---|---|---|---|---|---|
| S | Z | 0 | AC | 0 | P | 1 | CY |

0 (which is the case for the 8080A chip). We consider the SUB flag to be a useful feature of 8080A-type microprocessors, and hope that it becomes incorporated in future versions of the chip by manufacturers such as Texas Instruments, National Semiconductor, Intel, etc.

An example of the operation of the Stack is given in Fig. 18-17. The section of code employed is

```
            Subroutine
CALL        PUSH B
<B2>        PUSH D
<B3>        PUSH H
            PUSH PSW
```

The stack pointer originally was located at HI = 003 and LO = 303. After the CALL subroutine instruction, the two program counter bytes are pushed onto the stack and the stack pointer moves to HI = 003 and LO = 301. Note that the HI program counter byte goes on the stack first, but comes off the stack last. A succession of four PUSH

|  | Address | Contents | |
|---|---|---|---|
| SP − 1 | 003 270 | | |
| SP | 003 271 | Flags | Top of stack |
| SP + 1 | 003 272 | Accumulator | |
| | 003 273 | Register L | |
| | 003 274 | Register H | |
| | 003 275 | Register E | |
| | 003 276 | Register D | |
| | 003 277 | Register C | |
| | 003 300 | Register B | |
| | 003 301 | Program Counter (LO byte) | |
| | 003 302 | Program Counter (HI byte) | |
| | 003 303 | | Original SP location |
| | 003 304 | | |
| | 003 305 | | |
| | 003 306 | | |

Fig. 18-17. The "stack."

instructions load the stack with the contents of the six general-purpose registers, the accumlator, and the flag register. After all of this, stack pointer (SP) location is $HI = 003$ and $LO = 271$, the top filled location on the stack.

Once the subroutine has been executed, there is the problem of removing the contents of the stack and placing them back into the 8080A microprocessor chip. The section of code, located at the end of the subroutine, that accomplishes this is

```
POP PSW
POP H
POP D
POP B
RET
```

In each case, the LO byte comes off the stack first. Recall that in three-byte instructions, the LO byte is always the second byte of the instruction. Thus, the 8080A chip is consistent in its handling of 16-bit address words. Once the contents of the stack have been popped off, the stack pointer resumes its original location of HI = 003 and LO = 303.

Registers can be pushed and popped in any order. However, the program counter is almost always pushed first and popped last. The caution that you must observe is that you must pop registers in the reverse order with which you pushed them. For example, with the stack configuration shown in Fig. 18-17, if you executed the following section of code at the end of the subroutine,

```
POP PSW
POP B
POP H
POP D
RET
```

you would encounter problems with the execution of the main program. The original register contents would not be returned to their original locations. The chip would attempt to execute the program, but there is not much chance of a useful result.

If you do not need to push registers on a stack during a subroutine call, do not do so. Store only that information on the stack which is needed by the 8080A chip when it resumes the main program.

## XTHL

The XTHL instruction is used to exchange the contents of the HL register pair with the top pair of items on the stack. The contents of the top location, the one addressed by the stack pointer SP, are exchanged with the contents of register L. The stack pointer is incremented, and the contents of memory addressed by this new value of SP are exchanged with the contents of register H.

**XTHL** (Exchange stack top with H and L)

(L) ↔ (SP)
(H) ↔ [(SP) + 1]

The content of the L register is exchanged with the content of the memory location whose address is specified by the content of register SP. The content of the H register is exchanged with the content of the memory location whose address is one more than the content of register SP.

| 1 | 1 | 1 | 0 | 0 | 0 | 1 | 1 |
|---|---|---|---|---|---|---|---|

Cycles: 5
States: 18
Addressing: register indirect
Flags: none

## SPHL

The SPHL instruction is used to load the stack pointer register with the contents of the register pair H, L. The contents of L are placed in the LO eight bits of the stack pointer, and the contents of H are placed in the HI eight bits of the stack pointer. As pointed out in the NEC Microcomputers, Inc., manual:

"The SPHL instruction can be used to load the stack pointer with a value which has been computed using the double register arithmetic operations available with the HL register pair. This should always be done with care, since it is easy to lose track of where the stack pointer is pointing, with subsequent loss of stack content."

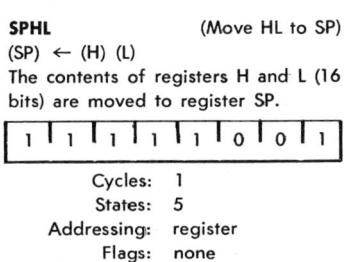

**SPHL** (Move HL to SP)

(SP) ← (H) (L)

The contents of registers H and L (16 bits) are moved to register SP.

| 1 | 1 | 1 | 1 | 1 | 1 | 0 | 0 | 1 |
|---|---|---|---|---|---|---|---|---|

Cycles: 1
States: 5
Addressing: register
Flags: none

## OUT port

The OUT port instruction moves the 8-bit contents of the accumulator to the output port specified by the second byte of the instruction. Two hundred and fifty-six unique output ports can be selected. During the third machine cycle of the instruction, the device code appears on the address bus, an $\overline{\text{OUT}}$ control pulse is generated, and the contents of the accumulator appear on the external bidirectional data bus.

**OUT port** (Output)
(data) ← (A)
The content of register A is placed on the 8-bit bidirectional data bus for transmission to the specified port.

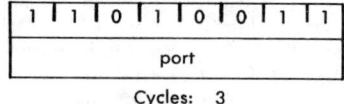

| 1 | 1 | 0 | 1 | 0 | 0 | 1 | 1 |
|---|---|---|---|---|---|---|---|
| port ||||||||

Cycles: 3
States: 10
Addressing: direct
Flags: none

## IN port

The IN port instruction permits the 8080A chip to read the data present at the input port given by the second byte of the instruction. Two hundred and fifty-six unique input ports can be addressed. During the third machine cycle of the instruction, the device code for the input device appears on the address bus, an $\overline{\text{IN}}$ control signal appears on the control bus, and information appearing on the bidirectional data bus also appears in the accumulator.

**IN port** (Input)
(A) ← (data)
The data placed on the 8-bit bidirectional data bus by the specified port is moved to register A.

| 1 | 1 | 0 | 1 | 1 | 0 | 1 | 1 |
|---|---|---|---|---|---|---|---|
| port ||||||||

States: 3
Cycles: 10
Addressing: direct
Flags: none

## EI and DI

To quote the NEC Microcomputers, Inc. *μCOM-8 Software Manual:*

> Whether the 8080A responds to an interrupt request is determined by the state of an internal interrupt flip-flop, INTE. When this flip-flop is set to one, the processor responds to interrupts. When it is reset to zero, the processor ignores interrupt requests. The INTE flip-flop is affected by both program control and system operation. System operations which affect INTE are a system reset and the acknowledgement of an interrupt. Both operations clear INTE and thus disable the interrupt facility. If further interrupts are to be acknowledged after a Reset or Acknowledge Interrupt, the program must re-enable the flip-flop. Two instructions, EI, Enable Interrupt, and DI, Disable Interrupt, provide programmed control of the INTE flip-flop. The EI in-

struction sets the INTE flip-flop to one, enabling the interrupt facility, while the DI instruction clears the INTE flip-flop to zero, disabling the interrupt facility. Thus, if it is desired that a section of the program be executed with high speed and without the possibility of being interrupted, the DI instruction may be used to disable interrupts for that section of code. After the section is complete, EI re-enables the interrupt facility. Since the acknowledgement of an interrupt request resets the INTE flip-flop to zero, an EI should be the first instruction in any routine that services interrupts. (This assumes that the interrupt acknowledgement resets the interrupt request. This must be done to prevent hanging up the 8080A processor.) An exception should be made when servicing the fastest I/O device. To avoid disturbing service to this I/O unit, the INTE flip-flop should be enabled at the end of the routine.

**EI** (Enable interrupts)
The interrupt system is enabled **following the execution of the next instruction.**

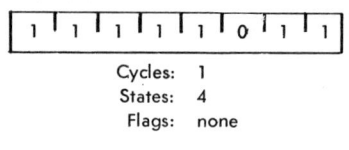

| 1 | 1 | 1 | 1 | 1 | 1 | 0 | 1 | 1 |

Cycles: 1
States: 4
Flags: none

**DI** (Disable interrupts)
The interrupt system is disabled **immediately following the execution of the DI instruction.**

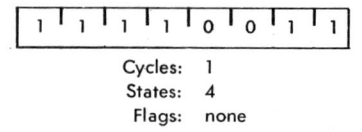

| 1 | 1 | 1 | 1 | 1 | 0 | 0 | 1 | 1 |

Cycles: 1
States: 4
Flags: none

## HLT

The HLT instruction causes the processor to suspend operation until the 8080A chip receives a RESET signal or receives an interrupt request signal (INT). The processor accepts the INT request only if the internal interrupt flip-flop is set (interrupt enabled). After processing the interrupt, instruction execution continues at the next location after the HLT instruction.

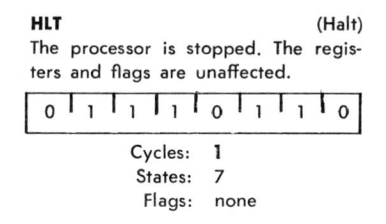

**HLT** (Halt)
The processor is stopped. The registers and flags are unaffected.

| 0 | 1 | 1 | 1 | 1 | 0 | 1 | 1 | 0 |

Cycles: 1
States: 7
Flags: none

## NOP

The NOP instruction does absolutely nothing except consume a location in memory and take up four states during program execution. It is used for program debugging, in which extra NOP instructions are placed in a program for subsequent modification. When deletions are made to a program, NOPs should be inserted in their place.

**NOP** (No op)
No operation is performed. The registers and flags are unaffected.

| 0 | 0 | 0 | 0 | 0 | 0 | 0 | 0 |
|---|---|---|---|---|---|---|---|

Cycles: 1
States: 4
Flags: none

## INSTRUCTION SET

Fig. 18-18 gives a chart of the instruction set. It provides a summary of the processor instructions.

| Mnemonic | Description | $D_7$ | $D_6$ | $D_5$ | $D_4$ | $D_3$ | $D_2$ | $D_1$ | $D_0$ | Clock[2] Cycles |
|---|---|---|---|---|---|---|---|---|---|---|
| MOV r1,r2 | Move register to register | 0 | 1 | D | D | D | S | S | S | 5 |
| MOV M,r | Move register to memory | 0 | 1 | 1 | 1 | 0 | S | S | S | 7 |
| MOV r,M | Move memory to register | 0 | 1 | D | D | D | 1 | 1 | 0 | 7 |
| HLT | Halt | 0 | 1 | 1 | 1 | 0 | 1 | 1 | 0 | 7 |
| MVI r | Move immediate register | 0 | 0 | D | D | D | 1 | 1 | 0 | 7 |
| MVI M | Move immediate memory | 0 | 0 | 1 | 1 | 0 | 1 | 1 | 0 | 10 |
| INR r | Increment register | 0 | 0 | D | D | D | 1 | 0 | 0 | 5 |
| DCR r | Decrement register | 0 | 0 | D | D | D | 1 | 0 | 1 | 5 |
| INR M | Increment memory | 0 | 0 | 1 | 1 | 0 | 1 | 0 | 0 | 10 |
| DCR M | Decrement memory | 0 | 0 | 1 | 1 | 0 | 1 | 0 | 1 | 10 |
| ADD r | Add register to A | 1 | 0 | 0 | 0 | 0 | S | S | S | 4 |
| ADC r | Add register to A with carry | 1 | 0 | 0 | 0 | 1 | S | S | S | 4 |
| SUB r | Subtract register from A | 1 | 0 | 0 | 1 | 0 | S | S | S | 4 |
| SBB r | Subtract register from A with borrow | 1 | 0 | 0 | 1 | 1 | S | S | S | 4 |
| ANA r | And register with A | 1 | 0 | 1 | 0 | 0 | S | S | S | 4 |
| XRA r | Exclusive Or register with A | 1 | 0 | 1 | 0 | 1 | S | S | S | 4 |
| ORA r | Or register with A | 1 | 0 | 1 | 1 | 0 | S | S | S | 4 |
| CMP r | Compare register with A | 1 | 0 | 1 | 1 | 1 | S | S | S | 4 |
| ADD M | Add memory to A | 1 | 0 | 0 | 0 | 0 | 1 | 1 | 0 | 7 |
| ADC M | Add memory to A with carry | 1 | 0 | 0 | 0 | 1 | 1 | 1 | 0 | 7 |
| SUB M | Subtract memory from A | 1 | 0 | 0 | 1 | 0 | 1 | 1 | 0 | 7 |
| SBB M | Subtract memory from A with borrow | 1 | 0 | 0 | 1 | 1 | 1 | 1 | 0 | 7 |
| ANA M | And memory with A | 1 | 0 | 1 | 0 | 0 | 1 | 1 | 0 | 7 |
| XRA M | Exclusive Or memory with A | 1 | 0 | 1 | 0 | 1 | 1 | 1 | 0 | 7 |
| ORA M | Or memory with A | 1 | 0 | 1 | 1 | 0 | 1 | 1 | 0 | 7 |
| CMP M | Compare memory with A | 1 | 0 | 1 | 1 | 1 | 1 | 1 | 0 | 7 |
| ADI | Add immediate to A | 1 | 1 | 0 | 0 | 0 | 1 | 1 | 0 | 7 |
| ACI | Add immediate to A with carry | 1 | 1 | 0 | 0 | 1 | 1 | 1 | 0 | 7 |
| SUI | Subtract immediate from A | 1 | 1 | 0 | 1 | 0 | 1 | 1 | 0 | 7 |
| SBI | Subtract immediate from A with borrow | 1 | 1 | 0 | 1 | 1 | 1 | 1 | 0 | 7 |
| ANI | And immediate with A | 1 | 1 | 1 | 0 | 0 | 1 | 1 | 0 | 7 |
| XRI | Exclusive Or immediate with A | 1 | 1 | 1 | 0 | 1 | 1 | 1 | 0 | 7 |
| ORI | Or immediate with A | 1 | 1 | 1 | 1 | 0 | 1 | 1 | 0 | 7 |
| CPI | Compare immediate with A | 1 | 1 | 1 | 1 | 1 | 1 | 1 | 0 | 7 |
| RLC | Rotate A left | 0 | 0 | 0 | 0 | 0 | 1 | 1 | 1 | 4 |
| RRC | Rotate A right | 0 | 0 | 0 | 0 | 1 | 1 | 1 | 1 | 4 |
| RAL | Rotate A left through carry | 0 | 0 | 0 | 1 | 0 | 1 | 1 | 1 | 4 |
| RAR | Rotate A right through carry | 0 | 0 | 0 | 1 | 1 | 1 | 1 | 1 | 4 |
| JMP | Jump unconditional | 1 | 1 | 0 | 0 | 0 | 0 | 1 | 1 | 10 |
| JC | Jump on carry | 1 | 1 | 0 | 1 | 1 | 0 | 1 | 0 | 10 |
| JNC | Jump on no carry | 1 | 1 | 0 | 1 | 0 | 0 | 1 | 0 | 10 |
| JZ | Jump on zero | 1 | 1 | 0 | 0 | 1 | 0 | 1 | 0 | 10 |
| JNZ | Jump on no zero | 1 | 1 | 0 | 0 | 0 | 0 | 1 | 0 | 10 |
| JP | Jump on positive | 1 | 1 | 1 | 1 | 0 | 0 | 1 | 0 | 10 |
| JM | Jump on minus | 1 | 1 | 1 | 1 | 1 | 0 | 1 | 0 | 10 |
| JPE | Jump on parity even | 1 | 1 | 1 | 0 | 1 | 0 | 1 | 0 | 10 |
| JPO | Jump on parity odd | 1 | 1 | 1 | 0 | 0 | 0 | 1 | 0 | 10 |
| CALL | Call unconditional | 1 | 1 | 0 | 0 | 1 | 1 | 0 | 1 | 17 |
| CC | Call on carry | 1 | 1 | 0 | 1 | 1 | 1 | 0 | 0 | 11/17 |
| CNC | Call on no carry | 1 | 1 | 0 | 1 | 0 | 1 | 0 | 0 | 11/17 |
| CZ | Call on zero | 1 | 1 | 0 | 0 | 1 | 1 | 0 | 0 | 11/17 |
| CNZ | Call on no zero | 1 | 1 | 0 | 0 | 0 | 1 | 0 | 0 | 11/17 |
| CP | Call on positive | 1 | 1 | 1 | 1 | 0 | 1 | 0 | 0 | 11/17 |
| CM | Call on minus | 1 | 1 | 1 | 1 | 1 | 1 | 0 | 0 | 11/17 |
| CPE | Call on parity even | 1 | 1 | 1 | 0 | 1 | 1 | 0 | 0 | 11/17 |
| CPO | Call on parity odd | 1 | 1 | 1 | 0 | 0 | 1 | 0 | 0 | 11/17 |
| RET | Return | 1 | 1 | 0 | 0 | 1 | 0 | 0 | 1 | 10 |
| RC | Return on carry | 1 | 1 | 0 | 1 | 1 | 0 | 0 | 0 | 5/11 |
| RNC | Return on no carry | 1 | 1 | 0 | 1 | 0 | 0 | 0 | 0 | 5/11 |
| RZ | Return on zero | 1 | 1 | 0 | 0 | 1 | 0 | 0 | 0 | 5/11 |
| RNZ | Return on no zero | 1 | 1 | 0 | 0 | 0 | 0 | 0 | 0 | 5/11 |
| RP | Return on positive | 1 | 1 | 1 | 1 | 0 | 0 | 0 | 0 | 5/11 |
| RM | Return on minus | 1 | 1 | 1 | 1 | 1 | 0 | 0 | 0 | 5/11 |
| RPE | Return on parity even | 1 | 1 | 1 | 0 | 1 | 0 | 0 | 0 | 5/11 |
| RPO | Return on parity odd | 1 | 1 | 1 | 0 | 0 | 0 | 0 | 0 | 5/11 |
| RST | Restart | 1 | 1 | A | A | A | 1 | 1 | 1 | 11 |
| IN | Input | 1 | 1 | 0 | 1 | 1 | 0 | 1 | 1 | 10 |
| OUT | Output | 1 | 1 | 0 | 1 | 0 | 0 | 1 | 1 | 10 |
| LXI B | Load immediate register Pair B & C | 0 | 0 | 0 | 0 | 0 | 0 | 0 | 1 | 10 |
| LXI D | Load immediate register Pair D & E | 0 | 0 | 0 | 1 | 0 | 0 | 0 | 1 | 10 |
| LXI H | Load immediate register Pair H & L | 0 | 0 | 1 | 0 | 0 | 0 | 0 | 1 | 10 |
| LXI SP | Load immediate stack pointer | 0 | 0 | 1 | 1 | 0 | 0 | 0 | 1 | 10 |
| PUSH B | Push register Pair B & C on stack | 1 | 1 | 0 | 0 | 0 | 1 | 0 | 1 | 11 |
| PUSH D | Push register Pair D & E on stack | 1 | 1 | 0 | 1 | 0 | 1 | 0 | 1 | 11 |
| PUSH H | Push register Pair H & L on stack | 1 | 1 | 1 | 0 | 0 | 1 | 0 | 1 | 11 |
| PUSH PSW | Push A and Flags on stack | 1 | 1 | 1 | 1 | 0 | 1 | 0 | 1 | 11 |
| POP B | Pop register pair B & C off stack | 1 | 1 | 0 | 0 | 0 | 0 | 0 | 1 | 10 |
| POP D | Pop register pair D & E off stack | 1 | 1 | 0 | 1 | 0 | 0 | 0 | 1 | 10 |
| POP H | Pop register pair H & L off stack | 1 | 1 | 1 | 0 | 0 | 0 | 0 | 1 | 10 |
| POP PSW | Pop A and Flags off stack | 1 | 1 | 1 | 1 | 0 | 0 | 0 | 1 | 10 |
| STA | Store A direct | 0 | 0 | 1 | 1 | 0 | 0 | 1 | 0 | 13 |
| LDA | Load A direct | 0 | 0 | 1 | 1 | 1 | 0 | 1 | 0 | 13 |
| XCHG | Exchange D & E, H & L Registers | 1 | 1 | 1 | 0 | 1 | 0 | 1 | 1 | 4 |
| XTHL | Exchange top of stack, H & L | 1 | 1 | 1 | 0 | 0 | 0 | 1 | 1 | 18 |
| SPHL | H & L to stack pointer | 1 | 1 | 1 | 1 | 1 | 0 | 0 | 1 | 5 |
| PCHL | H & L to program counter | 1 | 1 | 1 | 0 | 1 | 0 | 0 | 1 | 5 |
| DAD B | Add B & C to H & L | 0 | 0 | 0 | 0 | 1 | 0 | 0 | 1 | 10 |
| DAD D | Add D & E to H & L | 0 | 0 | 0 | 1 | 1 | 0 | 0 | 1 | 10 |
| DAD H | Add H & L to H & L | 0 | 0 | 1 | 0 | 1 | 0 | 0 | 1 | 10 |
| DAD SP | Add stack pointer to H & L | 0 | 0 | 1 | 1 | 1 | 0 | 0 | 1 | 10 |
| STAX B | Store A indirect | 0 | 0 | 0 | 0 | 0 | 0 | 1 | 0 | 7 |
| STAX D | Store A indirect | 0 | 0 | 0 | 1 | 0 | 0 | 1 | 0 | 7 |
| LDAX B | Load A indirect | 0 | 0 | 0 | 0 | 1 | 0 | 1 | 0 | 7 |
| LDAX D | Load A indirect | 0 | 0 | 0 | 1 | 1 | 0 | 1 | 0 | 7 |
| INX B | Increment B & C registers | 0 | 0 | 0 | 0 | 0 | 0 | 1 | 1 | 5 |
| INX D | Increment D & E registers | 0 | 0 | 0 | 1 | 0 | 0 | 1 | 1 | 5 |
| INX H | Increment H & L registers | 0 | 0 | 1 | 0 | 0 | 0 | 1 | 1 | 5 |
| INX SP | Increment stack pointer | 0 | 0 | 1 | 1 | 0 | 0 | 1 | 1 | 5 |
| DCX B | Decrement B & C | 0 | 0 | 0 | 0 | 1 | 0 | 1 | 1 | 5 |
| DCX D | Decrement D & E | 0 | 0 | 0 | 1 | 1 | 0 | 1 | 1 | 5 |
| DCX H | Decrement H & L | 0 | 0 | 1 | 0 | 1 | 0 | 1 | 1 | 5 |
| DCX SP | Decrement stack pointer | 0 | 0 | 1 | 1 | 1 | 0 | 1 | 1 | 5 |
| CMA | Complement A | 0 | 0 | 1 | 0 | 1 | 1 | 1 | 1 | 4 |
| STC | Set carry | 0 | 0 | 1 | 1 | 0 | 1 | 1 | 1 | 4 |
| CMC | Complement carry | 0 | 0 | 1 | 1 | 1 | 1 | 1 | 1 | 4 |
| DAA | Decimal adjust A | 0 | 0 | 1 | 0 | 0 | 1 | 1 | 1 | 4 |
| SHLD | Store H & L direct | 0 | 0 | 1 | 0 | 0 | 0 | 1 | 0 | 16 |
| LHLD | Load H & L direct | 0 | 0 | 1 | 0 | 1 | 0 | 1 | 0 | 16 |
| EI | Enable Interrupts | 1 | 1 | 1 | 1 | 1 | 0 | 1 | 1 | 4 |
| DI | Disable interrupt | 1 | 1 | 1 | 1 | 0 | 0 | 1 | 1 | 4 |
| NOP | No-operation | 0 | 0 | 0 | 0 | 0 | 0 | 0 | 0 | 4 |

NOTES: 1. DDD or SSS – 000,B – 001,C – 010,D – 011,E – 100,H – 101,L – 110,Memory – 111,A.
2. Two possible cycle times, (5/11) indicate instruction cycles dependent on condition flags.

Courtesy Intel Corp.

**Fig. 18-18. Summary of processor instructions.**

## INTRODUCTION TO THE EXPERIMENTS

The following experiments provide a number of interesting programs that you may need if you are working with digital instrumentation.

| Experiment No. | Comments |
|---|---|
| 1 | Demonstrates the execution of a routine that converts a two-digit BCD number into an 8-bit binary number. |
| 2 | Demonstrates the execution of a program that performs a sixteen-digit BCD addition of two numbers. The result must be less than or equal to 9,999,999,999,999,999. |
| 3 | Demonstrates the execution of a routine that converts a 16-bit binary number into a five-digit BCD number. |

### EXPERIMENT NO. 1

**Purpose**

The purpose of this experiment is to load and execute a BCD Input and Direct Conversion to Binary Routine, No. 80-147 in the Intel Microcomputer User's Library. The program was developed by M. H. Gansler.

**Program**

| LO Memory Address | Instruction Byte | Mnemonic | Description |
|---|---|---|---|
| 000 | 076 | MVI A | Move immediate byte to the accumulator |
| 001 | * | — | Two-BCD-digit data byte that is to be converted to an 8-bit binary number |
| 002 | 117 | MOV C,A | Move contents of accumulator to register C |
| 003 | 346 | ANI | AND immediate byte with contents of the accumulator |
| 004 | 017 | 017 | Mask byte that masks out the most significant BCD digit |
| 005 | 137 | MOV E,A | Move contents of accumulator to register E |
| 006 | 171 | MOV A,C | Move contents of register C to the accumulator |
| 007 | 346 | ANI | AND immediate byte with contents of the accumulator |
| 010 | 360 | 360 | Mask byte that masks out the least significant BCD digit |
| 011 | 017 | RRC | Rotate the accumulator contents one bit to the right and into the carry flag |
| 012 | 017 | RRC | Same |
| 013 | 117 | MOV C,A | Move contents of accumulator into register C |
| 014 | 017 | RRC | Rotate the accumulator contents one bit to the right and into the carry flag |

| LO Memory Address | Instruction Byte | Mnemonic | Description |
|---|---|---|---|
| 015 | 017 | RRC | Same |
| 016 | 201 | ADD C | Add contents of register C to the contents of the accumulator |
| 017 | 007 | RLC | Rotate the accumulator contents one bit to the left and into the carry flag |
| 020 | 203 | ADD E | Add contents of register E to the contents of the accumulator |
| 021 | 323 | OUT | Output contents of accumulator to output port given in the next instruction byte |
| 022 | 000 | 000 | Device code for output port 0 |
| 023 | 166 | HLT | Halt |

## Discussion

The program starts with the two-digit BCD number in the accumulator. The program can be a subroutine; substitute a RET instruction for the HLT instruction at memory address LO = 023. The above program can be located anywhere in memory. We located the original program at HI = 003 and LO = 000.

## Step 1

Load and execute the above program for the two-digit decimal number 56. The BCD equivalent is 01010110, or 56 in hexadecimal and 126 in octal. What binary number do you observe?

We observed 00111000 in binary, or 070 in octal.

## Step 2

Change the BCD number at memory address HI = 003 and LO = 001 to the numbers given in the following table. Compare your results with the results that we observed.

| Decimal number | Observed binary number | Predicted binary number |
|---|---|---|
| 1 | _____ | 00000001 |
| 10 | _____ | 00001010 |
| 20 | _____ | 00010100 |
| 50 | _____ | 00110010 |
| 75 | _____ | 01001011 |
| 80 | _____ | 01010000 |
| 90 | _____ | 01011010 |
| 99 | _____ | 01100011 |

To confirm the last BCD-to-binary conversions, 99 = 1 + 2 + 32 + 64. Yes, it works.

## EXPERIMENT NO. 2

### Purpose

The purpose of this experiment is to load and execute a 16-digit BCD addition subroutine, in which two BCD numbers are added together to produce a result that is less than or equal to 9,999,999,999,-999,999. This program is listed and described in considerable detail in the $\mu COM$-8 *Software Manual* and is given here courtesy of NEC Microcomputers, Inc. The program is started at memory location HI = 003 and LO = 024.

### Program

| LO Memory Address | Instruction Byte | Mnemonic | Description |
|---|---|---|---|
| 024 | 021 | LXI D | Load immediate two bytes into registers E and D, respectively |
| 025 | 347 | — | Registers D and E contain the 16-bit address of |
| 026 | 003 | — | the least significant digits in the augend |
| 027 | 041 | LXI H | Load immediate two bytes into registers L and H, respectively |
| 030 | 357 | — | Registers H and L contain the 16-bit address of |
| 031 | 003 | — | the least significant digits in the addend |
| ADD16: 032 | 365 | PUSH PSW | Push the program status word onto the stack [NOTE: Make certain that you have loaded the stack pointer before you execute this program.] |
| 033 | 305 | PUSH B | Push the contents of register pair B,C onto the stack. |
| 034 | 016 | MVI C | Move the immediate byte into register C |
| 035 | 010 | — | Binary number equal to one-half the number of BCD digits. Thus, for 16 BCD digits, the octal code would be 010. |
| 036 | 257 | XRA A | Clear the accumulator and carry flag |
| LOOP2: 037 | 032 | LDAX D | Load the accumulator from the memory location addressed by register pair D,E |
| 040 | 216 | ADC M | Add the contents of the memory location addressed by register pair H,L to the contents of the accumulator |
| 041 | 047 | DAA | Decimal adjust the contents of the accumulator |
| 042 | 022 | STAX D | Store the contents of the accumulator into the memory location addressed by register pair D,E |
| 043 | 015 | DCR C | Decrement contents of register C by 1 |
| 044 | 312 | JZ | Jump to the memory location DONE2 if the contents of register C are zero |
| 045 | 054 | — | LO address byte of DONE2 |
| 046 | 003 | — | HI address byte of DONE2 |
| 047 | 053 | DCX H | Decrement contents of register pair H,L by 1 |

| LO Memory Address | Instruction Byte | Mnemonic | Description |
|---|---|---|---|
| 050 | 033 | DCX D | Decrement contents of register pair D,E by 1 |
| 051 | 303 | JMP | Jump to the memory location LOOP2 |
| 052 | 037 | — | LO address byte of LOOP2 |
| 053 | 003 | — | HI address byte of LOOP2 |
| DONE2: 054 | 301 | POP B | Pop contents of register pair B,C off of stack |
| 055 | 361 | POP PSW | Pop the program status word off of stack |
| 056 | 172 | MOV A,D | Move contents of register D to accumulator |
| 057 | 323 | OUT | Output contents of accumulator |
| 060 | 001 | 001 | Device code of port 1 |
| 061 | 173 | MOV A,E | Move contents of register E to accumulator |
| 062 | 323 | OUT | Output contents of accumulator |
| 063 | 000 | 000 | Device code of port 0 |
| 064 | 166 | HLT | Halt |

**Discussion**

This program starts with a sixteen-digit BCD augend in LO memory addresses 340 through 347, with the least-significant BCD digit in location 347 and the most-significant BCD digit in location 340. The sixteen-digit BCD addend is initially in LO memory addresses 350 through 357, with the least-significant BCD digit in location 357 and the most-significant BCD digit in location 350. The terms *addend* and *augend* are defined as follows:[2]

*augend*—In an arithmetic addition, the number increased by having another number (called the addend) added to it.

*addend*—A quantity which, when added to another quantity (called the augend), produces a result called the sum.

Program execution starts at HI = 003 and LO = 024. The sum replaces the augend.

Consider an augend of 1,000,000,000,000,099 and an addend of 8,000,000,000,000,001. The memory map for these two sixteen-digit BCD numbers is as follows (all at HI = 003):

| LO Memory Address | BCD Digits | Octal Code | Binary Code |
|---|---|---|---|
| 340 | 1,0 | 020 | 00010000 |
| 341 | 0,0 | 000 | 00000000 |
| 342 | 0,0 | 000 | 00000000 |
| 343 | 0,0 | 000 | 00000000 |
| 344 | 0,0 | 000 | 00000000 |
| 345 | 0,0 | 000 | 00000000 |
| 346 | 0,0 | 000 | 00000000 |
| 347 | 9,9 | 231 | 10011001 |
| 350 | 8,0 | 200 | 10000000 |
| 351 | 0,0 | 000 | 00000000 |
| 352 | 0,0 | 000 | 00000000 |
| 353 | 0,0 | 000 | 00000000 |
| 354 | 0,0 | 000 | 00000000 |
| 355 | 0,0 | 000 | 00000000 |
| 356 | 0,0 | 000 | 00000000 |
| 357 | 0,1 | 001 | 00000001 |

When these two numbers are added, the sum—9,000,000,000,000,-100—replaces the augend in memory locations HI = 003 and LO = 340 to HI = 003 and LO = 347.

### Step 1

Load the above program into memory. Load the augend, 1,000,-000,000,000,099, and the addend, 8,000,000,000,000,001, into memory. Execute the program. What is the sum, which appears starting at LO memory address 340? *NOTE:* Don't forget to add a three-byte LXI SP instruction, perhaps at HI = 003 and LO = 021, to set the stack pointer.

We observed the following sequence of BCD numbers in successive memory locations starting at LO = 340:

$$90, 00, 00, 00, 00, 00, 01, 00$$

which correspond to the sixteen-digit BCD number, 9,000,000,000,000,100.

### Step 2

Add the following BCD numbers and compare your results with those that we observed.

| Augend | Addend | Sum |
|---|---|---|
| 3,000,000,000,000,100 | 1,000,000,000,000,001 | 4,000,000,000,000,101 |
| 0,000,000,000,123,456 | 0,000,000,000,240,833 | 0,000,000,000,364,289 |
| 0,000,000,000,927,928 | 0,000,000,000,844,992 | 0,000,000,001,772,920 |
| 9,999,999,999,999,999 | 0,000,000,000,000,001 | 0,000,000,000,000,000 |

### Step 3

Change the byte at LO = 035 to 002, which corresponds to the addition of two four-digit BCD numbers. Perform the following additions:

$$0099 + 0001 = 0100$$
$$9999 + 0001 = 0000$$
$$0001 + 0001 = 0002$$
$$0023 + 0077 = 0100$$

### EXPERIMENT NO. 3

#### Purpose

The purpose of this experiment is to load and execute a Binary to BCD subroutine, No. 80-67 in the Intel Microcomputer User's Library. The program was developed by Niels S. Gundestrup of the Geophysical Isotope Laboratory in Denmark.

| LO Memory Address | Instruction Byte | Mnemonic | Description |
|---|---|---|---|
| | 222 | 021 | LXI D | Move immediate two bytes into register pair D. This is the 16-bit binary number that will be converted to a 5-BCD digit number. |
| | 223 | * | — | Least significant 8 bits of 16-bit binary number |
| | 224 | * | — | Most significant 8 bits of 16-bit binary number |
| | 225 | 041 | LXI H | Move immediate two bytes into register pair H. This is the memory address of the most significant digit (MSD) of the 5-digit BCD number. The remaining four digits are stored in successive memory locations, one digit per location. |
| | 226 | 340 | — | L register byte |
| | 227 | 003 | — | H register byte |
| BNBCD: | 230 | 365 | PUSH PSW | Push contents of program status word on stack |
| | 231 | 305 | PUSH B | Push contents of register pair B on stack |
| | 232 | 325 | PUSH D | Push contents of register pair D on stack |
| | 233 | 345 | PUSH H | Push contents of register pair H on stack |
| | 234 | 353 | XCHG | Exchange the contents of register pair H with the contents of register pair D |
| | 235 | 001 | LXI B | Move immediate two bytes into register pair B. (10,000) |
| | 236 | 360 | — | C register byte |
| | 237 | 330 | — | B register byte |
| | 240 | 315 | CALL | Call subroutine DECNO, which performs the binary to BCD conversion. (MSD) |
| | 241 | 276 | — | LO address byte |
| | 242 | 003 | — | HI address byte |
| | 243 | 001 | LXI B | Move immediate two bytes into register pair B. (1000) |
| | 244 | 030 | — | C register byte |
| | 245 | 374 | — | B register byte |
| | 246 | 315 | CALL | Call subroutine DECNO |
| | 247 | 276 | — | LO address byte |
| | 250 | 003 | — | HI address byte |
| | 251 | 001 | LXI B | Move immediate two bytes into register pair B. (100) |
| | 252 | 234 | — | C register byte |
| | 253 | 377 | — | B register byte |
| | 254 | 315 | CALL | Call subroutine DECNO |
| | 255 | 276 | — | LO address byte |
| | 256 | 003 | — | HI address byte |
| | 257 | 001 | LXI B | Move immediate two bytes into register pair B. (10) |
| | 260 | 366 | — | C register byte |
| | 261 | 377 | — | B register byte |
| | 262 | 315 | CALL | Call subroutine DECNO |
| | 263 | 276 | — | LO address byte |
| | 264 | 003 | — | HI address byte |
| | 265 | 175 | MOV A,L | Move contents of register L to the accumulator |
| | 266 | 306 | ADI | Add immediate byte to contents of accumulator |
| | 267 | 000 | 000 | [NOTE: 260 if ASCII code is desired] |
| | 270 | 022 | STAX D | Store contents of accumulator in the memory location addressed by register pair D |

| LO Memory Address | Instruction Byte | Mnemonic | Description |
|---|---|---|---|
| 271 | 341 | POP H | Pop register pair H off stack |
| 272 | 321 | POP D | Pop register pair D off stack |
| 273 | 301 | POP B | Pop register pair B off stack |
| 274 | 361 | POP PSW | Pop program status word off stack |
| 275 | 311 | RET | Return from subroutine |
| DECNO: 276 | 076 | MVI A | Clear contents of accumulator |
| 277 | 000 | 000 | [NOTE: 260 if ASCII code is desired] |
| 300 | 325 | PUSH D | Push register D on stack |
| 301 | 135 | MOV E,L | Move contents of register L to register E |
| 302 | 124 | MOV D,H | Move contents of register H to register D |
| 303 | 074 | INR A | Increment contents of accumulator by 1 |
| 304 | 011 | DAD B | Add contents of register pair B to contents of register pair H and store in register pair H |
| 305 | 332 | JC | Jump if carry flag is at logic 1 |
| 306 | 301 | — | LO address byte |
| 307 | 003 | — | HI address byte |
| 310 | 075 | DCR A | Decrement contents of accumulator by 1 |
| 311 | 153 | MOV L,E | Move contents of register E to register L |
| 312 | 142 | MOV H,D | Move contents of register D to register H |
| 313 | 321 | POP D | Pop register pair D off stack |
| 314 | 022 | STAX D | Store contents of accumulator in the memory location addressed by register pair D |
| 315 | 023 | INX D | Increment contents of register pair D by 1 |
| 316 | 311 | RET | Return from subroutine DECNO |

## Discussion

This program starts with a 16-bit binary number in register pair D, E. The number is converted into a five-BCD digit number that is stored starting at HI = 003 and LO = 340. The most-significant BCD digit is stored at this location, and the remaining four digits in subsequent locations. The least-significant BCD digit is stored at LO = 344. The program BNBCD starts at HI = 003 and LO = 230; however, the 16-bit binary number must exist in register pair D, and the location of the most-significant digit in register pair H. We have used LXI instructions to set this information in the registers before BNBCD is executed. The output can either be as decimal numerals or as 8-bit ASCII code, with the most-significant bit (the parity bit) at logic 1.

The original contribution to the Intel User's library is shown in Step 4. It is presented here by courtesy of the Intel User's Library and the Intel Corporation. Note the style of the assembled program, the program comments (which follow the semicolon on each line), and the fact that both the memory addresses and the instruction bytes are listed in hexadecimal code. This listing gives you important clues concerning the operation of the program.

### Step 1

Load the program into memory. Set the bytes at LO = 267 and LO = 277 to 000. Load 377 into both LO = 223 and LO = 224. These two instruction bytes correspond to the 16-bit binary number, 1111111111111111, which has a decimal value of 65,535.

### Step 2

Execute the program. Reset the microcomputer and determine the contents of memory locations LO = 240 through LO = 244. What sequence of decimal numbers do you observe in these locations?

We observed 6 5 5 3 5, as expected.

### Step 3

Determine the 5-digit BCD equivalent of the following 16-bit binary numbers, which should be loaded at memory locations LO = 223 and LO = 224 before you execute the program. Check your results with ours.

| D register byte (LO = 224) | E register byte (LO = 223) | Observed 5-digit BCD number |
|---|---|---|
| 377 | 377 | 65535 |
| 200 | 000 | 32768 |
| 100 | 000 | 16384 |
| 040 | 000 | 08192 |
| 020 | 000 | 04096 |
| 010 | 000 | 02048 |
| 004 | 000 | 01024 |
| 002 | 000 | 00512 |
| 001 | 000 | 00256 |
| 000 | 200 | 00128 |
| 000 | 100 | 00064 |
| 000 | 040 | 00032 |
| 000 | 020 | 00016 |
| 000 | 010 | 00008 |
| 000 | 004 | 00004 |
| 000 | 002 | 00002 |
| 000 | 001 | 00001 |
| 000 | 000 | 00000 |

### Step 4

A program to convert a 16-bit binary word to a 5-digit BCD number is quite useful. Given below is a hexadecimal listing of a program that starts at HI = 001 and LO = 000. We have loaded it into EPROM and can use it as a subroutine.

| LO Memory Address | Instruction Byte | |
|---|---|---|
| BNBCD: 00 | F5 | |
| 01 | C5 | |
| 02 | D5 | |
| 03 | E5 | |
| 04 | EB | |
| 05 | 01 | |
| 06 | F0 | |
| 07 | D8 | |
| 08 | CD | |
| 09 | 24 | |
| 0A | 01 | |
| 0B | 01 | |
| 0C | 18 | |
| 0D | FC | |
| 0E | CD | |
| 0F | 24 | |
| 10 | 01 | |
| 11 | 01 | |
| 12 | 9C | |
| 13 | FF | |
| 14 | CD | |
| 15 | 24 | |
| 16 | 01 | |
| 17 | 01 | |
| 18 | F6 | |
| 19 | FF | |
| 1A | CD | |
| 1B | 24 | |
| 1C | 01 | |
| 1D | 7D | |
| 1E | 12 | |
| 1F | E1 | |
| 20 | D1 | |
| 21 | C1 | |
| 22 | F1 | |
| 23 | C9 | RET |
| DECNO: 24 | AF | |
| 25 | D5 | |
| 26 | 5D | |
| 27 | 54 | |
| 28 | 3C | |
| 29 | 09 | |
| 2A | DA | |
| 2B | 26 | |
| 2C | 01 | |
| 2D | 3D | |
| 2E | 6B | |
| 2F | 62 | |
| 30 | D1 | |
| 31 | 12 | |
| 32 | 13 | |
| 33 | C9 | RET |

```
                ;BINARY TO BCD SUBROUTINE
                ;INPUT:    UNSIGNED BINARY NUMBER IN D,E
                ;          POINTER TO LOWEST BUFFER LOC IN HL
                ;OUTPUT:   5 BCD-DIGITS, ONE DIGIT PER MEMORY LOC.
                ;          HL POINT TO MSD IN LOWEST LOCATION.

0100  F5        BNBCD:    PUSH PSW       ;SAVE VARIABLES
0101  C5                  PUSH B
0102  D5                  PUSH D
0103  E5                  PUSH H
0104  EB                  XCHG           ;GET NUMBER IN HL, ADDR IN DE
0105  01F0D8              LXI B,-10000
0108  CD2401              CALL DECNO     ;GET MSD
010B  0118FC              LXI B,-1000
010E  CD2401              CALL DECNO
0111  019CFF              LXI B,-100
0114  CD2401              CALL DECNO
0117  01F6FF              LXI B,-10
011A  CD2401              CALL DECNO
011D  7D                  MOV A,L        ;GET LSD
011E  12                  STAX D         ;STORE IT
011F  E1                  POP H
0120  D1                  POP D
0121  C1                  POP B
0122  F1                  POP PSW
0123  C9                  RET

0124  AF        DECNO:    XRA A          ;0 TO A. USE 30H IF ASCII
0125  D5                  PUSH D         ;SAVE ADDR
0126  5D                  MOV E,L        ;SAVE BINARY
0127  54                  MOV D,H
0128  3C                  INR A          ;INCREMENT DIGIT
0129  09                  DAD B          ;SUBTRACT
012A  DA2601              JC DECNO+2     ;RESULT NEGATIVE?
012D  3D                  DCR A          ;YES, RESTORE DIGIT COUNT
012E  6B                  MOV L,E        ;BINARY NUMBER
012F  62                  MOV H,D
0130  D1                  POP D          ;AND ADDRESS
0131  12                  STAX D         ;STORE DIGIT
0132  13                  INX D          ;INCREMENT POINTER
0133  C9                  RET
```

## OCTAL/HEXADECIMAL LISTING OF THE 8080 INSTRUCTION SET

On the following pages, we have provided an extensive listing of the 256 instruction codes in the 8080 microprocessor instruction set. This listing provides the following information:

- The instruction code, in octal.
- The instruction code, in hexadecimal.
- The instruction code, in the Intel Corporation Mnemonic code.
- A brief description of what the instruction code does.

You may wish to make a copy of this listing and keep it handy. We have found the listing to be of particular value when we attempt to convert an octal listing into hexadecimal, or vice versa.

We have also provided a summary of the 8080 instruction set following the octal/hexadecimal listing. The summary is arranged according to the number of bytes in the instruction.

## The Instruction Codes

&lt;B1&gt;

| OCTAL | HEX | MNEMONIC | DESCRIPTION |
|-------|-----|----------|-------------|
| 000 | 00 | NOP | No operation |
| 001 | 01 | LXI B &lt;B2&gt; &lt;B3&gt; | Load immediate into register pair B and C |
| 002 | 02 | STAX B | Store A into M addressed by B and C |
| 003 | 03 | INX B | Increment contents of register pair B and C by 1 |
| 004 | 04 | INR B | Increment register B by 1 |
| 005 | 05 | DCR B | Decrement register B by 1 |
| 006 | 06 | MVI B &lt;B2&gt; | Move immediate into register B |
| 007 | 07 | RLC | Rotate A left |
| 010 | 08 | — | — |
| 011 | 09 | DAD B | Add contents of B,C to H,L and store in H,L |
| 012 | 0A | LDAX B | Load A from M addressed by B and C |
| 013 | 0B | DCX B | Decrement contents of register pair B and C by 1 |
| 014 | 0C | INR C | Increment register C by 1 |
| 015 | 0D | DCR C | Decrement register C by 1 |
| 016 | 0E | MVI C &lt;B2&gt; | Move immediate into register C |
| 017 | 0F | RRC | Rotate A right |
| 020 | 10 | — | — |
| 021 | 11 | LXI D &lt;B2&gt; &lt;B3&gt; | Load immediate into register pair D and E |
| 022 | 12 | STAX D | Store A into M addressed by D and E |
| 023 | 13 | INX D | Increment contents of register pair D and E by 1 |
| 024 | 14 | INR D | Increment register D by 1 |
| 025 | 15 | DCR D | Decrement register D by 1 |
| 026 | 16 | MVI D &lt;B2&gt; | Move immediate into register D |
| 027 | 17 | RAL | Rotate A left through carry |
| 030 | 18 | — | — |
| 031 | 19 | DAD D | Add contents of D,E to H,L and store in H,L |
| 032 | 1A | LDAX D | Load A from M addressed by D and E |
| 033 | 1B | DCX D | Decrement contents of register pair D and E by 1 |
| 034 | 1C | INR E | Increment register E by 1 |
| 035 | 1D | DCR E | Decrement register E by 1 |
| 036 | 1E | MVI E &lt;B2&gt; | Move immediate into register E |
| 037 | 1F | RAR | Rotate A right through carry |
| 040 | 20 | — | — |
| 041 | 21 | LXI H &lt;B2&gt; &lt;B3&gt; | Load immediate into register pair H and L |
| 042 | 22 | SHLD &lt;B2&gt; &lt;B3&gt; | Store L and H into M and M+1, where M = &lt;B2&gt; &lt;B3&gt; |
| 043 | 23 | INX H | Increment contents of register pair H and L by 1 |
| 044 | 24 | INR H | Increment register H by 1 |
| 045 | 25 | DCR H | Decrement register H by 1 |
| 046 | 26 | MVI H &lt;B2&gt; | Move immediate into register H |
| 047 | 27 | DAA | Decimal adjust A |
| 050 | 28 | — | — |
| 051 | 29 | DAD H | Add contents of H,L to H,L and store in H,L |

<B1>

| OCTAL | HEX | MNEMONIC | DESCRIPTION |
|---|---|---|---|
| 052 | 2A | LHLD <B2> <B3> | Load L and H with contents of M and M+1, respectively |
| 053 | 2B | DCX H | Decrement contents of register pair H and L by 1 |
| 054 | 2C | INR L | Increment register L by 1 |
| 055 | 2D | DCR L | Decrement register L by 1 |
| 056 | 2E | MVI L <B2> | Move immediate into register L |
| 057 | 2F | CMA | Complement A |
| 060 | 30 | — | — |
| 061 | 31 | LXI SP <B2> <B3> | Load immediate into stack pointer |
| 062 | 32 | STA <B2> <B3> | Store A into M addressed by <B2> <B3> |
| 063 | 33 | INX SP | Increment register SP by 1 |
| 064 | 34 | INR M | Increment contents of M by 1 |
| 065 | 35 | DCR M | Decrement contents of M by 1 |
| 066 | 36 | MVI M <B2> | Move immediate into M addressed by H and L |
| 067 | 37 | STC | Set carry flip-flop to logic 1 |
| 070 | 38 | — | |
| 071 | 39 | DAD SP | Add stack pointer contents to H,L and store in H,L |
| 072 | 3A | LDA <B2> <B3> | Load A direct with contents of M addressed by <B2> <B3> |
| 073 | 3B | DCX SP | Decrement register SP by 1 |
| 074 | 3C | INR A | Increment register A by 1 |
| 075 | 3D | DCR A | Decrement register A by 1 |
| 076 | 3E | MVI A <B2> | Move immediate into register A |
| 077 | 3F | CMC | Complement carry flip-flop |
| 100 | 40 | MOV B,B | Move contents of register B to register B |
| 101 | 41 | MOV B,C | Move contents of register C to register B |
| 102 | 42 | MOV B,D | Move contents of register D to register B |
| 103 | 43 | MOV B,E | Move contents of register E to register B |
| 104 | 44 | MOV B,H | Move contents of register H to register B |
| 105 | 45 | MOV B,L | Move contents of register L to register B |
| 106 | 46 | MOV B,M | Move contents of M to register B |
| 107 | 47 | MOV B,A | Move contents of register A to register B |
| 110 | 48 | MOV C,B | Move contents of register B to register C |
| 111 | 49 | MOV C,C | Move contents of register C to register C |
| 112 | 4A | MOV C,D | Move contents of register D to register C |
| 113 | 4B | MOV C,E | Move contents of register E to register C |
| 114 | 4C | MOV C,H | Move contents of register H to register C |
| 115 | 4D | MOV C,L | Move contents of register L to register C |
| 116 | 4E | MOV C,M | Move contents of M to register C |
| 117 | 4F | MOV C,A | Move contents of register A to register C |
| 120 | 50 | MOV D,B | Move contents of register B to register D |
| 121 | 51 | MOV D,C | Move contents of register C to register D |
| 122 | 52 | MOV D,D | Move contents of register D to register D |
| 123 | 53 | MOV D,E | Move contents of register E to register D |
| 124 | 54 | MOV D,H | Move contents of register H to register D |
| 125 | 55 | MOV D,L | Move contents of register L to register D |
| 126 | 56 | MOV D,M | Move contents of M to register D |
| 127 | 57 | MOV D,A | Move contents of register A to register D |
| 130 | 58 | MOV E,B | Move contents of register B to register E |
| 131 | 59 | MOV E,C | Move contents of register C to register E |
| 132 | 5A | MOV E,D | Move contents of register D to register E |
| 133 | 5B | MOV E,E | Move contents of register E to register E |

<B1>

| OCTAL | HEX | MNEMONIC | DESCRIPTION |
|---|---|---|---|
| 134 | 5C | MOV E,H | Move contents of register H to register E |
| 135 | 5D | MOV E,L | Move contents of register L to register E |
| 136 | 5E | MOV E,M | Move contents of M to register E |
| 137 | 5F | MOV E,A | Move contents of register A to register E |
| 140 | 60 | MOV H,B | Move contents of register B to register H |
| 141 | 61 | MOV H,C | Move contents of register C to register H |
| 142 | 62 | MOV H,D | Move contents of register D to register H |
| 143 | 63 | MOV H,E | Move contents of register E to register H |
| 144 | 64 | MOV H,H | Move contents of register H to register H |
| 145 | 65 | MOV H,L | Move contents of register L to register H |
| 146 | 66 | MOV H,M | Move contents of M to register H |
| 147 | 67 | MOV H,A | Move contents of register A to register H |
| 150 | 68 | MOV L,B | Move contents of register B to register L |
| 151 | 69 | MOV L,C | Move contents of register C to register L |
| 152 | 6A | MOV L,D | Move contents of register D to register L |
| 153 | 6B | MOV L,E | Move contents of register E to register L |
| 154 | 6C | MOV L,H | Move contents of register H to register L |
| 155 | 6D | MOV L,L | Move contents of register L to register L |
| 156 | 6E | MOV L,M | Move contents of M to register L |
| 157 | 6F | MOV L,A | Move contents of register A to register L |
| 160 | 70 | MOV M,B | Move contents of register B to M |
| 161 | 71 | MOV M,C | Move contents of register C to M |
| 162 | 72 | MOV M,D | Move contents of register D to M |
| 163 | 73 | MOV M,E | Move contents of register E to M |
| 164 | 74 | MOV M,H | Move contents of register H to M |
| 165 | 75 | MOV M,L | Move contents of register L to M |
| 166 | 76 | HLT | Halt |
| 167 | 77 | MOV M,A | Move contents of register A to M |
| 170 | 78 | MOV A,B | Move contents of register B to register A |
| 171 | 79 | MOV A,C | Move contents of register C to register A |
| 172 | 7A | MOV A,D | Move contents of register D to register A |
| 173 | 7B | MOV A,E | Move contents of register E to register A |
| 174 | 7C | MOV A,H | Move contents of register H to register A |
| 175 | 7D | MOV A,L | Move contents of register L to register A |
| 176 | 7E | MOV A,M | Move contents of M to register A |
| 177 | 7F | MOV A,A | Move contents of register A to register A |
| 200 | 80 | ADD B | Add contents of register B to register A |
| 201 | 81 | ADD C | Add contents of register C to register A |
| 202 | 82 | ADD D | Add contents of register D to register A |
| 203 | 83 | ADD E | Add contents of register E to register A |
| 204 | 84 | ADD H | Add contents of register H to register A |
| 205 | 85 | ADD L | Add contents of register L to register A |
| 206 | 86 | ADD M | Add contents of M to register A |
| 207 | 87 | ADD A | Add contents of register A to register A |
| 210 | 88 | ADC B | Add carry and contents of register B to register A |
| 211 | 89 | ADC C | Add carry and contents of register C to register A |
| 212 | 8A | ADC D | Add carry and contents of register D to register A |
| 213 | 8B | ADC E | Add carry and contents of register E to register A |
| 214 | 8C | ADC H | Add carry and contents of register H to register A |
| 215 | 8D | ADC L | Add carry and contents of register L to register A |
| 216 | 8E | ADC M | Add carry and contents of M to register A |
| 217 | 8F | ADC A | Add carry and contents of register A to register A |
| 220 | 90 | SUB B | Subtract contents of register B from register A |

<B1>
| OCTAL | HEX | MNEMONIC | DESCRIPTION |
|---|---|---|---|
| 221 | 91 | SUB C | Subtract contents of register C from register A |
| 222 | 92 | SUB D | Subtract contents of register D from register A |
| 223 | 93 | SUB E | Subtract contents of register E from register A |
| 224 | 94 | SUB H | Subtract contents of register H from register A |
| 225 | 95 | SUB L | Subtract contents of register L from register A |
| 226 | 96 | SUB M | Subtract contents of M from register A |
| 227 | 97 | SUB A | Clear register A |
| 230 | 98 | SBB B | Subtract carry and contents of register B from A |
| 231 | 99 | SBB C | Subtract carry and contents of register C from A |
| 232 | 9A | SBB D | Subtract carry and contents of register D from A |
| 233 | 9B | SBB E | Subtract carry and contents of register E from A |
| 234 | 9C | SBB H | Subtract carry and contents of register H from A |
| 235 | 9D | SBB L | Subtract carry and contents of register L from A |
| 236 | 9E | SBB M | Subtract carry and contents of M from register A |
| 237 | 9F | SBB A | Subtract carry and contents of register A from A |
| 240 | A0 | ANA B | AND contents of register B with register A |
| 241 | A1 | ANA C | AND contents of register C with register A |
| 242 | A2 | ANA D | AND contents of register D with register A |
| 243 | A3 | ANA E | AND contents of register E with register A |
| 244 | A4 | ANA H | AND contents of register H with register A |
| 245 | A5 | ANA L | AND contents of register L with register A |
| 246 | A6 | ANA M | AND contents of M with register A |
| 247 | A7 | ANA A | AND contents of register A with register A |
| 250 | A8 | XRA B | Exclusive-OR contents of register B with register A |
| 251 | A9 | XRA C | Exclusive-OR contents of register C with register A |
| 252 | AA | XRA D | Exclusive-OR contents of register D with register A |
| 253 | AB | XRA E | Exclusive-OR contents of register E with register A |
| 254 | AC | XRA H | Exclusive-OR contents of register H with register A |
| 255 | AD | XRA L | Exclusive-OR contents of register L with register A |
| 256 | AE | XRA M | Exclusive-OR contents of M with register A |
| 257 | AF | XRA A | Clear register A |
| 260 | B0 | ORA B | OR contents of register B with register A |
| 261 | B1 | ORA C | OR contents of register C with register A |
| 262 | B2 | ORA D | OR contents of register D with register A |
| 263 | B3 | ORA E | OR contents of register E with register A |
| 264 | B4 | ORA H | OR contents of register H with register A |
| 265 | B5 | ORA L | OR contents of register L with register A |
| 266 | B6 | ORA M | OR contents of M with register A |
| 267 | B7 | ORA A | OR contents of register A with register A |
| 270 | B8 | CMP B | Compare contents of register B with register A |
| 271 | B9 | CMP C | Compare contents of register C with register A |
| 272 | BA | CMP D | Compare contents of register D with register A |
| 273 | BB | CMP E | Compare contents of register E with register A |
| 274 | BC | CMP H | Compare contents of register H with register A |
| 275 | BD | CMP L | Compare contents of register L with register A |
| 276 | BE | CMP M | Compare contents of M with register A |
| 277 | BF | CMP A | Compare contents of register A with register A |
| 300 | C0 | RNZ | Return from subroutine if zero flip-flop = logic 0 |
| 301 | C1 | POP B | Pop register pair B and C off the stack |
| 302 | C2 | JNZ <B2> <B3> | Jump if zero flip-flop = logic 0 |
| 303 | C3 | JMP <B2> <B3> | Jump unconditionally to M addressed by <B2> <B3> |
| 304 | C4 | CNZ <B2> <B3> | Call subroutine if zero flip-flop = logic 0 |

<B1>

| OCTAL | HEX | MNEMONIC | DESCRIPTION |
|---|---|---|---|
| 305 | C5 | PUSH B | Push contents of register pair B and C on stack |
| 306 | C6 | ADI <B2> | Add immediate to register A |
| 307 | C7 | RST 0 | Call subroutine at address $000_8$ |
| 310 | C8 | RZ | Return from subroutine if zero flip-flop = logic 1 |
| 311 | C9 | RET | Return from subroutine |
| 312 | CA | JZ <B2> <B3> | Jump if zero flip-flop = logic 1 |
| 313 | CB | — | — |
| 314 | CC | CZ <B2> <B3> | Call subroutine if zero flip-flop = logic 1 |
| 315 | CD | CALL <B2> <B3> | Call subroutine located at M = <B2> <B3> |
| 316 | CE | ACI <B2> | Add immediate byte and carry to register A |
| 317 | CF | RST 1 | Call subroutine at address $000_8/010_8$ |
| 320 | D0 | RNC | Return from subroutine if carry flip-flop = logic 0 |
| 321 | D1 | POP D | Pop register pair D and E off the stack |
| 322 | D2 | JNC <B2> <B3> | Jump if carry flip-flop = logic 0 |
| 323 | D3 | OUT <B2> | Output to device addressed by <B2> |
| 324 | D4 | CNC <B2> <B3> | Call subroutine if carry flip-flop = logic 0 |
| 325 | D5 | PUSH D | Push contents of register pair D and E on stack |
| 326 | D6 | SUI <B2> | Subtract immediate from register A |
| 327 | D7 | RST 2 | Call subroutine at address $000_8/020_8$ |
| 330 | D8 | RC | Return from subroutine if carry flip-flop = logic 1 |
| 331 | D9 | — | — |
| 332 | DA | JC <B2> <B3> | Jump if carry flip-flop = logic 1 |
| 333 | DB | IN <B2> | Input from device addressed by <B2> |
| 334 | DC | CC <B2> <B3> | Call subroutine if carry flip-flop = logic 1 |
| 335 | DD | — | — |
| 336 | DE | SBI <B2> | Subtract immediate byte and carry from register A |
| 337 | DF | RST 3 | Call subroutine at address $000_8/030_8$ |
| 340 | E0 | RPO | Return from subroutine if parity flip-flop = logic 0 |
| 341 | E1 | POP H | Pop register pair H and L off the stack |
| 342 | E2 | JPO <B2> <B3> | Jump if parity flip-flop = logic 0 |
| 343 | E3 | XTHL | Exchange top of stack with contents of H and L |
| 344 | E4 | CPO <B2> <B3> | Call subroutine if parity flip-flop = logic 0 |
| 345 | E5 | PUSH H | Push contents of register pair H and L on stack |
| 346 | E6 | ANI <B2> | AND immediate with contents of register A |
| 347 | E7 | RST 4 | Call subroutine at address $000_8/040_8$ |
| 350 | E8 | RPE | Return from subroutine if parity flip-flop = logic 1 |
| 351 | E9 | PCHL | Jump to M addressed by register pair H and L |
| 352 | EA | JPE <B2> <B3> | Jump if parity flip-flop = logic 1 |
| 353 | EB | XCHG | Exchange contents of registers H,L with registers D,E |
| 354 | EC | CPE <B2> <B3> | Call subroutine if parity flip-flop = logic 1 |
| 355 | ED | — | — |
| 356 | EE | XRI <B2> | Exclusive-OR immediate with contents of register A |
| 357 | EF | RST 5 | Call subroutine at address $000_8/050_8$ |
| 360 | F0 | RP | Return from subroutine if sign flip-flop = logic 0 |
| 361 | F1 | POP PSW | Pop register A and flags off the stack |
| 362 | F2 | JP <B2> <B3> | Jump if sign flip-flop = logic 0 [positive sign] |
| 363 | F3 | DI | Disable interrupt |
| 364 | F4 | CP <B2> <B3> | Call subroutine if sign flip-flop = logic 0 |
| 365 | F5 | PUSH PSW | Push contents of register A and flags on stack |
| 366 | F6 | ORI <B2> | OR immediate with contents of register A |
| 367 | F7 | RST 6 | Call subroutine at address $000_8/060_8$ |

<B1>

| OCTAL | HEX | MNEMONIC | DESCRIPTION |
|---|---|---|---|
| 370 | F8 | RM | Return from subroutine if sign flip-flop = logic 1 |
| 371 | F9 | SPHL | Transfer contents of registers H,L to stack pointer |
| 372 | FA | JM <B2> <B3> | Jump if sign flip-flop = logic 1 [minus sign] |
| 373 | FB | EI | Enable interrupt |
| 374 | FC | CM <B2> <B3> | Call subroutine if sign flip-flop = logic 1 |
| 375 | FD | — | — |
| 376 | FE | CPI <B2> | Compare immediate with contents of register A |
| 377 | FF | RST 7 | Call subroutine at address $000_8/070_8$ |

## 8080 INSTRUCTION SET SUMMARY

### Single-Byte Instructions

| | | | | | | | | | |
|---|---|---|---|---|---|---|---|---|---|
| INR r | 0S4 | INX B | 003 | POP B | 301 | RNZ | 300 | XCHG | 353 |
| DCR r | 0S5 | INX D | 023 | POP D | 321 | RZ | 310 | XTHL | 343 |
| | | INX H | 043 | POP H | 341 | RNC | 320 | SPHL | 371 |
| MOV $r_1r_2$ | 1DS | INX SP | 063 | POP PSW | 361 | RC | 330 | PCHL | 351 |
| | | | | | | RPO | 340 | HLT | 166 |
| ADD r | 20S | DCX B | 013 | PUSH B | 305 | RPE | 350 | NOP | 000 |
| ADC r | 21S | DCX D | 033 | PUSH D | 325 | RP | 360 | DI | 363 |
| SUB r | 22S | DCX H | 053 | PUSH H | 345 | RM | 370 | EI | 373 |
| SBB r | 23S | DCX SP | 073 | PUSH PSW | 365 | RET | 311 | | |
| ANA r | 24S | | | | | | | DAA | 047 |
| XRA r | 25S | DAD B | 011 | STAX B | 002 | RLC | 007 | CMA | 057 |
| ORA r | 26S | DAD D | 031 | STAX D | 022 | RRC | 017 | STC | 067 |
| CMP r | 27S | DAD H | 051 | LDAX B | 012 | RAL | 027 | CMC | 077 |
| | | DAD SP | 071 | LDAX D | 032 | RAR | 037 | RST | 3A7 |

*S and D:* B = 0, C = 1, D = 2, E = 3, H = 4, L = 5, M = 6, accumulator = 7

*A:* 0 through 7

### Two-Byte Instructions

| | | | | | |
|---|---|---|---|---|---|
| ADI <B2> | 306 | IN <B2> | 333 | MVI B <B2> | 006 |
| ACI <B2> | 316 | OUT <B2> | 323 | MVI C <B2> | 016 |
| SUI <B2> | 326 | | | MVI D <B2> | 026 |
| SBI <B2> | 336 | | | MVI E <B2> | 036 |
| ANI <B2> | 346 | | | MVI H <B2> | 046 |
| XRI <B2> | 356 | | | MVI L <B2> | 056 |
| ORI <B2> | 366 | | | MVI M <B2> | 066 |
| CPI <B2> | 376 | | | MVI A <B2> | 076 |

### Three-Byte Instructions

| | | | | | |
|---|---|---|---|---|---|
| JNZ <B2> <B3> | 302 | CNZ <B2> <B3> | 304 | LXI B <B2> <B3> | 001 |
| JZ <B2> <B3> | 312 | CZ <B2> <B3> | 314 | LXI D <B2> <B3> | 021 |
| JNC <B2> <B3> | 322 | CNC <B2> <B3> | 324 | LHI H <B2> <B3> | 041 |
| JC <B2> <B3> | 332 | CC <B2> <B3> | 334 | LXI SP <B2> <B3> | 061 |
| JPO <B2> <B3> | 342 | CPO <B2> <B3> | 344 | | |
| JPE <B2> <B3> | 352 | CPE <B2> <B3> | 354 | STA <B2> <B3> | 062 |
| JP <B2> <B3> | 362 | CP <B2> <B3> | 364 | LDA <B2> <B3> | 072 |
| JM <B2> <B3> | 372 | CM <B2> <B3> | 374 | SHLD <B2> <B3> | 042 |
| JMP <B2> <B3> | 303 | CALL <B2> <B3> | 315 | LHLD <B2> <B3> | 052 |

## REVIEW QUESTIONS

The following questions will help you review the 8080A instruction set.

1. Which flags do the following instructions influence when they are executed? Use the following abbreviations: Zero flag = Z, Carry flag = C, Parity flag = P, Sign flag = S, and Auxiliary Carry flag = AC.
   a. JMP
   b. POP B
   c. INX D
   d. STAX D
   e. RST n
   f. LXI SP
   g. RET
   h. SHLD
   i. LHLD
   j. DAA
   k. EI
   l. XTHL
   m. PCHL
   n. DAD B
   o. CMC
   p. CMP r
2. The stack pointer is initially HI = 004 and LO = 000. You call a subroutine and then execute the following instructions in the order given:
   PUSH D
   PUSH H
   PUSH PSW
   PUSH B

   At what memory locations in the stack do the contents of the internal registers appear? Answer this question for each register.
3. What instructions can you use to control the location of the stack? List them.
4. Explain the similarities and/or differences between the following pairs of concepts. (This is a review question that contains material from other units.)
   a. Register vs register pair
   b. Byte vs bit
   c. Byte vs word
   d. Word vs memory address
   e. HI address byte vs LO address byte
   f. Jump vs call

g. Conditional vs unconditional instruction
h. OR vs Exclusive-OR for a 2-input gate
i. Zero flag vs sign flag
j. Carry flag vs auxiliary carry flag
k. PUSH vs POP
l. Accumulator vs ALU
m. Data byte vs address byte
n. Octal code vs hexadecimal code
o. Increment vs decrement
p. Increment vs ADD
q. IN vs OUT instructions
r. ADD vs ADC instructions
s. MOV vs MVI instructions
t. MVI vs LXI instructions
u. EI vs DI instructions
v. SUB A vs XRA A instructions
w. Carry vs borrow
x. Machine code vs mnemonic code
y. B register pair vs the H register pair
z. Instruction register vs instruction decoder

# 19

# Data Bus Techniques Using Three-State Devices

### INTRODUCTION

A bus is a set of common conducting paths over which digital information is transferred, from any of several sources to any of several destinations. The fundamental objective of a bus is to minimize the number of interconnections required to transfer information between digital devices. In this unit, we shall describe three-state bussing, the bussing technique that is currently used in microprocessor chips and microprocessor systems.

### OBJECTIVES

At the completion of this unit, you will be able to do the following:

- Define bus and the verb, to bus.
- Describe the characteristics of a TRI-STATE, or three-state, buffer, including the data and enable/disable inputs as well as the three-state output.
- Write a truth table for a three-state device.
- Provide one or two examples of simple bus systems.
- Describe the general characteristics of three-state chips such as the 74125, 74126, and 8095.
- Write a truth table for a three-state latch/buffer.

## WHAT IS A BUS?

A digital *bus* is a path over which digital information is transferred, from any of several sources to any of several destinations. Only one transfer of information can take place at any one time. While such a transfer is taking place, all other sources that are tied to the bus must be disabled. The verb *to bus* means to interconnect several digital devices, which either receive or transmit digital information, by a common set of conducting paths (called a bus) over which all information between such devices is transferred.

The fundamental purpose of a bus is to minimize the number of interconnections required to transfer information between digital devices. Busses are present within integrated-circuit chips, e.g., the internal data bus within an 8080A microprocessor chip; between integrated-circuit chips, e.g., the address, control, and bidirectional data busses present in an 8080A-based microcomputer; and between digital systems and instruments, e.g., the Hewlett-Packard interface bus that is now a standard interface between digital instruments.

Though not discussed much in textbooks on digital electronics, the concept of a bus is probably one of the most important concepts in digital electronics. Without the ability to share information paths, most digital devices would probably require three to four times the number of wire connections that they presently have. Printed-circuit boards for microcomputers and minicomputers would be considerably more complex and expensive.

## THREE-STATE BUSSING

In a bus system, the optimum gate should have two digital output states (logic 0 and logic 1) and a third disconnected or isolated state. That such should be the case can be easily seen from the following truth table:

| Input Data | Gating Signal | Output |
|---|---|---|
| 0 | enable | 0 |
| 1 | enable | 1 |
| 0 | disable | Disconnected from bus |
| 1 | disable | Disconnected from bus |

In other words, the third state is a condition in which the gate is "disconnected" from the bus and no input data appears on the bus *from this specific gate.*

The solution pioneered by National Semiconductor Corporation is the TRI-STATE®, or three-state, output. It is appropriate to quote from their catalogue, *Digital Integrated Circuits,* a description of the TRI-STATE concept.

This unique TRI-STATE concept allows outputs to be tied together and then connected to a common bus line. Normal TTL outputs cannot to be connected owing to the low-impedance logical "1" output current which one device would have to sink from the other. If, however, on all but one of the connected devices, both the upper and lower output transistors are turned off, then the one remaining device in the normal low-impedance state will have to supply to or sink from the other devices only a small amount of leakage current. . . .

. . . While true that in a TTL system open-collector gates could be used to perform the logic function of these TRI-STATE elements, neither waveform integrity nor optimum speed would be achieved. The low output impedance of TRI-STATE devices provides good capacitance drive capability and rapid transition from the logical "0" to logical "1" level, thus assuring both speed and waveform integrity.

It is possible to connect as many as 128 devices to a common bus line and still have adequate drive capability to allow fan-out from the bus. . . .

Another advantage of these buffers is that in the high-impedance state, their inputs do not present the normal loading to the driving device. This is significant when it is desirable to transmit in both directions over a common line. . . .

To summarize the above, a TRI-STATE device has three possible output states: (1) A logical "0" state, (2) A logical "1" state, and (3) A high-impedance output state that is, in effect, disconnected from the bus line. All three-state devices have an input pin called an *enable/disable* input, which permits the logic devices either to behave normally or to exist in the high-impedance state. When enabled, a TRI-STATE device behaves as a normal TTL device; when disabled, a TRI-STATE device behaves as if it is, in effect, disconnected from the circuit.

The truth table for a typical three-state device, as shown in Fig. 19-1, is as follows:

| Input Data | Gating Signal | Output Data |
|---|---|---|
| 0 | enable | 0 |
| 1 | enable | 1 |
| X | disable | High impedance |

X = irrelevant

**Fig. 19-1. Diagram of a three-state buffer.**

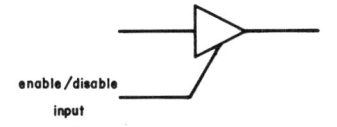

enable/disable input

## EXAMPLES OF SIMPLE BUS SYSTEMS

In Fig. 19-2, we show a simple four-device one-line bus system that is based upon the use of a single 74126 quad three-state buffer chip. We recognize the circuit as a bus system because the outputs of gates A through D are connected together. With standard 7400-series TTL chips, it is not possible to do so unless the chips have special output circuits that permit bussing, such as either *three-state* or *open-collector* outputs.

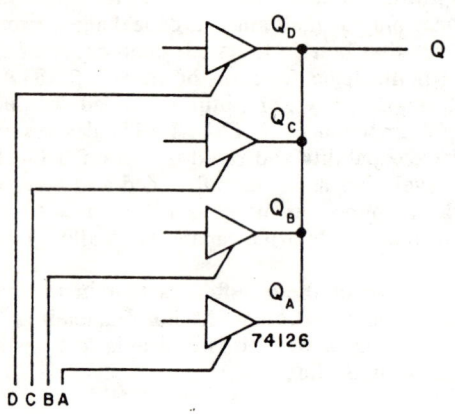

Fig. 19-2. A simple four-device, one-line bus system.

If we assume that gates A through D in Fig. 19-2 are enabled by a logic 1 input, the operation of the circuit should be clear. *Only one buffer gate may be enabled at any instant of time; the remaining buffer gates must be disabled.* Thus, digital information from only one of the four buffers appears on the single-line bus at any given instant of time. Information from the remaining three buffers is blocked since the corresponding buffers are disabled. The following truth table applies to the operation of this circuit:

| D C B A | Output | Comments |
|---|---|---|
| 0 0 0 1 | $Q_A$ | Buffers B, C, and D are "disconnected" from the bus |
| 0 0 1 0 | $Q_B$ | Buffers A, C, and D are "disconnected" from the bus |
| 0 1 0 0 | $Q_C$ | Buffers A, B, and D are "disconnected" from the bus |
| 1 0 0 0 | $Q_D$ | Buffers A, B, and C are "disconnected" from the bus |

It is important to note that all other input conditions are considered to be "illegal" for this circuit since they permit information from more than one buffer to appear on the single-line bus. In addition, if you attempt to implement any of these "illegal" input conditions, you will most likely burn out the three-state chip!

Typical bus systems consist of multiline busses, such as shown in Fig. 19-3, rather than single-line busses. Other than the fact that the gating inputs enable or disable four buffer gates at a time, this circuit is identical to the one shown in Fig. 19-2. For example, the truth table is essentially the same:

| D C B A | Output | Comments |
|---------|--------|----------|
| 0 0 0 1 | Device A | Devices B, C, and D are "disconnected" from the bus |
| 0 0 1 0 | Device B | Devices A, C, and D are "disconnected" from the bus |
| 0 1 0 0 | Device C | Devices A, B, and D are "disconnected" from the bus |
| 1 0 0 0 | Device D | Devices A, B, and C are "disconnected" from the bus |

As was the case previously, all other input conditions are "illegal" since they permit information from more than one device to appear on the bus.

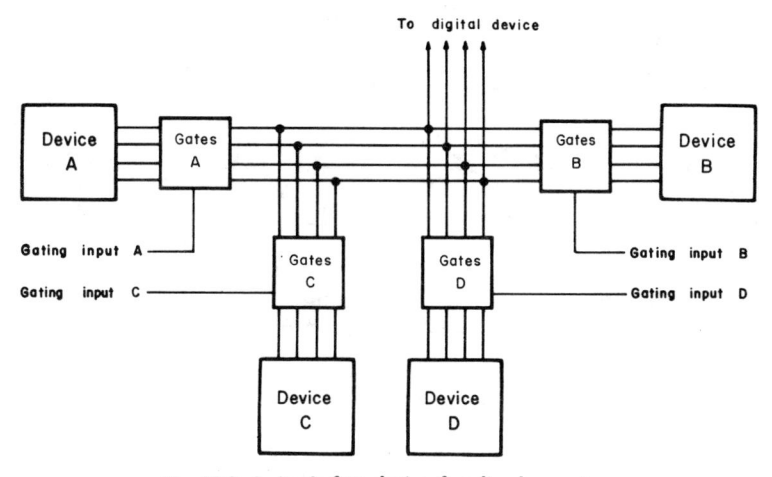

Fig. 19-3. A simple four-device, four-line bus system.

#### 74125 THREE-STATE BUFFER

A typical 74125 three-state buffer contains a separate enable/disable input in addition to the normal input and output pins,

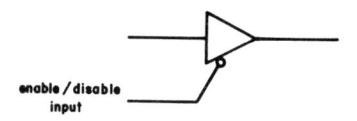

The pin configuration for a 74125 chip, as given in *The TTL Data Book for Design Engineers,* by Texas Instruments Incorporated, is shown in Fig. 19-4.

The four independent buffers can be identified as follows:

First buffer: Input 1A, output 1Y, and enable/disable input 1C

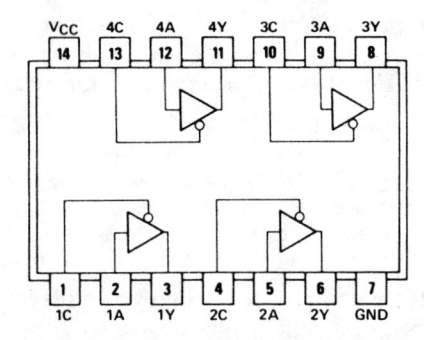

**Fig. 19-4.** Pin configuration diagram for the SN74125.

Second buffer: Input 2A, output 2Y, and enable/disable input 2C
Third buffer: Input 3A, output 3Y, and enable/disable input 3C
Fourth buffer: Input 4A, output 4Y, and enable/disable input 4C

These four buffers can be schematically represented as

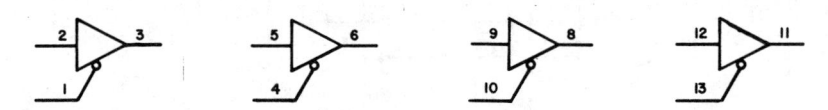

Based upon our experience, we recommend the use of this chip in preference to the 74126. When an enable/disable input is not connected, the corresponding 74125 buffer is disabled.

### 74126 THREE-STATE BUFFER

The pin configuration for the 74126 three-state quad buffer chip is shown in Fig. 19-5. The power inputs are at pins 7 and 14. There are four independent buffers on the chip:

First buffer: Input 1A, output 1Y, and enable/disable input 1C
Second buffer: Input 2A, output 2Y, and enable/disable input 2C

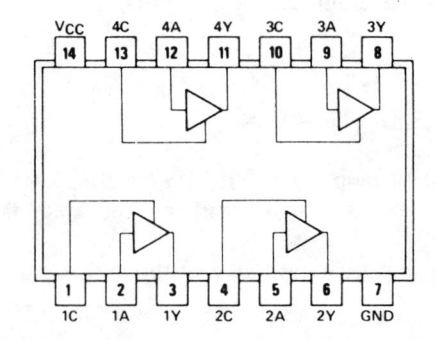

**Fig. 19-5.** Pin configuration diagram for the SN74126.

Third buffer: Input 3A, output 3Y, and enable/disable input 3C
Fourth buffer: Input 4A, output 4Y, and enable/disable input 4C

which can be schematically represented as follows:

## 8095 THREE-STATE BUFFER

The 8095 three-state hex buffer chip contains six buffers that are enabled simultaneously from the output of a 2-input NOR gate. The truth table is given below. The pin configuration diagram is given in Fig. 19-6.

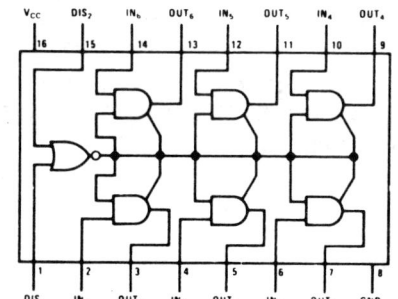

Fig. 19-6. Pin configuration diagram for the DM8095 hex buffer.

| Enable/Disable Inputs | | Input Data | Buffer Output |
|:---:|:---:|:---:|:---:|
| $DIS_1$ | $DIS_2$ | | |
| 0 | 0 | 0 | 0 |
| 0 | 0 | 1 | 1 |
| 0 | 1 | X | High impedance |
| 1 | 0 | X | High impedance |
| 1 | 1 | X | High impedance |

X = irrelevant

This is an excellent chip that is frequently used in microcomputer input buffer circuits. For example, a single 8095 chip will permit six separate input lines to be connected to an 8080A microprocessor chip via the bidirectional data bus.

## OTHER THREE-STATE DEVICES

Currently, the dominant bussing technology, which is used in most microprocessor chips, is three-state. A very common circuit configuration that is found within microprocessor chips is the three-state latch/buffer (shown in Fig. 19-7). It consists of a 7475 D-type latch

**Fig. 19-7. A three-state latch/buffer circuit configuration.**

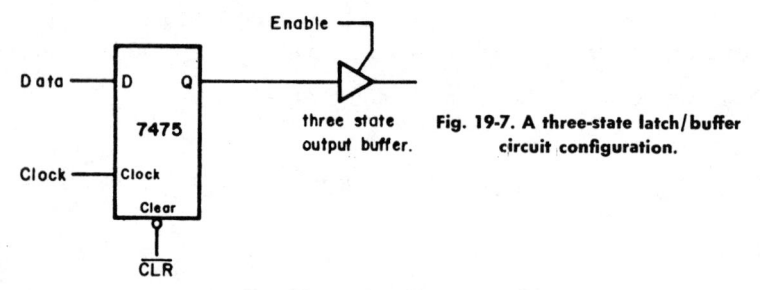

and a three-state output buffer that requires a logic 1 enable input. The following truth table applies:

| Clock | Enable | Output Condition |
|---|---|---|
| 0 | 0 | Previous data is latched; three-state output is disabled |
| 0 | 1 | Previous data is latched and output to bus |
| 1 | 0 | Latch follows Data input; three-state output is disabled |
| 1 | 1 | Behaves as simple three-state buffer |

Many of the new programmable interface chips, such as the Intel Corporation 8251, 8253, 8255, 8257, and 8259 employ the three-state latch/buffer circuit in 8-bit internal programmable registers.

Although several chips in the 7400 series, including the 74125, 74126, and 74200, have three-state outputs, most of the three-state devices are available from National Semiconductor Corporation, with second-sourcing by Texas Instruments and others. The following is a partial listing of the TRI-STATE® devices that are currently available. Note that TRI-STATE is a registered trademark of National Semiconductor.

| | |
|---|---|
| 74200 | Three-state 256-bit read/write memory |
| 74251 | Three-state 8-channel multiplexer |
| 74284 | Three-state 4-bit multiplier |
| 74285 | Three-state 4-bit multiplier |
| 74365 | Three-state hex buffer (same as 8065) |
| 8093 | Three-state quad buffer (same as 74125) |
| 8094 | Three-state quad buffer (same as 74126) |
| 8095 | Three-state hex buffer |
| 8096 | Three-state hex inverter |
| 8097 | Three-state hex buffer |
| 8098 | Three-state hex inverter |
| 8123 | Three-state quad 2-input multiplexer |
| 8214 | Three-state dual 4:1 multiplexer |
| 8219 | Three-state 16-line-to-1-line multiplexer |
| 8230 | Three-state demultiplexer |
| 8542 | Three-state quad I/O register |
| 8544 | Three-state quad switch debouncer |
| 8551 | Three-state quad D flip-flop |
| 8552 | Three-state decade counter/latch |
| 8553 | Three-state 8-bit latch |
| 8554 | Three-state binary counter/latch |

| | |
|---|---|
| 8555 | Three-state programmable decade counter |
| 8556 | Three-state programmable binary counter |
| 8598 | Three-state 256-bit read-only memory |
| 8599 | Three-state 64-bit read/write memory (same as 74189) |
| 8831 | Three-state line driver |
| 8832 | Three-state line driver |
| 8833 | Three-state quad transceiver |
| 8834 | Three-state quad transceiver |
| 8835 | Three-state quad transceiver |
| 8875 | Three-state 4-bit multiplier |

If you have trouble locating a parts outlet, many of the above chips are available from James Electronics, 1021A Howard Avenue, San Carlos, California, 94070.

## INTRODUCTION TO THE EXPERIMENTS

The following experiments demonstrate the use of three-state bussing techniques and three-state buffers.

| *Experiment No.* | *Comments* |
|---|---|
| 1 | Demonstrates the operation of a single 74125 buffer with three-state output. |
| 2 | Demonstrates how you create a four-source single-line bus system using a single 74125 three-state buffer chip. |
| 3 | Demonstrates how you create a two-source four-line bus using a pair of 74126 three-state buffer chips. The sources of digital information are a 7490 decade counter and a 7493 binary counter. |
| 4 | Demonstrates the operation of a simple latch/buffer circuit that is based upon a 7475 D-type latch and a 74125 three-state buffer. This type of one-bit circuit is widely used in registers within microprocessor chips such as the 8080A. |

## EXPERIMENT NO. 1

**Purpose**

The purpose of this experiment is to demonstrate the operation of a single 74125 bus buffer with three-state output.

## Pin Configuration of Integrated-Circuit Chip (Fig. 19-8)

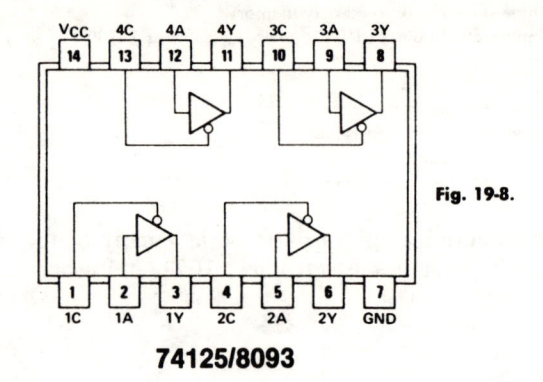

Fig. 19-8.

**74125/8093**

## Schematic Diagram of Circuit (Fig. 19-9)

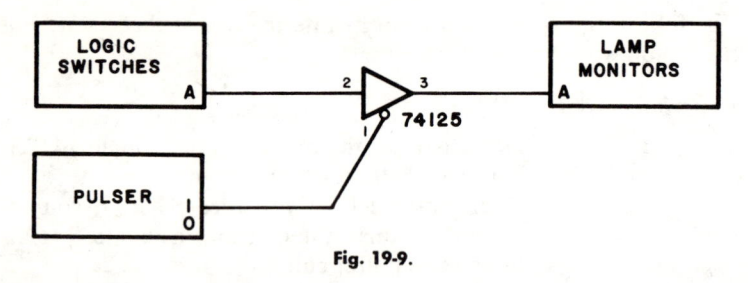

Fig. 19-9.

### Step 1

Wire the circuit shown in Fig. 19-9. The 74125 chip contains four independent bus buffers. You will use only one of them.

### Step 2

Set logic switch A to logic 1 state. Apply power to the breadboard. Is the lamp monitor lit or unlit?

The lamp monitor is unlit, which indicates that the buffer is disabled or burned out.

### Step 3

Now press the pulser button in. Does the lamp monitor become lighted?

Yes. The buffer is now enabled with a logic 0 state.

### Step 4

With the pulser pressed in, vary the logic switch setting between logic 1 and logic 0. What do you observe on the lamp monitor?

The lamp monitor indicates the state of the logic switch as long as the buffer is enabled.

### Step 5

Is the following truth table the correct one for the operation of the 74125 buffer? If not, write the correct truth table.

| A | Pulser | Lamp Monitor |
|---|--------|--------------|
| 0 | 0      | 0            |
| 0 | 1      | 0            |
| 1 | 0      | 0            |
| 1 | 1      | 1            |

No, the table is not correct. The correct truth table is:

| A | Pulser | Lamp Monitor |
|---|--------|--------------|
| 0 | 0      | 0            |
| 0 | 1      | 0            |
| 1 | 0      | 1            |
| 1 | 1      | 0            |

## EXPERIMENT NO. 2

### Purpose

The purpose of this experiment is to bus four different sources of data onto a single-line bus.

## Pin Configurations of Integrated-Circuit Chips (Fig. 19-10)

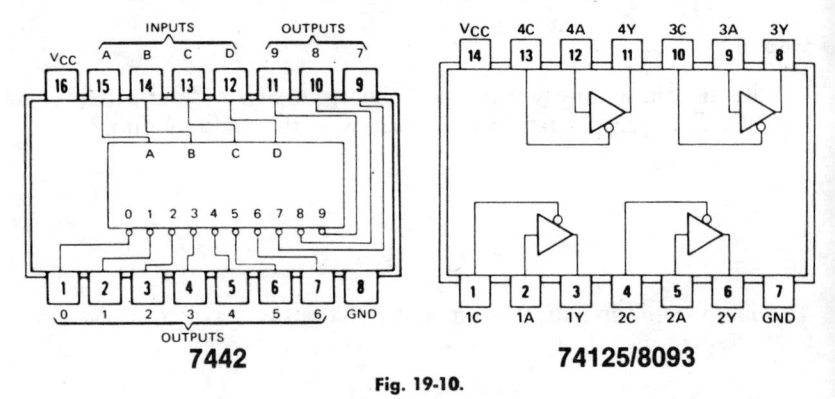

**Fig. 19-10.**

## Schematic Diagram of Circuit (Fig. 19-11)

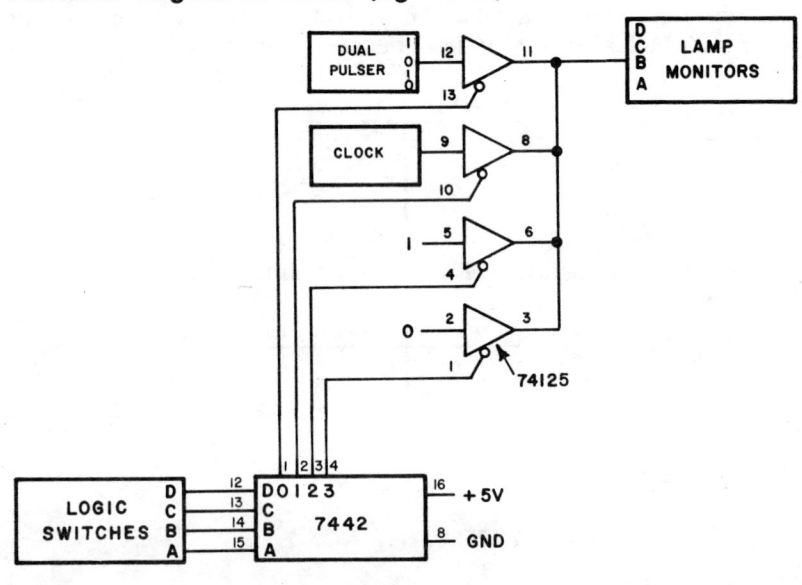

**Fig. 19-11.**

## Step 1

Wire the circuit shown in Fig. 19-11. What is the purpose of the 7442 decoder chip?

The purpose of the 7442 chip in the circuit is to enable only one buffer at a time.

**Step 2**

Select, in turn, output channels 0, 1, 2, and 3 from the 7442 decoder and write down what you observe the lamp monitor output to be in each case.

We observed the following results:

| Channel | Lamp Monitor Output |
|---|---|
| 0 | Output from pulser |
| 1 | Clock output |
| 2 | 1 (lighted lamp monitor) |
| 3 | 0 (unlighted lamp monitor) |

**Step 3**

What occurs when you choose channels 4 through 9 on the 7442 chip? Do you observe any lamp monitor output?

We observed that the lamp monitor output remained at logic 0. The reason was that all four 74125 buffers were disabled.

**Step 4**

Keep in mind that in any three-state bus system, *only a single data input to the bus must be enabled at any given instant of time.* In this experiment, the 7442 decoder ensures the fact that only one 74125 buffer is enabled at a time. The use of decoders for such a purpose is common in three-state bus systems.

## EXPERIMENT NO. 3

**Purpose**

The purpose of this experiment is to bus two different digital devices, a 7490 counter and a 7493 counter, to a single 7-segment LED display using a pair of 74126 bus buffer chips.

## Pin Configurations of Integrated-Circuit Chips (Fig. 19-12)

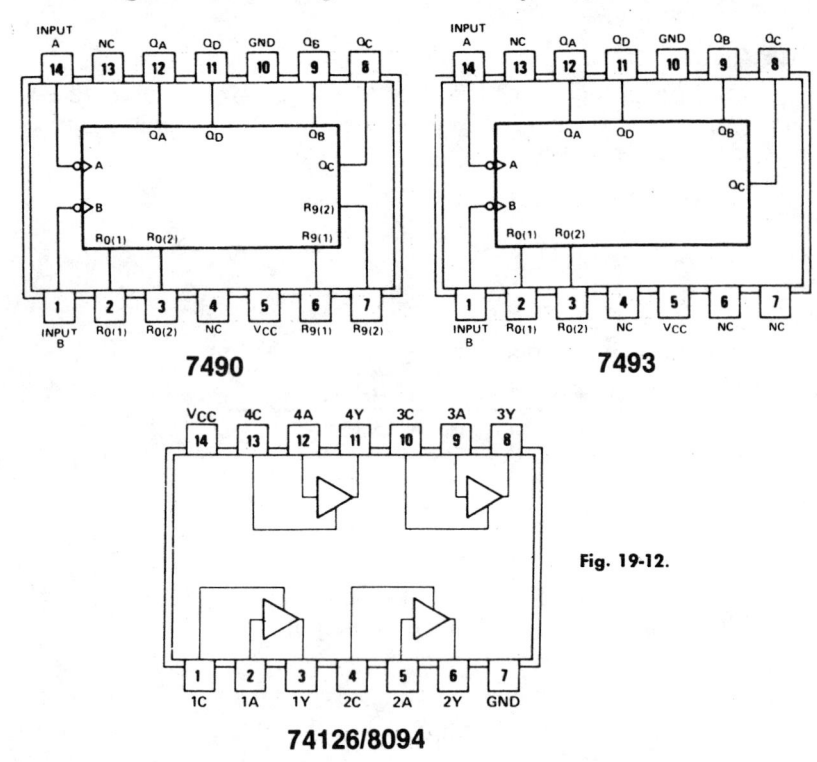

**7490**

**7493**

**74126/8094**

Fig. 19-12.

## Schematic Diagram of Circuit (Fig. 19-13)

### Step 1

Study the circuit diagram in Fig. 19-13 carefully. Observe that two 74126 buffer chips are required. The enable/disable inputs from each chip are tied either to the "0" or "1" output on a single pulser. Why is only a single pulser used?

A single pulser is used to ensure the fact that only one input device is enabled at any given instant of time.

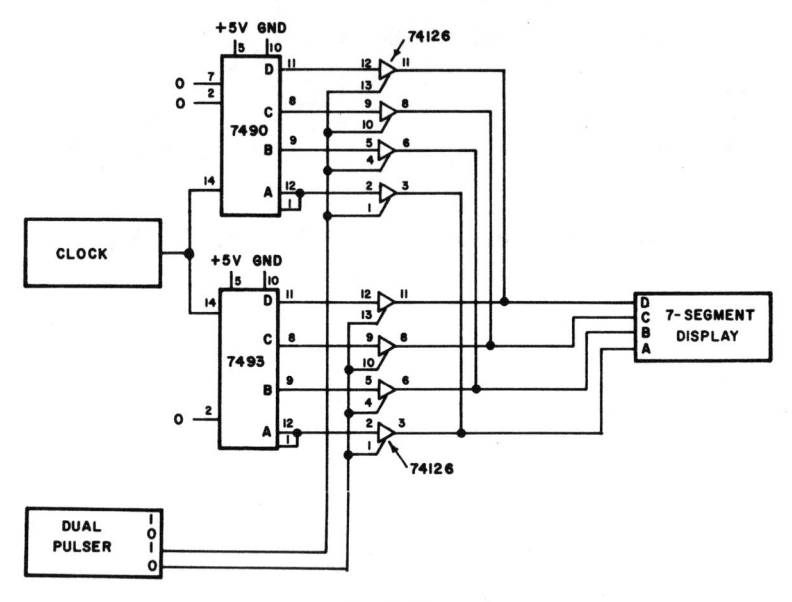

**Fig. 19-13.**

## Step 2

Wire the circuit and then apply power to the breadboard. Which counter, the 7490 decade counter or the 7493 binary counter, is enabled?

The 7490 decade counter is enabled, since a logic 1 enable input is required to enable the 74126 buffers.

## Step 3

Press the pulser button in. Which counter is now enabled?

The 7493 binary counter is enabled. Is there any possible way in which both counters can be simultaneously enabled in the above circuit?

No.

# EXPERIMENT NO. 4

## Purpose

The purpose of this experiment is to test a simple latch/buffer circuit based upon a 7475 D-type latch and a 74125 three-state buffer.

## Pin Configurations of Integrated-Circuit Chips (Fig. 19-14)

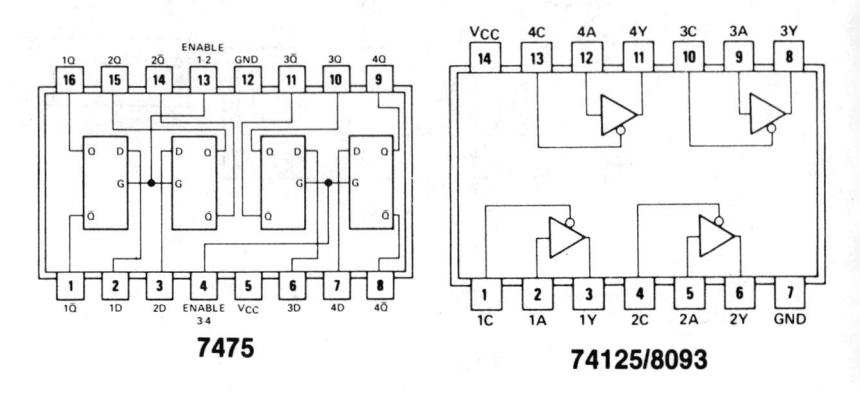

Fig. 19-14.

## Schematic Diagram of Circuit (Fig. 19-15)

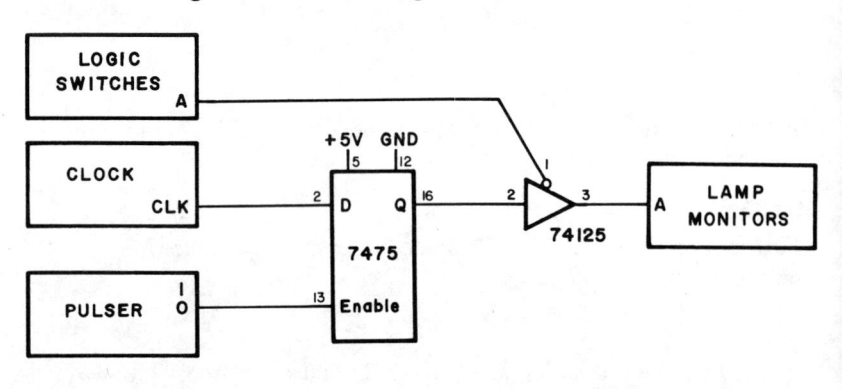

Fig. 19-15.

## Step 1

Wire the circuit shown in Fig. 19-15. Adjust the clock output so that the clock frequency is approximately 1 Hz.

### Step 2

The truth table for the operation of the circuit can be summarized as follows:

| 7475 Enable | 74125 Enable | D | Three-State Buffer Output |
|---|---|---|---|
| 0 | 0 | 0 | Previously latched D |
| 0 | 0 | 1 | Previously latched D |
| 1 | 0 | 0 | 0 |
| 1 | 0 | 1 | 1 |
| 0 | 1 | X | High-impedance state |
| 1 | 1 | X | High-impedance state |

X = irrelevant

What logic state disables the 74125 buffer?

A logic 1 state.

### Step 3

Enable both the 7475 latch and the 74125 buffer. What do you observe on the lamp monitor?

A train of clock pulses that appear at a rate of approximately 1 Hz.

### Step 4

Now disable the 74125 buffer. What happens to the lamp monitor output?

It becomes logic 0. The 74125 buffer output is now in its high impedance state.

### Step 5

Enable the 74125 buffer, but this time enable or disable the 7475 latch. Is it possible to latch either the logic 0 or logic 1 data input at D (pin 2)?

Yes, it is possible to operate the 7475 latch independent of the 74125 buffer.

## REVIEW QUESTIONS

The following questions will help you review three-state devices and three-state bussing techniques.

1. What is a digital bus and why is it used?
2. In what types of digital devices might you find a bus? List at least three such devices.
3. Write a truth table for a simple three-state device.
4. Why is a three-state latch/buffer such a useful circuit?
5. List different types of digital devices, e.g., gates, latches, etc., that are available with three-state outputs.
6. What happens in a three-state bus system when two or more sources of bus data are simultaneously enabled?

# 20

# An Introduction to Accumulator Input/Output Techniques

### INTRODUCTION

The objective of a microcomputer input/output operation in an 8080A-based microcomputer is to transfer data between an input/output device and one of the internal registers within the 8080A chip. In accumulator I/O, you employ the IN and OUT instructions and data transfer occurs between the accumulator and the I/O device. In this unit, you will learn how to write simple I/O programs and wire simple interface circuits that, when working together, permit you to transfer input/output data to and from the 8080A chip.

### OBJECTIVES

At the completion of this unit, you will be able to do the following:

- State the objective of a microcomputer input/output operation.
- Distinguish between accumulator I/O and memory-mapped I/O in 8080A-based microcomputers.
- Sketch several simple latch circuits that are useful for accumulator output in an 8080A-based microcomputer.
- Sketch one or two simple three-state buffer circuits that are useful for accumulator input in an 8080A-based microcomputer.
- Explain the significance of output drive capability in microcomputer output circuits.
- Explain how device select pulses are employed to achieve the objective of accumulator input/output.

- Summarize the accumulator I/O instructions in the 8080A instruction set.
- Wire a simple microcomputer output circuit.
- Wire a simple microcomputer input circuit.

## WHAT IS INPUT/OUTPUT?

When the term *input/output,* or *I/O,* is employed, we usually mean that one or more data bytes are transferred between an input/output device and the microprocessor chip. The important concepts associated with this data transfer are summarized in Figs. 20-1 through 20-4.

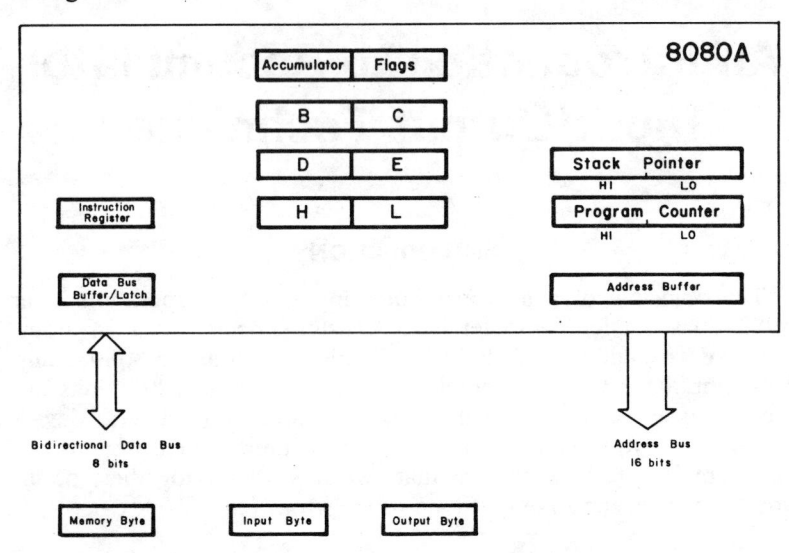

**Fig. 20-1. Diagram of the register architecture within the 8080A microprocessor chip.**

The objective of a microcomputer input/output operation in an 8080A-based microcomputer is to transfer data between an input/output device and one of the internal registers within the 8080A chip. As shown in Figs. 20-1 and 20-2, available registers include the accumulator and general-purpose registers B, C, D, E, H, and L. It is not possible to load either the stack pointer or the program counter registers directly from an external I/O device. If the I/O instructions IN and OUT are used, the data transfer occurs between the external I/O device and the accumulator within the 8080A chip. This type of input-output operation is called *isolated I/O* by the Intel Corporation and *accumulator I/O* by others. If memory reference instructions such as MOV M,r or MOV r,M are used, the data transfer can

Courtesy Intel Corp.

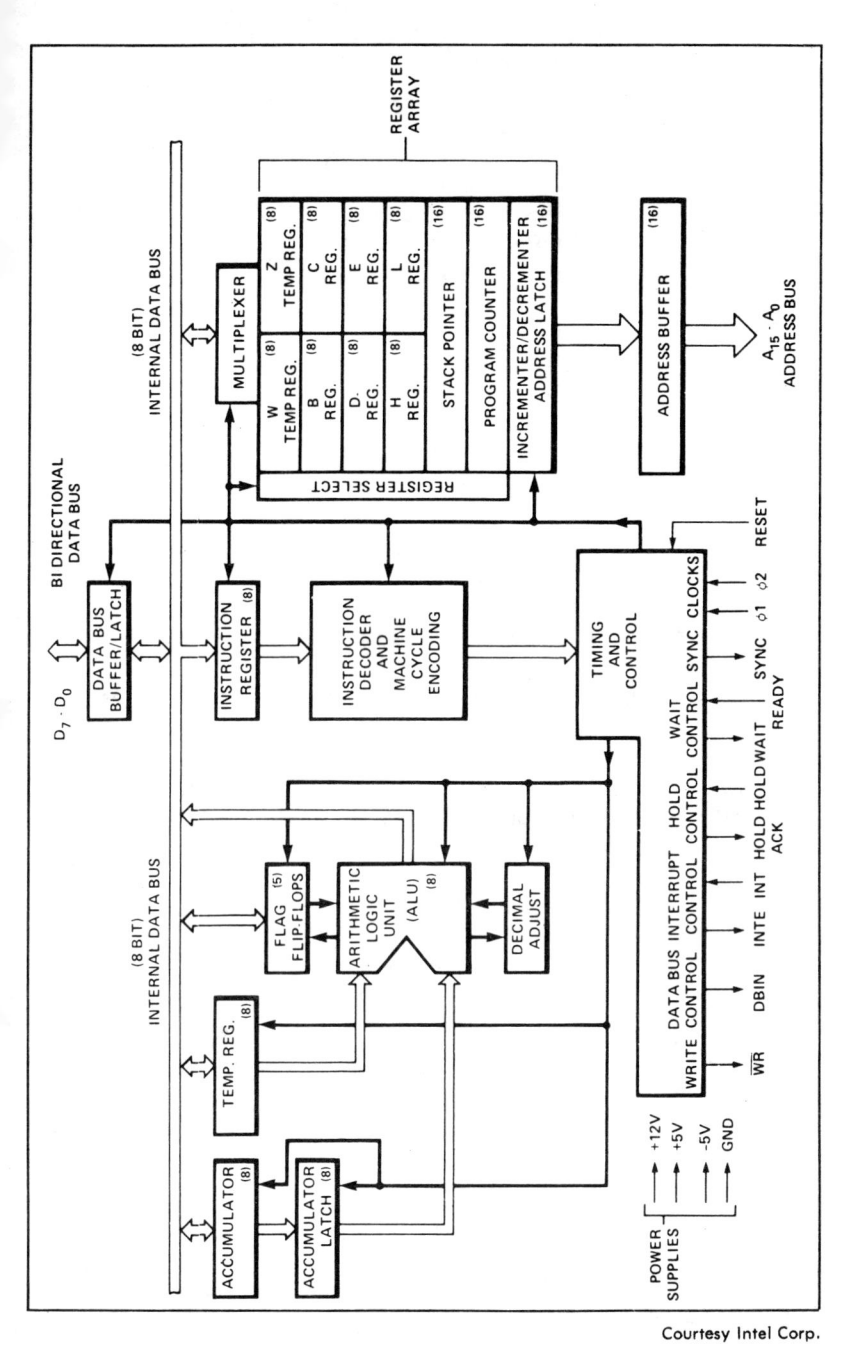

Fig. 20-2. A more detailed diagram of the register architecture shown in Fig. 20-1.

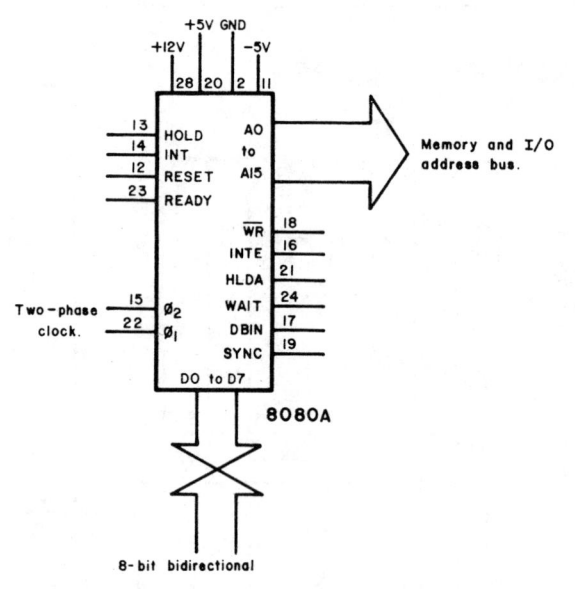

Fig. 20-3. Block diagram of the 8080 microprocessor chip.

occur between the external I/O device and *any* of the seven general-purpose registers. This second type of input-output operation is called *memory-mapped I/O* or, simply, *memory I/O*.

All microcomputer input/output occurs eight bits at a time over the 8080A bidirectional data bus. As shown in Fig. 20-3, the only

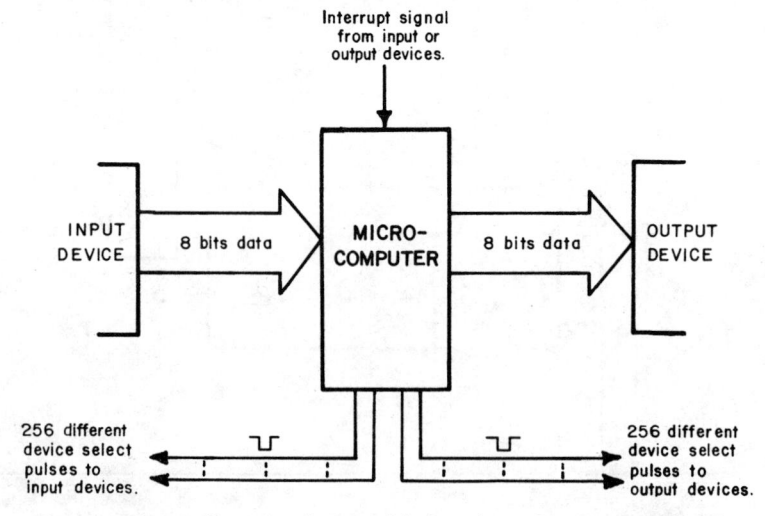

Fig. 20-4. Diagram illustrating the role of device select pulses in accumulator I/O.

path over which data can be transferred into the 8080A chip is the bidirectional data bus, D0 through D7, which is an 8-bit three-state bus.

*Synchronization pulses* are required to control all data transfers to and from the 8080A microprocessor chip. For accumulator I/O, they are called *device select pulses,* while for memory I/O, they are called *address select pulses.* In accumulator I/O, the two-byte IN and OUT instructions permit you to select and synchronize the operation of 256 different input "devices" and 256 different output "devices," as shown in Fig. 20-4. For an 8080A-based microcomputer, such pulses have a duration of about one clock period and can be either positive or negative pulses depending upon the decoding scheme used. Typically, such pulses are generated as negative device select pulses from a decoder chip and must be inverted if positive device select pulses are required.

The reference point for the terms "input" and "output" is the microprocessor chip. The 8080A chip outputs data to an "output" device, and inputs data from an "input" device. This rule holds in all cases.

## MICROCOMPUTER OUTPUT

The basic technique that you use to output data from the accumulator to an output device is quite simple. *You generate, via software and hardware, a single output device select pulse and use it to enable*

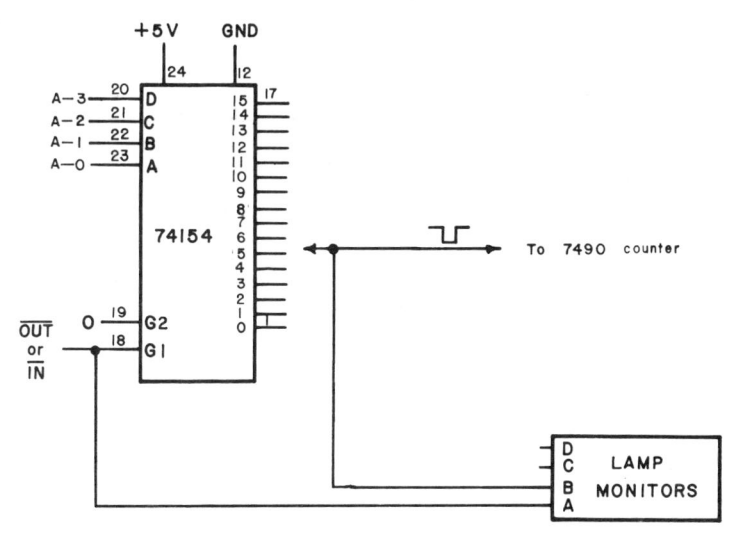

Fig. 20-5. Circuit diagram using a 74154 decoder chip to generate sixteen different negative device select pulses.

*a latch chip at the instant when the accumulator data appears on the bidirectional data bus.* The 8080A microprocessor chip is responsible for the entire synchronization process. The latch chip plays a passive role in the data transfer process and latches data only when instructed to do so by a device select pulse. Depending upon the type of latch chip used, either a positive or a negative device select pulse is used to latch the data. Recall Experiment No. 4 in Unit 17, in which you used a 74154 decoder chip to generate sixteen different negative device select pulses (Fig. 20-5). It is such pulses that you use, either directly or with inversion, to latch microcomputer data into chips such as the 8212, 74100, 7475, 74198, 74175, or 74193.

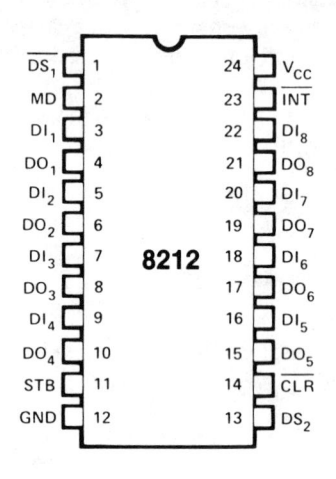

Fig. 20-6. Pin configuration of the 8212 latch/buffer chip.

## SOME OUTPUT LATCH CIRCUITS

Typical microcomputer output circuits are varied. They include those based upon the 8212 chip, the 74100 8-bit D-type latch, a pair of 7475 D-type latch chips, the 74198 8-bit shift register, the 74175 eight-bit latch, and a pair of 74193 up/down counters.

The pin configuration of the 8212 8-bit latch/buffer chip is shown in Fig. 20-6, while Fig. 20-7 is the schematic diagram of a circuit in which an 8212 chip serves as an output latch. When using the 8212 chip as an output latch, you should make certain that the clear input, $\overline{CLR}$, at pin 14 is tied to logic 1 when not in use.

Figs. 20-8 and 20-9 show the pin configuration of the 74100 8-bit D-type latch chip and a microcomputer output latch circuit that is based upon the use of the 74100 D-type latch. In the circuit of Fig. 20-9, the output is provided as a three-digit octal word.

Fig. 20-10 illustrates the pin configuration of the 7475 4-bit D-type latch chip. A microcomputer output latch circuit which is

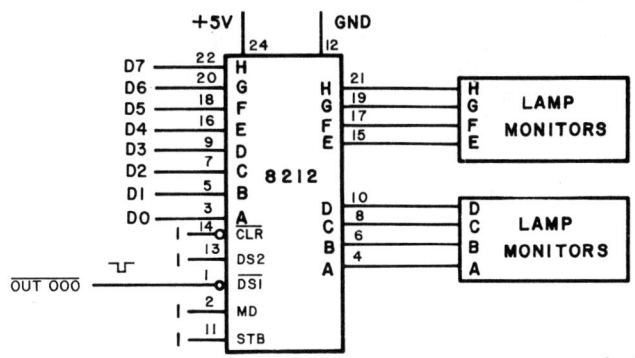

Fig. 20-7. Schematic diagram of a circuit where the 8212 chip is an output latch chip.

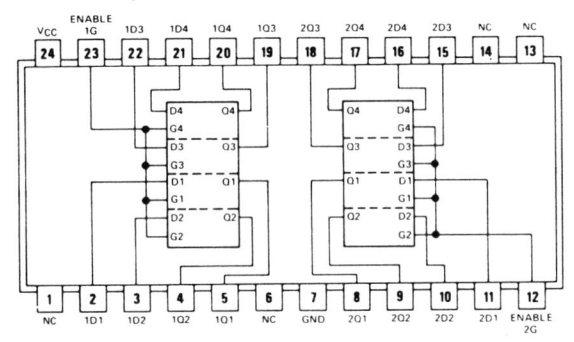

Fig. 20-8. Pin configuration of the 74100 D-type latch.

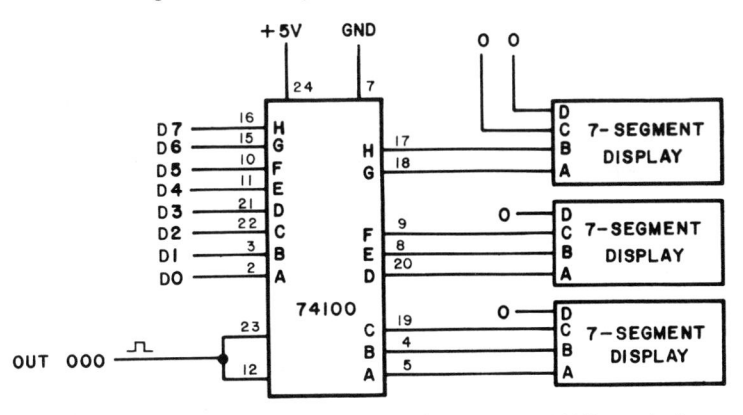

Fig. 20-9. Output latch circuit based upon the use of a 74100 D-type latch.

based upon the use of a pair of 7475 4-bit D-type latches is given in Fig. 20-11.

The pin configurations of both the 74198 8-bit shift register and the 74175 4-bit latch are shown in Fig. 20-12. Both chips

Fig. 20-10. Pin configuration of a 7475 D-type latch chip.

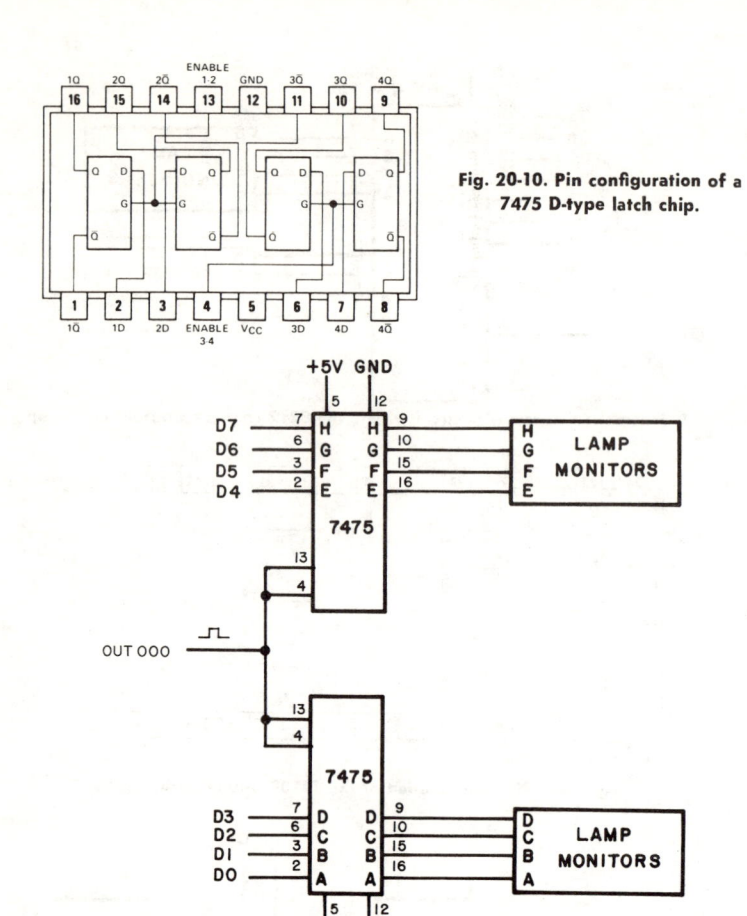

Fig. 20-11. Output latch circuit using a pair of 7475 D-type latches.

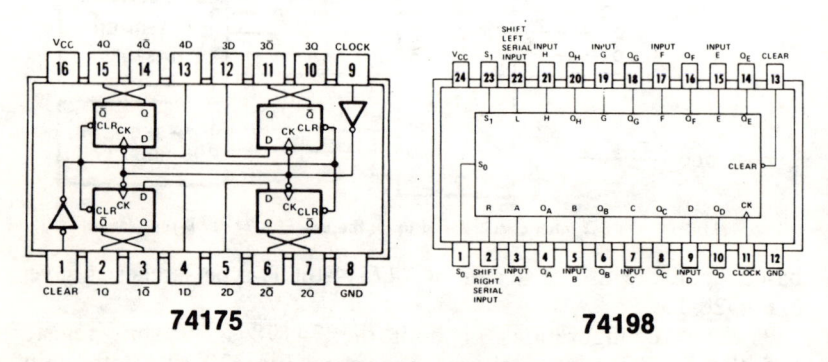

**74175**

**74198**

Fig. 20-12. Pin configuration drawings.

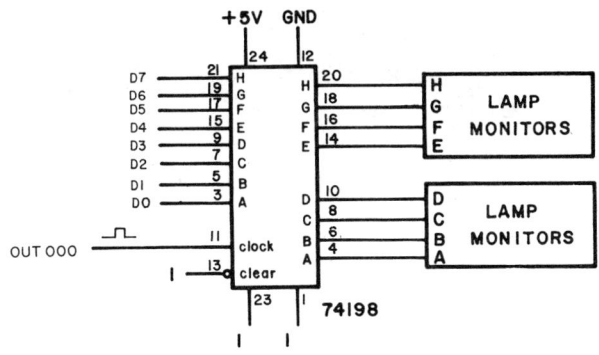

**Fig. 20-13.** Output latch circuit based upon a 74198 8-bit shift register.

contain positive-edge-triggered flip-flops of the 7474 type. A microcomputer output latch circuit based upon the use of a 74198 8-bit shift register is shown in Fig. 20-13. The same type of output circuit, but based upon a pair of 74175 latch chips, is given in Fig. 20-14.

The 74193 up/down counter is illustrated in Figs. 20-15 and 20-16. Each 74193 counter chip package contains an internal 4-bit latch of the 7475 type. Fig. 20-15 shows the pin configuration dia-

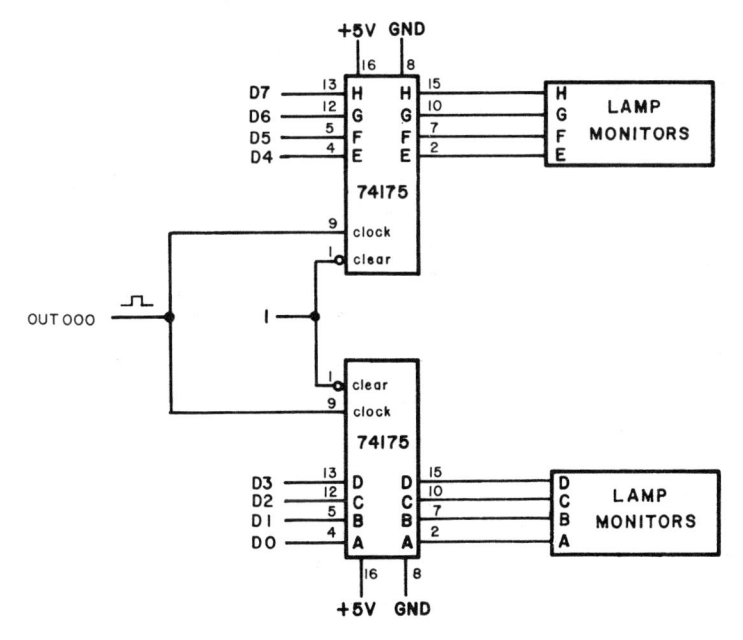

**Fig. 20-14.** Microcomputer output latch circuit using two 74175 latch chips.

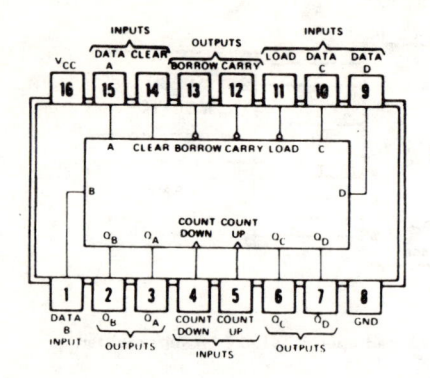

Fig. 20-15. Pin configuration for the 74193 up/down counter chip.

gram of the 74193 chip, while Fig. 20-16 illustrates a microcomputer output latch circuit based upon the use of a pair of 74193 chips. The circuit of Fig. 20-16 demonstrates how you could preload a count into an up/down counter directly from the accumulator in an 8080A-based microcomputer. You would not normally use such a chip as a general-purpose latch.

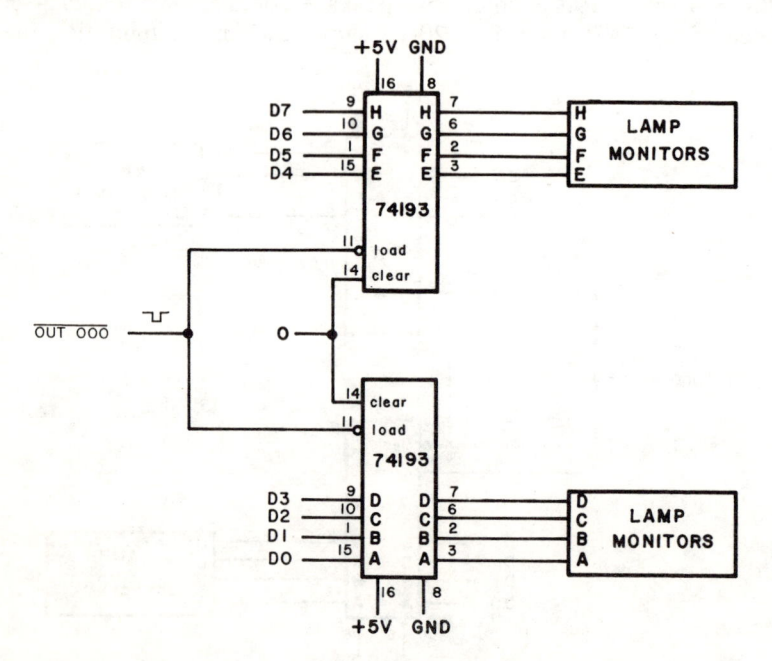

Fig. 20-16. Output latch circuit based upon the use of a pair of 74193 up/down counter chips.

## OUTPUT DRIVE CAPABILITY

The characteristics of different TTL subfamilies and the concepts of *fan-in* and *fan-out* have been previously described in Unit 10. Such considerations are extremely important when you construct microcomputer output circuits. Thus, the fan-out of an 8080A microprocessor chip is 1.2, which means that a single output pin can drive (sink) a maximum current of only 1.9 mA. The fan-in of a normal 7400-series input is 1.6 mA, so it should be clear that *you should always employ output devices that have a much lower fan-in whenever you make a direct connection to an 8080A output pin.*

What chips should you use? We recommend the following:

1. Chips in the 74LS subfamily, in which the fan-in of an input is only 0.2, or 0.32 mA.
2. Chips in the 74L subfamily, in which the fan-in of an input is only 0.1, or 0.16 mA.
3. Microprocessor-compatible chips such as the 8205 decoder, 8212 8-bit I/O port, 8111-2 static read/write memory, and related chips, in which the fan-in is only 0.15, or 0.25 mA. Such chips are manufactured specifically for interfacing to an 8080A chip.

Many low-power chips can be tied to the output busses from a microprocessor chip provided only that such busses are not overloaded.

You should also concern yourself with the fan-out of a low-power chip that is connected to an 8080A bus. If output signals must travel over a distance that is greater than several inches, it is good policy to buffer the latch outputs. Latch and flip-flop outputs are inherently sensitive to drive problems. The fan-out of a 74LS output is only 5 while that for a 74L output is 2.25.

In constructing a microcomputer, it is common to have bus runs that are as long as nine to twelve inches. Do not make a bus over one foot long without using special bus drivers and a termination network.

## MICROCOMPUTER INPUT

The technique that you use to input data from an external device into the accumulator is analogous to the technique used for microcomputer output. You generate, via software and hardware, a single input device select pulse and use it to enable a three-state buffer at the instant when a direct path is opened up between the bidirectional data bus and the accumulator. As with microcomputer output, the 8080A chip is responsible for the entire synchronization process. The three-state buffer chip plays a passive role in the data transfer process and applies data to the data bus only when instructed to do so by a device select pulse. Either a positive or negative device select pulse is used to enable the buffer, depending upon the type of buffer used. Typical three-state buffer chips are the 8212 and 8095. The 8255 programmable peripheral interface chip has become popular as an input buffer.

## SOME THREE-STATE BUFFER INPUT CIRCUITS

Typical microcomputer input circuits include those based upon the 8095 or 8212 chips, as shown in Figs. 20-17 through 20-19. The

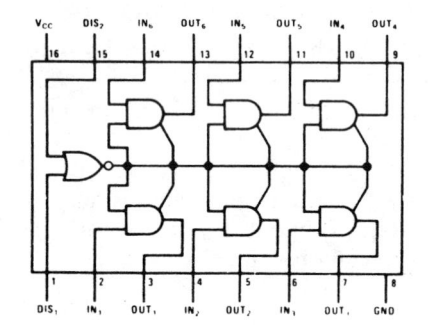

Fig. 20-17. Pin configuration of the 8095 buffer chip.

8212 8-bit latch/buffer chip was shown previously in Fig. 20-6 as an output latch. The 8095 three-state buffer chip is inexpensive and widely used in microcomputer interface circuits. It should be emphasized that *only one* three-state buffer input to an 8080A-based microcomputer must be enabled at any given time. All input device select pulses should be absolutely decoded, which means, for accumulator I/O, that all eight bits of the device code should be used to uniquely identify the desired input device. If a nonexistent device is called to input data, usually the byte $377_8$ will be input to the accumulator.

In Fig. 20-18, the enabled 8095 three-state buffers permit data to be transferred to the bidirectional data bus lines, D0 through D7, connected to the outputs of the 8095 chips. The accumulator acquires the logic switch data during the clock period of the input device select pulse which, for the 8095 chips, is a negative pulse. The inputs of the 8095 chips can be connected to any source of digital data, such as a laboratory instrument. This data is transferred through the 8095 buffers, placed on the bidirectional data bus lines, and *copied* or *jammed* into the accumulator during an IN microcomputer instruction. Data is input to the accumulator each time that the IN instruc-

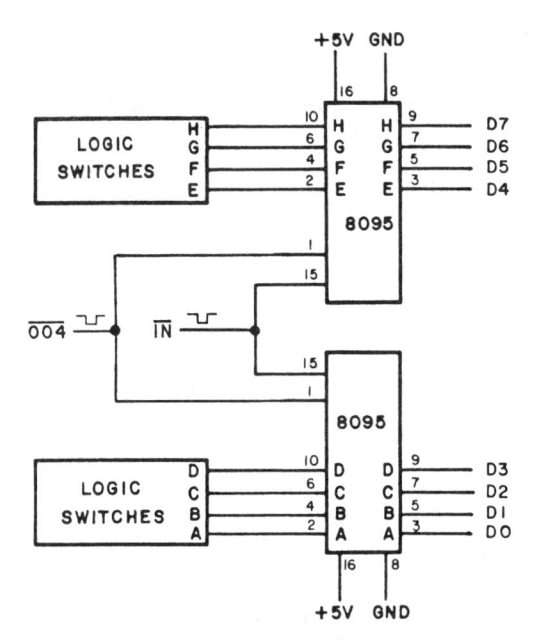

**Fig. 20-18.** Microcomputer input circuit based upon the use of a pair of 8095 buffer chips.

tion and a device code are executed. The accumulator need not be cleared before the IN instruction, since both a logic 0 and a logic 1 are jammed into the appropriate bit positions during the device select pulse clock period.

## ACCUMULATOR I/O INSTRUCTIONS

There are only two 8080A accumulator I/O instructions which transfer data between the accumulator and external I/O devices con-

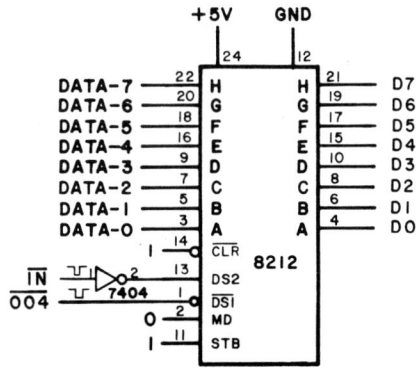

**Fig. 20-19.** Microcomputer input circuit using an 8212 latch/buffer chip.

current with the generation of the $\overline{\text{IN}}$ and $\overline{\text{OUT}}$ synchronization pulses:

323 &lt;B2&gt; OUT  Output the accumulator contents to the output latch selected by the device code in the second byte. This instruction is executed in ten clock cycles, or 13.33 µs for an 8080A-based microcomputer operating at 750 kHz.

333 &lt;B2&gt; IN  Input into the accumulator the contents of the digital device and three-state buffer circuit selected by the device code in the second byte. This instruction is executed in ten clock cycles, or 13.33 µs for a 750-kHz clock rate.

In this unit, in contrast to Unit 17, the device select pulses generated by the above instructions are used to transfer information to and from the accumulator.

## FIRST INPUT/OUTPUT PROGRAM

A simple program used to input the logic switch data into the accumulator (in Fig. 20-18) and then to immediately output it to Output Latch 000 (shown in Fig. 20-11) is as follows:

| LO Memory Address | Instruction Byte | Mnemonic | Description |
|---|---|---|---|
| 000 | 333 | START, IN | Input logic switch data associated with input device 004, a pair of 8095 three-state buffers |
| 001 | 004 | 004 | Device code 004 |
| 002 | 323 | OUT | Output accumulator data to output latch 000, a pair of 7475 latch chips |
| 003 | 000 | 000 | Device code 000 |
| 004 | 166 | HLT | Halt |

This program will input the logic switch data into the accumulator, then output the accumulator data to an output latch, and finally halt.

## SECOND PROGRAM

To continuously input and output the data acquired by input device 004, change the HALT instruction to a JMP instruction that loops back to HI = 003 and LO = 000.

| LO Memory Address | Instruction Byte | Mnemonic | Description |
|---|---|---|---|
| 000 | 333 | START, IN | Input logic switch data from input device 004 |
| 001 | 004 | 004 | Device code 004 |
| 002 | 323 | OUT | Output data to output device 000 |
| 003 | 000 | 000 | Device code 000 |
| 004 | 303 | JMP | Unconditional jump to memory location START |
| 005 | 000 | START | LO address byte of START |
| 006 | 003 | — | HI address byte of START |

## THIRD PROGRAM

To store the input data into a memory location and update the memory contents each time a new 8-bit data point is input, you would use the following program:

| LO Memory Address | Instruction Byte | Mnemonic | Description |
|---|---|---|---|
| 000 | 333 | START, IN | Input logic switch data from input device 004 |
| 001 | 004 | 004 | Device code 004 |
| 002 | 323 | OUT | Output data to output device 000 |
| 003 | 000 | 000 | Device code 000 |
| 004 | 062 | STA | Store the accumulator contents in the memory location given by the following two bytes |
| 005 | 200 | STORE | LO address byte of STORE |
| 006 | 003 | — | HI address byte of STORE |
| 007 | 166 | HLT | Halt |

This program is similar to the second program, but this time a STA <B2> <B3> instruction has been added to permit you to store the accumulator contents into memory location STORE, which is at HI = 003 and LO = 200. After the program comes to a halt, examine location STORE to see if the input logic switch data from device 004 is present. Change the switch settings, execute the program again, and again, and once more examine memory location STORE. You may ask, "How can data by stored when it has previously been sent out to Output Latch 000?" At first glance, it appears that the input data has been "used up" when it is output to Latch 000. The answer is that when a data byte is transferred from one location to another, it is *copied* to the new location. The original data is still present and is not "used up." This general rule holds for almost all data transfers in a microcomputer system—from register to register, register to memory, memory to register, accumulator to output device, etc.

## FOURTH PROGRAM

This program is especially interesting if you have an MMD-1 (Dyna-Micro) microcomputer in which the keyboard is Input Port 000 (see Unit 4).

| LO Memory Address | Instruction Byte | Mnemonic | | Description |
|---|---|---|---|---|
| 000 | 333 | START, | IN | Input data from keyboard on MMD-1 microcomputer |
| 001 | 000 | | 000 | Device code 000 |
| 002 | 323 | | OUT | Output data to output port 000 on MMD-1 microcomputer |
| 003 | 000 | | 000 | Device code 000 |
| 004 | 303 | | JMP | Unconditional jump to memory location START |
| 005 | 000 | | START | LO address byte of START |
| 006 | 003 | | — | HI address byte of START |

When you execute this program, you will be able to determine the encoding for each of the fifteen keys on the MMD-1 microcomputer. The sixteenth key is RESET, which is hardwired directly to the 8224 chip. The encoding of the keys can be summarized as follows:

| Key Heading | D7 | D6 | D5 | D4 | D3 | D2 | D1 | D0 | Octal Code |
|---|---|---|---|---|---|---|---|---|---|
| No key | 0 | 1 | 1 | 1 | 0 | 0 | 0 | 0 | 160 (irrelevant) |
| 0 | 1 | 1 | 1 | 1 | 0 | 0 | 0 | 0 | 360 |
| 1 | 1 | 1 | 1 | 1 | 0 | 0 | 0 | 1 | 361 |
| 2 | 1 | 1 | 1 | 1 | 0 | 0 | 1 | 0 | 362 |
| 3 | 1 | 1 | 1 | 1 | 0 | 0 | 1 | 1 | 363 |
| 4 | 1 | 1 | 1 | 1 | 0 | 1 | 0 | 0 | 364 |
| 5 | 1 | 1 | 1 | 1 | 0 | 1 | 0 | 1 | 365 |
| 6 | 1 | 1 | 1 | 1 | 0 | 1 | 1 | 0 | 366 |
| 7 | 1 | 1 | 1 | 1 | 0 | 1 | 1 | 1 | 367 |
| S | 1 | 1 | 1 | 1 | 1 | 0 | 0 | 0 | 370 |
| C | 1 | 1 | 1 | 1 | 1 | 0 | 1 | 0 | 372 |
| G | 1 | 1 | 1 | 1 | 1 | 0 | 1 | 1 | 373 |
| H | 1 | 1 | 1 | 1 | 1 | 1 | 0 | 0 | 374 |
| L | 1 | 1 | 1 | 1 | 1 | 1 | 0 | 1 | 375 |
| A | 1 | 1 | 1 | 1 | 1 | 1 | 1 | 0 | 376 |
| B | 1 | 1 | 1 | 1 | 1 | 1 | 1 | 1 | 377 |

With respect to the above table, you should observe that: (a) Bits D4, D5, and D6 are always at a logic 1 state since they are unconnected data bus bits; (b) When any of the fifteen keys are pressed, Bit D7 always goes to logic 1, indicating key closure and serving as a flag bit; and (c) Bits D0, D1, and D2 correspond to the octal code for the octal digit key, provided that Bit D3 is at logic 0.

### FIFTH PROGRAM

This program adds "one" to the contents of the accumulator, decimal adjusts the accumulator contents, and then outputs the binary coded decimal result, i.e., two BCD digits packed in an 8-bit data byte, to Output Port 002.

| LO Memory Address | Instruction Byte | | Mnemonic | Description |
|---|---|---|---|---|
| 000 | 257 | | XRA A | Clear the accumulator |
| 001 | 306 | REPEAT, | ADI | Add the immediate byte to the accumulator |
| 002 | 001 | | 001 | Immediate byte |
| 003 | 047 | | DAA | Decimal adjust the resulting accumulator contents |
| 004 | 006 | | MVI B | Move the following timing byte to the B register |
| 005 | 040 | | 040 | Timing byte |
| 006 | 315 | LOOP, | CALL | Call 10 ms time delay loop DELAY located in KEX |
| 007 | 277 | | DELAY | LO address byte of DELAY |
| 010 | 000 | | — | HI address byte of DELAY |
| 011 | 005 | | DCR B | Decrement B register |
| 012 | 302 | | JNZ | If B register is *not* equal to 000, jump to memory location LOOP; otherwise, continue to next instruction |
| 013 | 006 | | LOOP | LO address byte of LOOP |
| 014 | 003 | | — | HI address byte of LOOP |
| 015 | 323 | | OUT | Output BCD digits to output port 002 |
| 016 | 002 | | 002 | Device code 002 |
| 017 | 303 | | JMP | Jump to memory location REPEAT |
| 020 | 001 | | REPEAT | LO address byte of REPEAT |
| 021 | 003 | | — | HI address byte of REPEAT |

When you execute this program, you will observe the BCD numbers 00 through 99 at Output Port 002. Once the port reaches $99_{10}$, it returns to 00 and repeats the slow counting process. As a programming tip, we would like to point out that you should not use the INR A instruction to increment the accumulator immediately before a DAA instruction. The ADI 001 instruction accomplishes the same result, and properly adjusts the carry and auxiliary carry bits so that the DAA operation can be properly performed.

## INTRODUCTION TO THE EXPERIMENTS

The following simple experiments illustrate accumulator I/O techniques. More extensive accumulator I/O experiments are provided in Unit 22.

| Experiment No. | Comments |
|---|---|
| 1 | A simple microcomputer input-output circuit. Demonstrates the use of 7475 latches and 8095 three-state buffers in accumulator I/O. |
| 2 | Microcomputer input-output on the MMD-1 microcomputer. Demonstrates the operation of the keyboard on the MMD-1 microcomputer. |

3   Characteristics of the DAA instruction. Demonstrates that the use of a DAA instruction permits you to add two 8-bit packed BCD numbers.

## EXPERIMENT NO. 1
## A SIMPLE MICROCOMPUTER INPUT-OUTPUT CIRCUIT

### Purpose

The purpose of this experiment is to test the behavior of a simple microcomputer input-output circuit based upon the 8095 three-state buffer and the 7475 latch.

### Pin Configurations of Integrated-Circuit Chips (Fig. 20-20)

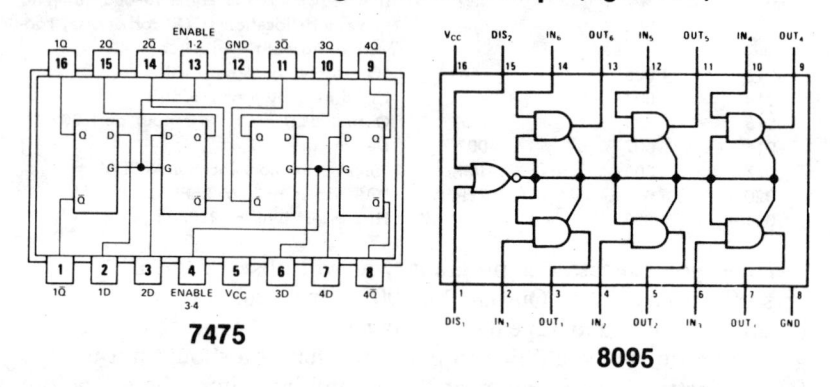

Fig. 20-20.

### Schematic Diagrams of Circuits (Figs. 20-21 and 20-22)

### Program

| LO Memory Address | Instruction Byte | Mnemonic | Description |
|---|---|---|---|
| 000 | 333 | START, IN | Input logic switch data from input port 004 |
| 001 | 004 | 004 | Device code 004 |
| 002 | 323 | OUT | Output accumulator to output port 003 |
| 003 | 003 | 003 | Device code 003 |
| 004 | 303 | JMP | Unconditional jump to memory location START |
| 005 | 000 | START | LO address byte of START |
| 006 | 003 | — | HI address byte of START |

### Step 1

This experiment provides you with experience in the wiring of both an input port and an output port. If you wish only to gain ex-

Fig. 20-21.

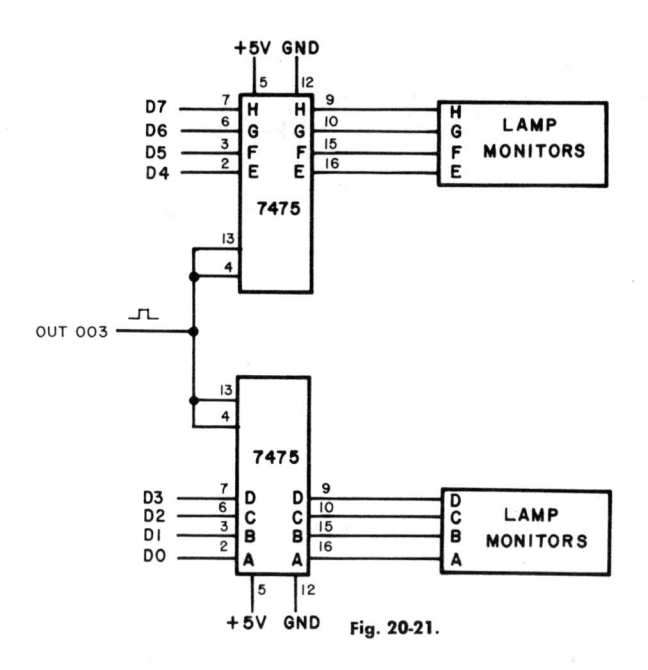

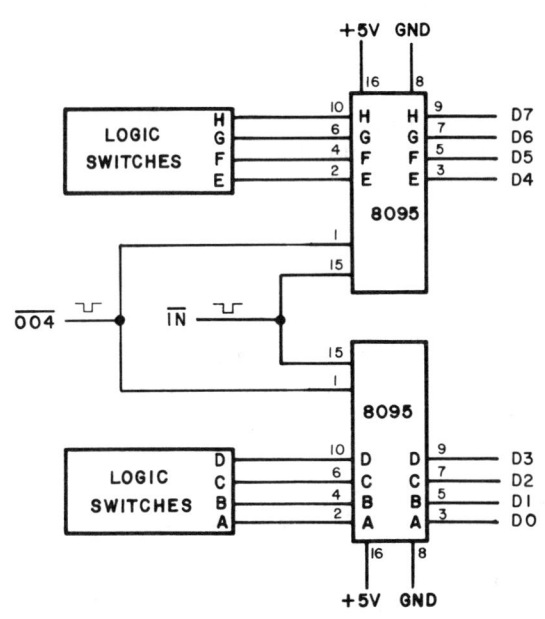

Fig. 20-22.

perience in wiring the input port, *do not* wire the 7475 latch circuit. Skip to Step 7 below.

On a breadboard that has sufficient room for four 16-pin integrated-circuit chips, wire the 7475 output port and the 8095 input port. Note the use of the 2-input 7402 NOR gate contained within the 8095 chip as a device select pulse generator. Use the circuit in Experiment No. 6 in Unit 17 to generate the OUT 003 pulse. The input port can also be wired with the aid of an LR-29 Outboard.

### Step 2

Load the program into memory starting at HI = 003 and LO = 000.

### Step 3

Set the logic switches all to logic 1. Execute the program at the full microcomputer speed. What do you observe at Output Port 003?


All of the lamp monitors of Output Port 003 are lighted.

### Step 4

With the microcomputer operating at full speed, return each logic switch, one at a time, to logic 0. While doing so, explain what you observe on the output port lamp monitors.




As soon as a logic switch is return to logic 0, the corresponding output lamp monitor also returns to logic 0. There is a one-to-one correspondence between the logic switches and the lamp monitors.

### Step 5

Set the eight logic switches to HGFEDCBA = 11110101, or 365 in octal code. Execute the program at the full microcomputer speed, and then, switch to single-step operation using a circuit such as that described in Experiment No. 2 in Unit 17. Wire a bus monitor circuit such as that described in Experiment No. 1 in Unit 17. The latch enable input should be at logic 0.

## Step 6

Single step through the execution of the program and verify the following sequence of bytes that should appear on the bus monitor. Note that you are executing a continuous loop, so you should always be able to start at the beginning through the application of several single-step pulses.

| Data Bus Byte That Appears on the Bus Monitor | Comments |
|---|---|
| 333 | FETCH machine cycle for IN instruction code |
| 004 | FETCH machine cycle for byte <B2> of the IN instruction that is the device code of the input port |
| 365 | INPUT machine cycle, during which information present on the external bidirectional data bus is transferred directly to the accumulator *and* the device code appears on the address bus. An $\overline{\text{IN}}$ control signal is also generated during this machine cycle. [NOTE: The input device select pulse 004 enables the pair of 8095 chips and permits logic switch data to appear on the data bus. In this case, the logic switches have been set to the octal byte, 365.] |
| 323 | FETCH machine cycle for OUT instruction code |
| 003 | FETCH machine cycle for byte <B2> of the OUT instruction that is the device code of the output port |
| 365 | OUTPUT machine cycle, during which the accumulator contents are made available on the directional data bus *and* the device code appears on the address bus. An $\overline{\text{OUT}}$ control signal is also generated during this machine cycle. [NOTE: The output device select pulse 003 enables the pair of 7475 latches and permits them to latch the octal byte, 365, that appears on the data bus. It is this data byte that was originally input during the IN instruction above.] |
| 303 | FETCH machine cycle for JMP instruction code |
| 000 | FETCH machine cycle for byte <B2> of the JMP instruction. This is the LO address byte of memory location START. |
| 003 | FETCH machine cycle for byte <B3> of the JMP instruction. This is the HI address byte of memory location START. |

As you continue to single step the microcomputer, the above sequence of bytes on the data bus will be repeated. The important point here is the fact that you can actually observe the transfer of data between the accumulator and an input or output device. When you work with more complex interface circuits, you may wish to have a bus monitor and single step circuit to verify that the proper data is being transferred at the proper time.

## Step 7

If you do not wish to wire a 7475 latch circuit and if you have an MMD-1 microcomputer, you can take advantage of the fact that there are three 7475-based output ports on the board. The device codes for these ports are 000, 001, and 002. We recommend that you use Output Port 002, which requires a change in the instruction byte at LO = 003 to 002. Make this change in the program.

### Step 8

Set the eight logic switches to logic 1. Execute the program at the full microcomputer speed. What do you observe at Output Port 002?

All of the lamp monitors on Output Port 002 are lighted.

### Step 9

With the microcomputer operating at full speed, return each logic switch, one at a time, to logic 0. While doing so, explain what you observe on the output port.

As soon as a logic switch is returned to logic 0, the corresponding output lamp monitor also returns to logic 0. There is a one-to-one correspondence between the logic switches and the lamp monitors.

### Step 10

Set the eight logic switches to HGFEDCBA = 11110101, or 365 in octal code. Execute the program at the full microcompter speed, and then switch to single-step operation using a circuit such as that described in Experiment No. 2 in Unit 17. Wire a bus monitor circuit similar to that described in Experiment No. 1 in Unit 17. The latch enable point should be at logic 0.

### Step 11

Single step through the execution of the program and verify the sequence of bytes given in Step 6 of this experiment. Keep in mind that the FETCH machine cycle for the output instruction device code places the byte 002 on the data bus instead of 003. Why?

You changed the output port device code from 003 to 002 in Step 7. Therefore, device code 002 must appear on the data bus during the FETCH machine cycle.

### Discussion

This experiment integrates much of what you have learned so far—the generation and use of device select pulses, accumulator input-output, and 8080A programming for input-output operation. A simple three-state input port and latch output port have been constructed and used under software control.

The accumulator I/O ports that you wired in this experiment will be used, with very little modification, in Experiments No. 1 through 3 in Unit 21. Do not remove the 7475 and 8095 I/O port circuits from your breadboard.

## EXPERIMENT NO. 2
## MICROCOMPUTER INPUT-OUTPUT ON THE MMD-1 MICROCOMPUTER

### Purpose

The purpose of this experiment is to demonstrate the operation of the keyboard on the MMD-1 microcomputer.

### Pin Configurations of Integrated-Circuit Chips

No interface circuitry is necessary since all of the necessary chips —the 7475 latch, the 8095 (SN74365) three-state buffer, and the 74148 priority encoder—are already present on the MMD-1 microcomputer. The pin configurations of the 7475 and 8095 chips have been given in the preceding experiment. The pin configuration and truth table for the 74148 chip are given in Fig. 20-23.

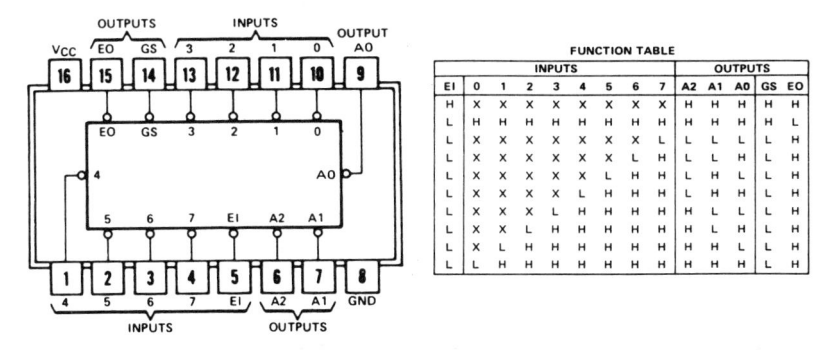

**FUNCTION TABLE**

| EI | 0 | 1 | 2 | 3 | 4 | 5 | 6 | 7 | A2 | A1 | A0 | GS | EO |
|----|---|---|---|---|---|---|---|---|----|----|----|----|----|
| H  | X | X | X | X | X | X | X | X | H  | H  | H  | H  | H  |
| L  | H | H | H | H | H | H | H | H | H  | H  | H  | H  | L  |
| L  | X | X | X | X | X | X | X | L | L  | L  | L  | L  | H  |
| L  | X | X | X | X | X | X | L | H | L  | L  | H  | L  | H  |
| L  | X | X | X | X | X | L | H | H | L  | H  | L  | L  | H  |
| L  | X | X | X | X | L | H | H | H | L  | H  | H  | L  | H  |
| L  | X | X | X | L | H | H | H | H | H  | L  | L  | L  | H  |
| L  | X | X | L | H | H | H | H | H | H  | L  | H  | L  | H  |
| L  | X | L | H | H | H | H | H | H | H  | H  | L  | L  | H  |
| L  | L | H | H | H | H | H | H | H | H  | H  | H  | L  | H  |

Fig. 20-23. Pin configuration and truth table for the 74148 priority encoder chip.

## Schematic Diagram of Circuit

The schematic diagram of the input/output section of the MMD-1 microcomputer is given in Fig. 20-24.

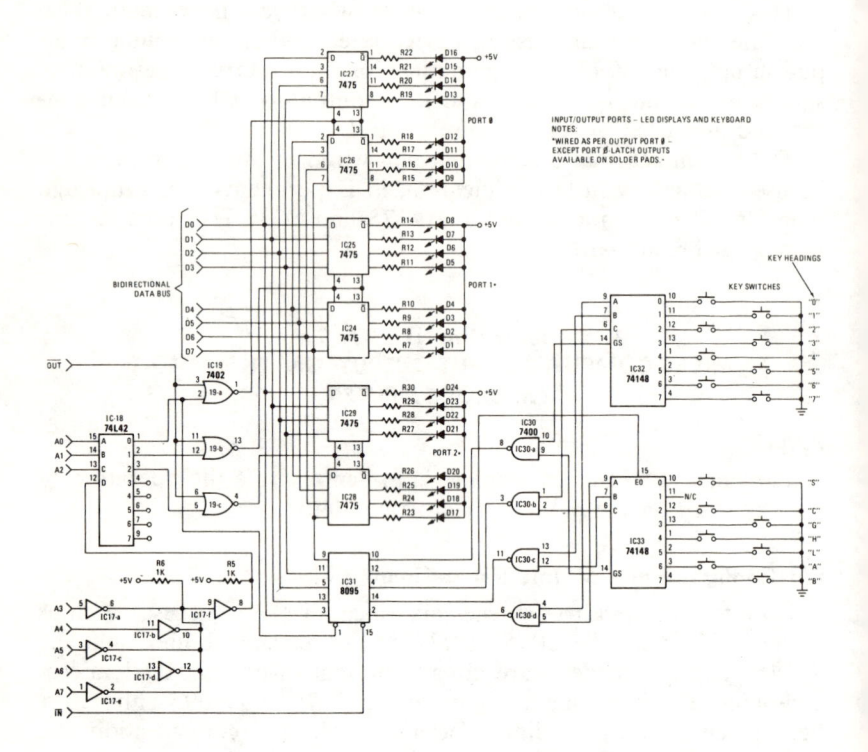

**Fig. 20-24. Input/output section of the MMD-1 microcomputer.**

Courtesy Gernsback Publications, Inc.

## Program

| LO Memory Address | Instruction Byte | Mnemonic | Description |
|---|---|---|---|
| 000 | 333 | START, IN | Input data from keyboard on MMD-1 microcomputer |
| 001 | 000 | 000 | Device code 000 |
| 002 | 323 | OUT | Output data to output port 002 on MMD-1 microcomputer |
| 003 | 002 | 002 | Device code 002 |
| 004 | 303 | JMP | Unconditional jump to memory location START |
| 005 | 000 | START | LO address byte of START |
| 006 | 003 | — | HI address byte of START |

## Step 1

Study the schematic diagram for the input/output section of the MMD-1 microcomputer that is shown in Fig. 20-24.
Note the following:

1. The input/output section is an example of accumulator I/O. The clues that we used to reach this conclusion were the $\overline{\text{IN}}$ and $\overline{\text{OUT}}$ control signals on the left-hand side of the diagram.
2. The I/O decoder circuit consists of a 74L42 decoder chip and three 2-input NOR gates. For further details, refer to Fig. 17-7 and the associated text as well as Experiment No. 5 in Unit 17.
3. There are three 8-bit output ports:
   a. Output Port 000, which consists of 7475 latches IC26 and IC27.
   b. Output Port 001, which consists of 7475 latches IC24 and IC25.
   c. Output Port 002, which consists of 7475 latches IC28 and IC29.
4. There is one 5-bit input port, which is ultimately connected to the keyboard. This is Input Port 000, which consists of an 8095 chip, IC31.
5. The pair of 74148 priority encoder chips and the 7400 2-input NAND gate provide the circuitry to encode fifteen keys on the keyboard. When a single key is pressed, the input to the 74148 corresponding to that key goes to a logic 0 and output GS on the 74148 chip goes to a logic 0. The three-bit octal code for the depressed key appears at output pins A, B, and C on the 74148 chip. Remember that the sixteenth key, RESET, is a hardwired function and generates no code.

## Step 2

Load the program into memory starting at HI = 003 and LO = 000.

## Step 3

Execute the program at the full microcomputer speed. When you do not press any key on the keyboard, which bits are lighted on Output Port 002?


Bits D4, D5 and D6 are at logic 1. The remaining five bits are at logic 0.

### Step 4

Press any key on the keyboard except the RESET key. Is it true that Bit D7 on Output Port 002 always goes to a logic 1 when any of the remaining fifteen keys is pressed?

Yes. The reason is that this is the bit that indicates to the KEX monitor program that a key has been depressed. Once the microcomputer detects that a key has been depressed, it then proceeds to determine which key was pressed. In the routine that accomplishes this task, there is a 10-ms time delay to eliminate the common problem of contact bounce. The keyboard input program from KEX is listed at the end of this experiment.

### Step 5

So far, you have observed that bits D4, D5, and D6 remain at logic 1 no matter which key is pressed and that Bit D7 goes to logic 1 whenever a key is pressed. You now must determine the function of the remaining four bits, D0 through D3. Press keys 0 through 7 in sequence and observe the bit pattern at the Output Port 002, bits D0 through D2. What do you observe?

The bit pattern at D0 through D2 corresponds to the 3-bit code for the keys 0 through 7. Thus, Key 5 will have the 3-bit code, 101.

### Step 6

Press keys H, L, G, S, A, B, and C and explain what you observe at Bit D3 on the output latch. Also explain the significance of bits D0 through D2 when these keys are pressed.

When any one of keys H, L, G, S, A, B, or C is pressed, Bit D3 always goes to a logic 1. The remaining three bits, D0 through D2, decode which key is pressed. You have now verified the encoding of the keyboard. Such encoding can be summarized by the following truth table, which gives the logic states of the bits when the key is depressed.

| Key heading | D7 | D6 | D5 | D4 | D3 | D2 | D1 | D0 |
|---|---|---|---|---|---|---|---|---|
| 0 | 1 | 1 | 1 | 1 | 0 | 0 | 0 | 0 |
| 1 | 1 | 1 | 1 | 1 | 0 | 0 | 0 | 1 |
| 2 | 1 | 1 | 1 | 1 | 0 | 0 | 1 | 0 |
| 3 | 1 | 1 | 1 | 1 | 0 | 0 | 1 | 1 |
| 4 | 1 | 1 | 1 | 1 | 0 | 1 | 0 | 0 |
| 5 | 1 | 1 | 1 | 1 | 0 | 1 | 0 | 1 |
| 6 | 1 | 1 | 1 | 1 | 0 | 1 | 1 | 0 |
| 7 | 1 | 1 | 1 | 1 | 0 | 1 | 1 | 1 |
| S | 1 | 1 | 1 | 1 | 1 | 0 | 0 | 0 |
| C | 1 | 1 | 1 | 1 | 1 | 0 | 1 | 0 |
| G | 1 | 1 | 1 | 1 | 1 | 0 | 1 | 1 |
| H | 1 | 1 | 1 | 1 | 1 | 1 | 0 | 0 |
| L | 1 | 1 | 1 | 1 | 1 | 1 | 0 | 1 |
| A | 1 | 1 | 1 | 1 | 1 | 1 | 1 | 0 |
| B | 1 | 1 | 1 | 1 | 1 | 1 | 1 | 1 |

**Step 7**

When keyboard keys are activated, the code is inputted and then outputted from the 8080A chip to the latched LEDs. It is important to remember that this data transfer process is under software control.

Change the device code byte at LO = 001 to 005 in the program, and then execute it at 750 kHz. Does it operate as it had previously? Why?

No. The device address for the input instruction has been changed so that the keyboard is no longer selected.

### LISTING OF SUBROUTINE KBRD

The instantaneous input of keyboard data may not match the codes shown previously because of contact bounce in the nonideal mechanical switches. A switch bounce "filter" program is available in the KEX programmable read-only memory (PROM) starting at address 000 315. This program, called KBRD, may be called with a subroutine call, and returns with the key code value located in the accumulator. If you wish to use this subroutine be sure that you have set up a stack area. A listing for the keyboard input subroutine is shown below.

| Memory Address | Instruction Byte | | Mnemonic | Comments |
|---|---|---|---|---|
| 000 315 | 333 | KBRD, | IN | /Input from keyboard encoders |
| 000 316 | 000 | | 000 | |
| 000 317 | 267 | | ORA A | /Set flags |
| 000 320 | 372 | | JM | /Jump back if last key not released |
| 000 321 | 315 | | KBRD | |
| 000 322 | 000 | | 0 | |
| 000 323 | 315 | | CALL | /Wait 10 msec |
| 000 324 | 277 | | TIMOUT | |
| 000 325 | 000 | | 0 | |
| 000 326 | 333 | FLAGCK, | IN | |
| 000 327 | 000 | | 000 | |
| 000 330 | 267 | | ORA A | |
| 000 331 | 362 | | JP | /Jump back to wait for a new key to |
| 000 332 | 326 | | FLAGCK | /be pressed |
| 000 333 | 000 | | 0 | |
| 000 334 | 315 | | CALL | /Wait 10 msec for bouncing |
| 000 335 | 277 | | TIMOUT | |
| 000 336 | 000 | | 0 | |
| 000 337 | 333 | | IN | |
| 000 340 | 000 | | 000 | |
| 000 341 | 267 | | ORA A | |
| 000 342 | 362 | | JP | /Jump back if new key not still |
| 000 343 | 326 | | FLAGCK | /pressed (false alarm) |
| 000 344 | 000 | | 0 | |
| 000 345 | 346 | | ANI | /Mask out all bits but key code |
| 000 346 | 017 | | 017 | |
| 000 347 | 345 | | PUSH H | /Save H and L registers |
| 000 350 | 046 | | MVI H | /Zero H register |
| 000 351 | 000 | | 000 | |
| 000 352 | 306 | | ADI | /Add the address of the beginning of |
| 000 353 | 360 | | 360 | /the table to the key code |
| 000 354 | 157 | | MOV L,A | |
| 000 355 | 176 | | MOV A,M | /Fetch new value from table |
| 000 356 | 341 | | POP H | /Restore H and L registers |
| 000 357 | 311 | | RET | |

The following translation table converts the code generated by key closures to the code used by the main KEX program.

| | | | |
|---|---|---|---|
| 000 360 | 000 | TABLE, 000 | |
| 000 361 | 001 | 001 | |
| 000 362 | 002 | 002 | |
| 000 363 | 003 | 003 | |
| 000 364 | 004 | 004 | |
| 000 365 | 005 | 005 | |
| 000 366 | 006 | 006 | |
| 000 367 | 007 | 007 | |
| 000 370 | 013 | 013 | /S |
| 000 371 | 000 | 000 | /This code cannot be generated |
| 000 372 | 017 | 017 | /C |
| 000 373 | 012 | 012 | /G |
| 000 374 | 010 | 010 | /H |
| 000 375 | 011 | 011 | /L |
| 000 376 | 015 | 015 | /A |
| 000 377 | 016 | 016 | /B |

## LISTING OF SUBROUTINE TIMOUT

| Memory Address | Instruction Byte | Mnemonic | Comments |
|---|---|---|---|
| 000 277 | 365 | TIMOUT, PUSH PSW | /Save accumulator and flags |
| 000 300 | 325 | PUSH D | /Save register pair D |
| 000 301 | 021 | LXI D | /Load D and E with value to be |
| 000 302 | 046 | 046 | /decremented |
| 000 303 | 001 | 001 | |
| 000 304 | 033 | MORE, DCX D | /Decrement register pair D |
| 000 305 | 172 | MOV A,D | /Move D to A |
| 000 306 | 263 | ORA E | /OR E with A |
| 000 307 | 302 | JNZ | /Is register A = 000? If not, jump |
| 000 310 | 304 | MORE | /to MORE at LO = 304 and |
| 000 311 | 000 | 0 | /HI = 000. Otherwise, continue to /next instruction. |
| 000 312 | 321 | POP D | /Restore register pair D |
| 000 313 | 361 | POP PSW | /Restore accumulator and flags |
| 000 314 | 311 | RET | /Return from subroutine TIMOUT |

The above programs have been written in the way they would appear as output from the Tychon, Inc. 8080A resident assembler/editor program.

## EXPERIMENT NO. 3
## CHARACTERISTICS OF THE DAA INSTRUCTION

**Purpose**

The purpose of this experiment is to explore some of the characteristics of the DAA instruction.

**Program**

| LO Memory Address | Instruction Byte | Mnemonic | | Description |
|---|---|---|---|---|
| 000 | 000 | BEGIN, | NOP | No operation |
| 001 | 257 | | XRA A | Clear the accumulator |
| 002 | 306 | ADD, | ADI | Add the following byte to the accumulator |
| 003 | 001 | | 001 | Data byte |
| 004 | 047 | | DAA | Decimal adjust the resulting accumulator contents |
| 005 | 006 | | MVI B | Move the following timing byte to register B |
| 006 | 040 | | 040 | Timing byte |
| 007 | 315 | REPEAT, | CALL | Call 10 ms time delay routine DELAY located in KEX EPROM |
| 010 | 277 | | DELAY | LO address byte of DELAY |
| 011 | 000 | | — | HI address byte of DELAY |
| 012 | 005 | | DCR B | Decrement register B |
| 013 | 302 | | JNZ | If register B is *not* equal to 000, jump to memory location REPEAT; otherwise, continue to next instruction |
| 014 | 007 | | REPEAT | LO address byte of REPEAT |
| 015 | 003 | | — | HI address byte of REPEAT |
| 016 | 323 | | OUT | Output pair of packed BCD digits to output port 002 |
| 017 | 002 | | 002 | Device code 002 |
| 020 | 303 | | JMP | Jump to memory location ADD |
| 021 | 002 | | ADD | LO address byte of ADD |
| 022 | 003 | | — | HI address byte of ADD |

## Step 1

This experiment uses Output Port 002 on the MMD-1 microcomputer. Our objective is to demonstrate how the DAA instruction is used to facilitate BCD arithmetic operations.

What do we mean by a pair of "packed BCD digits"?

We mean an 8-bit number that is composed of two 4-bit BCD digits, one occupying the four most-significant bits and the other occupying the four least-significant bits.

## Step 2

Load the program into read/write memory starting at HI= 003 and LO = 000. Execute the program. Observe what happens immediately after the BCD count in Output Port 002 reaches $99_{10}$. Summarize your observations in the space below.

We observed that the counts immediately following 99 were 00, 01, 02, 03, 04, 05, 06, 07, 08, 09, 10, . . . . In other words, the output port "rolled over" and started to count up from 00 once again. This type of behavior is what we would expect.

**Step 3**

Now change the following instruction bytes in the program:

| 002 | 000 | NOP | No operation |
|-----|-----|-----|--------------|
| 003 | 074 | INR A | Increment accumulator by one |
| 006 | 200 | 200 | Timing byte |

Execute the program, as modified above, and explain what you observe on Output Port 002 immediately following $99_{10}$.

We observed a strange sequence of digits that are best listed in hexadecimal code: 00, 61, C2, 23, 84, E5, 46, A7, 08, 69, D0, 31, 92, etc. The least significant BCD digit was correct, but the most significant BCD digit varied in an unpredictable manner.

### Step 4

Why did you suspect a problem when the INR A instruction replaced the ADI 001 instruction? After all, both increment the accumulator contents by one.

The ADI 001 instruction affects both the auxiliary carry and carry flags, depending upon the result of the operation. In contrast, the INR A instruction affects *only* the auxiliary carry flag. In order for the DAA instruction to operate properly, *both* the carry and auxiliary carry must respond properly to an arithmetic instruction that immediately precedes the DAA instruction. Such is not the case for the INR A instruction.

### Step 5

Consider the following inital steps in the program:

| | | | | |
|---|---|---|---|---|
| 000 | 076 | MVI A | | Move following byte into accumulator |
| 001 | AUG | AUG | | Data byte AUG that serves as the augend for an addition |
| 002 | 306 | ADI | | Add the following byte to the accumulator |
| 003 | ADD | ADD | | Data byte ADD that serves as the addend for an addition |

Once this addition has been performed, the time-delay routine executed, and the sum output to Port 002, the microcomputer execution is halted through the use of the following instruction byte,

| | | | |
|---|---|---|---|
| 020 | 166 | HLT | Halt |

The terms, augend and addend, have been defined by Graf[2] in the following manner:

> *addend*—A quantity which, when added to another quantity (called the augend), produces a result called the sum.
> *augend*—In an arithmetic addition, the number increased by having another number (called the addend) added to it.

222

In the above program section in this step, must the data bytes AUG and ADD be binary numbers or packed BCD numbers? The answer to this question is crucial to the operation of the DAA instruction, so consider it carefully. Please write your answer in the space below.

The correct answer is as follows:

- If the four steps *are not* followed by a DAA instruction, the data bytes AUG and ADD are treated as 8-bit binary numbers and a straight binary addition is performed to yield a binary number as a sum.
- If the four steps *are* followed by a DAA instruction, the data bytes AUG and ADD are treated as 8-bit packed BCD numbers and a BCD correction is automatically performed within the 8080A to yield a packed BCD number as a sum.

This is a very important distinction and one that you must remember. With the 8080A microprocessor chip, you can, in essence, perform simple BCD arithmetic, which means that the augend, addend and sum are all considered to be packed BCD numbers. You should remember, though, that the 8080A and almost all other computers are binary processors. All data is treated as binary 1s and 0s. Codes such as ASCII, EBCDIC, and BCD are transparent to the computer. To perform BCD math requires careful attention to programming and the careful use of the Decimal Adjust Accumulator (DAA) instruction. It should *not* be considered as a binary-to-decimal converter!

## Step 6

Change the instruction bytes at LO memory addresses 000, 002, and 020 to those indicated in Step 5. Perform the following additions between the BCD data bytes AUG and ADD, and compare the SUM that you observe on Output Port 002 with that given in the following table.

| AUG<br>(Packed BCD) | ADD<br>(Packed BCD) | SUM<br>(Packed BCD) | Output Port 002<br>(Packed BCD) |
|---|---|---|---|
| 10 | 00 | 10 | _____ |
| 10 | 10 | 20 | _____ |
| 10 | 15 | 25 | _____ |
| 15 | 19 | 34 | _____ |
| 27 | 28 | 55 | _____ |
| 33 | 48 | 81 | _____ |
| 38 | 75 | 13* | _____ |
| 99 | 99 | 98* | _____ |

* Carry = 1 for these sums.

What do you conclude?

You should conclude that when an addition operation is immediately followed by a DAA instruction, the data bytes added must be considered to be packed BCD quantities and the SUM is also a packed BCD quantity with or without carry. True BCD additions are being performed by the microcomputer through a binary addition and a BCD correction called Decimal Adjust Accumulator (DAA). The 8080A chip from the Intel Corporation can only perform BCD additions. Subtractions require tricks since the auxiliary carry is not affected by the SUB and SBB instructions. The NEC 8080A chip has an extra flag bit, SUB, that permits you to perform packed BCD subtractions using SUB, SBB, and SBI.

## Step 7

Eliminate the DAA instruction by substituting the following NOP instruction byte:

| 004 | 000 | NOP | No operation |

Perform the following additions between AUG and ADD and compare the SUM that you observe on Output Port 002 with that given in the following table.

| AUG (Hex) | ADD (Hex) | SUM (Hex) | Output Port 002 (Hex) |
|---|---|---|---|
| 10 | 00 | 10 | _____ |
| 10 | 10 | 20 | _____ |
| 10 | 15 | 25 | _____ |
| 15 | 19 | 2E | _____ |
| 27 | 28 | 4F | _____ |
| 33 | 48 | 7B | _____ |
| 38 | 75 | AD | _____ |
| 99 | 99 | 32* | _____ |

* Carry = 1 for this sum.

Note that the AUG and ADD column entries are given in hexadecimal code (hex) rather than in octal or decimal code. One of the objectives of this experiment is to give you some practice in converting decimal and hexadecimal quantities into octal code, and vice versa.

What do you conclude from the above table?

You should conclude that when an addition of two numbers is *not* immediately followed by a DAA instruction, the data bytes AUG and ADD as well as the SUM must be considered as regular 8-bit binary numbers. The microcomputer is now performing binary arithmetic, but the sum has not been corrected to give a BCD answer.

This experiment demonstrates that you can input and output either binary or packed BCD data and perform either binary or BCD additions using such data. We have simulated the input of BCD data through the use of the MVI A instruction.

## REVIEW QUESTIONS

The following questions will help you review accumulator input/output techniques.

1. What is meant by the term "accumulator input/output"?
2. What differences exist between chips used for microcomputer output and those used for microcomputer input? Explain why the chips for the two different uses must differ in function.

3. Why is the output drive capability of a microprocessor chip and the fan-in characteristics of interface chips important in microcomputer output circuits?
4. Based upon the information presented in the Fifth Program and Experiment No. 3, explain the characteristics of the DAA instruction in the addition of a pair of 8-bit numbers.

# 21

# An Introduction to Memory-Mapped Input/Output Techniques

### INTRODUCTION

In memory-mapped I/O, you treat any input/output device as if it were a memory location and use memory transfer instructions such as MOV, STAX, LDAX, STA, LDA, SHLD, and LHLD to input and output data. Any of the general-purpose registers can be the source or destination of memory-mapped I/O data. In this unit, you will use simple memory-mapped I/O programs and wire simple input/output interface circuits that permit you to transfer data with the aid of memory-reference instructions. In this unit and the one that follows, the term "memory I/O" will be used as a synonym for memory-mapped I/O.

### OBJECTIVES

At the completion of this unit, you will be able to do the following:

- Summarize the differences between accumulator I/O and memory-mapped I/O techniques.
- Sketch a circuit that can be used to generate memory address select pulses.
- Explain how memory address select pulses are employed to achieve the objective of memory-mapped input/output.
- Wire a simple memory input circuit.

- Wire a simple memory output circuit.
- Compare the memory input characteristics of the following three instructions: LDA, LDAX, and MOV A,M.
- Compare the memory output characteristics of the following three instructions: STA, STAX, and MOV M,A.

### MEMORY-MAPPED I/O VS ACCUMULATOR I/O

An input/output device can be a teletypewriter, cathode ray tube (CRT) display, laboratory instrument, minicomputer, another microcomputer, or a small digital device such as an integrated-circuit chip. All I/O devices can exchange data between the 8080A microprocessor chip via either *accumulator I/O* or *memory-mapped I/O* techniques, which are similar to each other in basic concept. We compare these two I/O techniques in Tables 21-1 and 21-2.

**Table 21-1. Summary of Characteristics of Accumulator I/O**

| 8080A instructions | OUT <B2><br>IN  <B2> |
|---|---|
| Control signals | $\overline{\text{OUT}}$<br>$\overline{\text{IN}}$ |
| Data transfer | Between accumulator and I/O device |
| Device decoding | An 8-bit device code, A0 to A7 or A8 to A15, that is byte <B2> in the IN or OUT instruction. We recommend that it be absolutely decoded, i.e., that all eight bits be used to designate, or decode, a specific I/O device. |
| Terminology | The I/O processes will be called **input** and **output**. The decoded signal that strobes an I/O device will be called a *device select pulse*. |

The advantages of memory I/O techniques can be clearly seen from a comparison of Tables 21-1 and 21-2. Data transfer can be between the I/O device and any of the seven general-purpose registers within the 8080A chip. If the 16-bit memory address has been previously stored in register pair H, then the data transfer can be quicker if either a MOV r,M or MOV M,r instruction is used. In principle, many more devices can be addressed by memory I/O techniques than by accumulator I/O techniques. Finally, two-byte data transfers in a single instruction are possible using the SHLD and LHLD instructions.

An important point is that memory I/O and accumulator I/O techniques are not fundamentally different from each other. In each case, a control signal indicates whether the operation is one of input or output. Also, in each case the address bus must be decoded to identify

### Table 21-2. Summary of Characteristics of Memory-Mapped I/O

| 8080A instructions | MOV B,M | MOV M,H | ANA M |
|---|---|---|---|
| | MOV C,M | MOV M,L | XRA M |
| | MOV D,M | MOV M,A | ORA M |
| | MOV E,M | STAX B | CMP M |
| | MOV H,M | STAX D | INR M |
| | MOV L,M | LDAX B | DCR M |
| | MOV A,M | LDAX D | MVI M |
| | MOV M,B | ADD M | STA $<$B2$>$ $<$B3$>$ |
| | MOV M,C | ADC M | LDA $<$B2$>$ $<$B3$>$ |
| | MOV M,D | SUB M | SHLD $<$B2$>$ $<$B3$>$ |
| | MOV M,E | SBB M | LHLD $<$B2$>$ $<$B3$>$ |
| Control signals | $\overline{\text{MEMR}}$ $\overline{\text{MEMW}}$ | | |
| Data transfer | Between memory I/O device and registers B, C, D, E, H, L, or the accumulator (register A) | | |
| Device decoding | A 16-bit device code, A0 to A15, that is contained either in register pair H; register pair B; register pair D; or in bytes $<$B2$>$ and $<$B3$>$ for the STA, LDA, SHLD, or LHLD instructions. In some instances, it is useful and convenient to reserve the upper 32K memory area for memory I/O addresses; when A15 on the address bus is at logic 1, memory I/0 exists. Bits A0 through A7 can be used to decode a specific I/O device when A15 = 1. The I/O device is made to look like a unique 8-bit memory location and the memory-reference instructions are used in their normal manner to read from or write into the specific memory I/O device. | | |
| Terminoloy | The memory I/O processes will be called **read** and **write** rather than input and output. The decoded signal that strobes a memory I/O device will be called an **address select pulse** rather than a device select pulse. | | |

a specific I/O device. Finally, in each case the actual data transfer occurs in a machine cycle during the execution of the instruction. For accumulator I/O, this machine cycle generates with the aid of a status latch the control signals $\overline{\text{IN}}$ and $\overline{\text{OUT}}$; for memory I/O, the signals $\overline{\text{MEMR}}$ and $\overline{\text{MEMW}}$ are generated instead. You can observe the data transfer over the bidirectional data bus with the aid of a bus monitor (see Experiment No. 1 in Unit 17).

We have observed during our work with memory I/O techniques that it is easy to be careless when they are used. Most problems can be attributed to the lack of absolute decoding of the entire 16-bit address bus. When addressing a memory I/O device, it is not sufficient to decode bits A0 to A9, since these same bits are used in any 8080A-based microcomputer system that has at least 1K of addressable memory. Therefore, if you wish to use memory I/O techniques, you should plan to decode some of the highest bits on the address bus, especially bits A13 to A15.

## GENERATING MEMORY-MAPPED I/O ADDRESS SELECT PULSES

To generate a memory I/O address select pulse, you need two types of information from the 8080A microcomputer:
1. A multibit identification code, called a *memory address,* for the external I/O device.
2. A single-bit synchronization pulse, either $\overline{\text{MEMR}}$ or $\overline{\text{MEMW}}$, that synchronizes the decoding of the device code.

The origin of both types of information is in software, i.e., in "memory-reference instructions" such as MOV r,M, STAX B, MOV M,r, STAX D, ADD M, MVI M, LDAX D, CMP M, etc. Such instructions cause the 8080A-microprocessor chip to place a 16-bit address on the address bus and also to generate either a memory read, $\overline{\text{MEMR}}$ (for memory data that is input to the 8080A chip), or a memory write, $\overline{\text{MEMW}}$ (for memory data that is output from the 8080A chip), control signal.

In other words, as with accumulator I/O, during the generation of a memory I/O address select pulse, both the address bus and the control bus are active. It is your responsibility to properly decode the signals on these two busses to produce unique address select pulses that can be used to transfer data between the external memory I/O device and the internal registers of the 8080A.

Fig. 21-1 provides a commonly used decoding technique for memory I/O address select pulses. This is a decoder circuit for the genera-

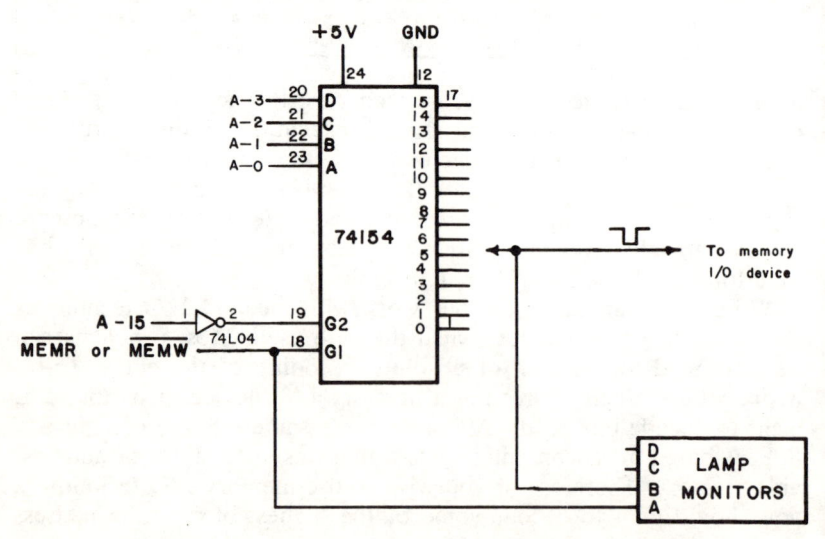

Fig. 21-1. A decoder circuit for memory I/O address select pulses.

tion of sixteen different memory-mapped I/O address select pulses. The HI address byte is 200 and the LO address byte ranges from 000 to 017, in octal code. (This is not an absolute decoder circuit for the 16-bit address bus.) Note the resemblance between this figure and Fig. 17-2 in Unit 17. The 74154 decoder is enabled by two signals—the complement of the A-15 address bus bit, and either $\overline{\text{MEMR}}$ or $\overline{\text{MEMW}}$. The following truth tables show the chip enable process.

| A-15 | $\overline{\text{MEMR}}$ | Decoder Behavior | A-15 | $\overline{\text{MEMW}}$ | Decoder Behavior |
|---|---|---|---|---|---|
| 0 | 0 | disabled | 0 | 0 | disabled |
| 0 | 1 | disabled | 0 | 1 | disabled |
| 1 | 0 | generate read pulse | 1 | 0 | generate write pulse |
| 1 | 1 | disabled | 1 | 1 | disabled |

Observe that the 74154 decoder is enabled only when address bus bit A-15 is at logic 1. Data is input only when $\overline{\text{MEMR}}$ is at logic 0 and output only when $\overline{\text{MEMW}}$ is at logic 0. When no memory data is being transferred, both $\overline{\text{MEMW}}$ and $\overline{\text{MEMR}}$ are at logic 1.

In Fig. 21-2, ten of the sixteen address bus bits are decoded by a pair of 74154 decoders. For the circuit shown, sixteen memory I/O address select pulses are produced, starting at HI = 300 and LO =

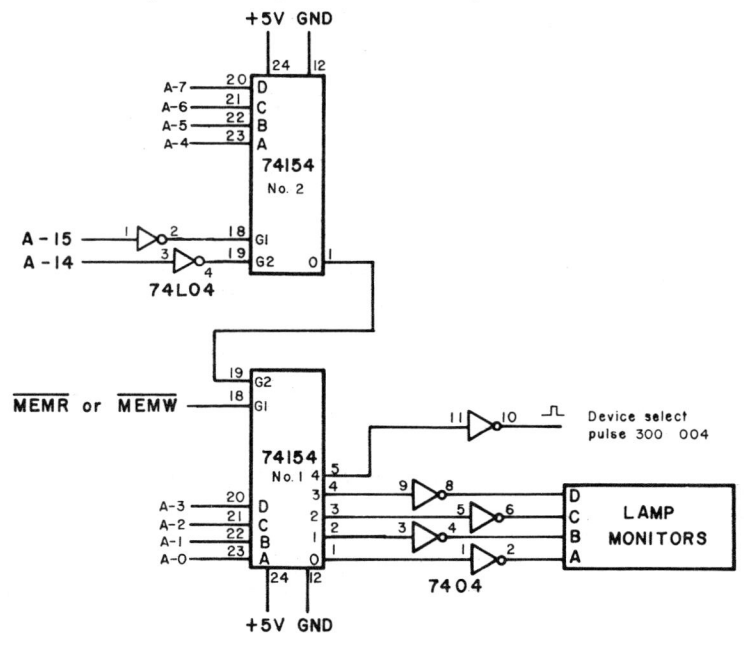

Fig. 21-2. Decoder circuit that generates sixteen different memory I/O address select pulses.

000 and terminating at HI = 300 and LO = 017. The 74154 decoder No. 2 is enabled only when both A-14 and A-15 are at logic 1. Any one of the sixteen output pins on decoder No. 2 can be used to enable the G2 input of decoder No. 1.

A final decoding technique is similar to that employed in Fig. 17-8 in Unit 17. The high address bits are input into a 74L30 eight-input NAND gate, which decodes the eight bits into a single unique logic 0 state. For example, if A8 through A15 are input into a 74L30 gate, a logic 0 output will be produced only when $A8 = A9 = A10 = A11 = A12 = A13 = A14 = A15 = 1$. This output can then be used to enable other decoder chips such as the 74154 or 7442.

### MEMORY-MAPPED I/O: USE OF ADDRESS BIT A-15

Intel Corporation's *8080 Microcomputer Systems User's Manual* provides interesting diagrams that demonstrate how to use address bus bit A-15 to distinguish between memory and a memory I/O device. In accumulator I/O, four control signals are generated either by the 8228 chip (Fig. 21-3) or by equivalent circuitry—$\overline{\text{MEMR}}$,

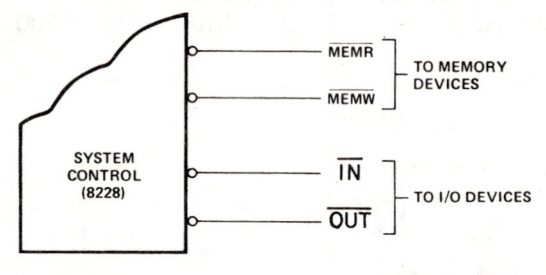

Fig. 21-3. Control signals used in accumulator I/O.

$\overline{\text{MEMW}}$, $\overline{\text{IN}}$, and $\overline{\text{OUT}}$. These signals permit you to distinguish between a memory location and an I/O device. In memory I/O, it can be observed from Fig. 21-4 that only $\overline{\text{MEMR}}$ and $\overline{\text{MEMW}}$ are used to address both memory and memory I/O devices. In the drawing of Fig. 21-4, address bit A-15 is gated with these two control signals to produce two new control signals, $\overline{\text{MEMIOR}}$ and $\overline{\text{MEMIOW}}$, that are used only with I/O devices.

The effect of the use of address bit A-15 is to subdivide the 64K of memory into two 32K blocks, one for memory and the other for memory I/O devices. This is shown in Fig. 21-5. In contrast, in normal accumulator I/O, only 256 input or 256 output devices can be addressed, but the maximum size of the memory can be as large as 64K.

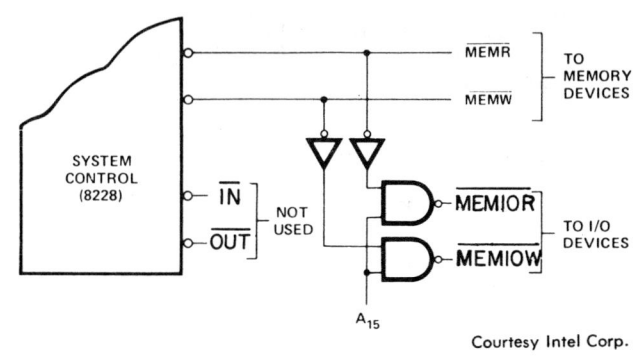

**Fig. 21-4. Control signals used in one type of memory I/O.**

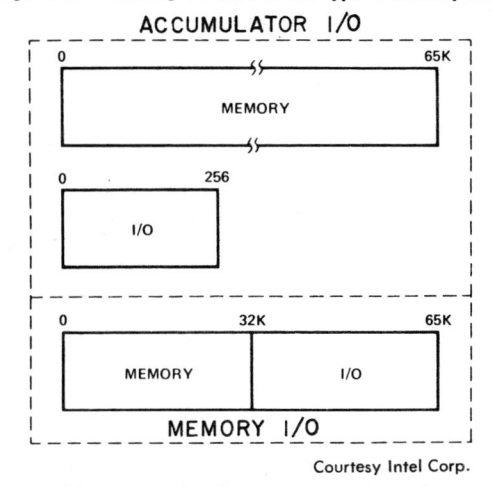

**Fig. 21-5. Memory block comparison between accumulator I/O and memory I/O.**

The accumulator I/O and memory I/O techniques discussed in this and previous units do not exhaust the available possibilities. Rather than use a 74154 decoder, the individual bits in the 8-bit device code could be decoded directly to select six I/O devices, each of which requires a 2-bit port select code (Fig. 21-6). This type of accumulator I/O is very useful when you have only a few I/O devices. The same technique can be applied in memory I/O, as shown in Fig. 21-7. In the drawing of Fig. 21-7, you can select up to thirteen different memory I/O devices, each of which requires a 2-bit port select code. Bit A-15, as in Fig. 21-4, is used to distinguish between memory and a memory I/O device. Finally, if you wish to restrict the 32K memory I/O block in Fig. 21-5 to a smaller region of memory, you can simultaneously decode several of the higher bits on the address bus. For example, if you use a pair of 7420 4-input NAND

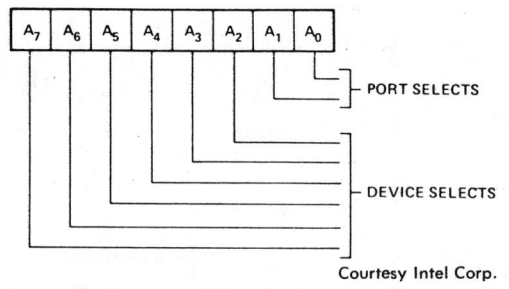

**Fig. 21-6. Example of the use of the 8-bit device code in accumulator I/O to address six devices.** *Courtesy Intel Corp.*

gates in Fig. 21-4 rather than the 2-input NAND gates shown, you can decode address bits A-13, A-14, and A-15 and restrict the memory I/O memory block to 8K and expand the memory block to 56K.

## MEMORY-MAPPED I/O INSTRUCTIONS

There are twenty-two 8080A instructions that permit you to transfer data between the internal registers and external memory devices.

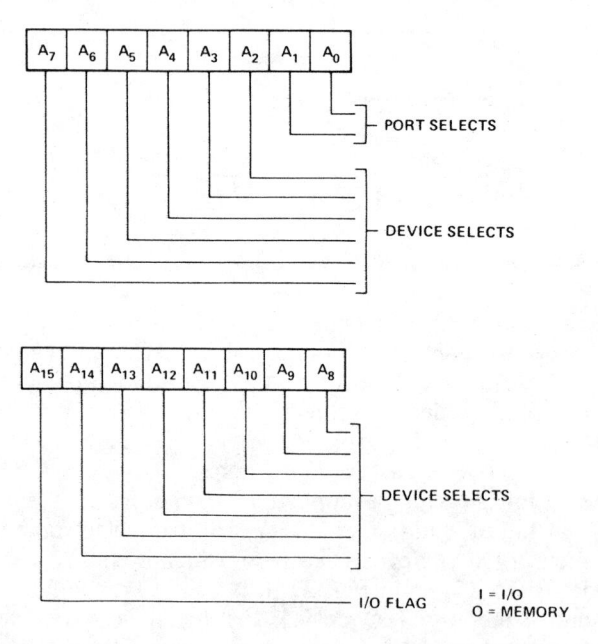

**Fig. 21-7. Example of the use of the 16-bit memory address word in memory I/O to address thirteen different memory I/O devices.** *Courtesy Intel Corp.*

These external devices can be either semiconductor memory, i.e., read/write memory, ROMs, EPROMs, etc., or else they can be input-output devices that are addressed *as if they were memory locations*. Implied in any memory-reference instruction is a 16-bit memory address that uniquely identifies a memory byte. This memory address is contained either in register pair H, register pair B, register pair D, or else in bytes <B2> and <B3> of the instruction itself.

In addition to the twenty-two data transfer instructions, there exist eleven other memory-reference instructions. One such instruction permits you to move an immediate byte in a program to a memory location. Two other instructions permit you to increment or decrement the contents of a specific memory location. Finally, eight other instructions permit you to perform logical or arithmetic operations between the contents of a memory location and the contents of the accumulator.

All thirty-three memory-reference instructions are summarized below. They are subdivided according to the location of the memory address word.

### Address of Memory Location M is Contained in Register Pair H

Twenty-five of the thirty-three 8080A memory-reference instructions are contained in this group. All require the address of the memory location to be stored in register pair H before the memory-reference instruction is executed.

| | | |
|---|---|---|
| MOV B,M | 106 | Move contents of memory location M to register B |
| MOV C,M | 116 | Move contents of memory location M to register C |
| MOV D,M | 126 | Move contents of memory location M to register D |
| MOV E,M | 136 | Move contents of memory location M to register E |
| MOV H,M | 146 | Move contents of memory location M to register H |
| MOV L,M | 156 | Move contents of memory location M to register L |
| MOV A,M | 176 | Move contents of memory location M to register A |
| MOV M,B | 160 | Move contents of register B to memory location M |
| MOV M,C | 161 | Move contents of register C to memory location M |
| MOV M,D | 162 | Move contents of register D to memory location M |
| MOV M,E | 163 | Move contents of register E to memory location M |
| MOV M,H | 164 | Move contents of register H to memory location M |
| MOV M,L | 165 | Move contents of register L to memory location M |
| MOV M,A | 167 | Move contents of register A to memory location M |
| MVI M <B2> | 066 <B2> | Move immediate byte <B2> to memory location M |
| INR M | 064 | Increment contents of memory location M |
| DCR M | 065 | Decrement contents of memory location M |
| ADD M | 206 | Add contents of memory location M to contents of accumulator and store result in accumulator |

| ADC M | 216 | Add with carry contents of memory location M to contents of accumulator and store result in accumulator |
| SUB M | 226 | Subtract contents of memory location M from contents of accumulator and store result in accumulator |
| SBB M | 236 | Subtract with borrow contents of memory location M from contents of accumulator and store result in accumulator |
| ANA M | 246 | AND contents of memory location M with contents of accumulator and store result in accumulator |
| XRA M | 256 | Exclusive-OR contents of memory location M with contents of accumulator and store result in accumulator |
| ORA M | 266 | OR contents of memory location M with contents of accumulator and store result in accumulator |
| CMP M | 276 | Compare contents of memory location M with contents of accumulator. Leave accumulator unchanged and alter the flag bits to correspond to the results of the compare operation. |

The MOV H,M, MOV L,M, MOV M,H and MOV M,L instructions are not particularly useful and the INR M and DCR M instructions are only useful when the memory I/O device can be read from and written to.

### Address of Memory Location M is Contained in Register Pair B

Only two of the thirty-three memory-reference instructions are contained in this group, STAX B and LDAX B.

| STAX B | 002 | Store contents of accumulator at memory location M given by the contents of register pair B |
| LDAX B | 012 | Load the accumulator with the contents of memory location M given by the contents of register pair B |

### Address of Memory Location M is Contained in Register Pair D

Only two of the thirty-three memory-reference instructions are contained in this group, STAX D and LDAX D.

| STAX D | 022 | Store contents of accumulator at memory location M given by the contents of register pair D |
| LDAX D | 032 | Load the accumulator with the contents of memory location M given by the contents of register pair D |

### Address of Memory Location M is Contained in Second and Third Instruction Bytes

| STA <br> <B2> <br> <B3> | 062 <br> <B2> <br> <B3> | Store contents of accumulator at memory location M defined by instruction bytes <B2> and <B3> |
| LDA <br> <B2> <br> <B3> | 072 <br> <B2> <br> <B3> | Load the accumulator with the contents of memory location M defined by instruction bytes <B2> and <B3> |
| SHLD <br> <B2> <br> <B3> | 042 <br> <B2> <br> <B3> | Store contents of register L into memory location M defined by instruction bytes <B2> and <B3>; store contents of register H into succeeding memory location, M+1. [NOTE: This is a two-byte data transfer in a single instruction.] |

| LHLD | 052 | Load register L with the contents of memory location M defined |
| <B2> | <B2> | by instruction bytes <B2> and <B3>; load register H with |
| <B3> | <B3> | the contents of the succeeding memory location, M+1. |
| | | [NOTE: This is a two-byte data transfer in a single instruction.] |

The SHLD and LHLD instructions differ from the remaining thirty-one memory-reference instructions in the fact that two data bytes are transferred.

## THE MEMORY READ AND MEMORY WRITE MACHINE CYCLES

As with the IN and OUT instructions, the 8080A microprocessor has a machine cycle during which data transfer occurs between the memory location and the internal registers. The machine cycle is called either a MEMORY READ or a MEMORY WRITE cycle, during which the following occurs:

- Either an $\overline{\text{MEMR}}$ or an $\overline{\text{MEMW}}$ pulse is generated on the control bus.
- A unique 16-bit memory address appears on the address bus.
- The external bidirectional data bus and the internal data bus within the microprocessor chip are opened to permit direct data communication between one of the internal general-purpose registers and the I/O device, whether input or output.

The SHLD instruction differs from the others in the fact that two successive MEMORY WRITE machine cycles are executed by the 8080A. With the LHLD instruction, two successive MEMORY READ machine cycles are executed. In all other cases, only one machine cycle, either a MEMORY READ or a MEMORY WRITE, is executed.

### FIRST PROGRAM

Consider the following program:

| LO Memory Address | Instruction Byte | Mnemonic | Description |
|---|---|---|---|
| 000 | 062 | START, STA | Write contents of accumulator into the memory output device that has the following memory address |
| 001 | 000 | 000 | LO address byte of output device |
| 002 | 200 | 200 | HI address byte of output device |
| 003 | 074 | INR A | Increment accumulator |
| 004 | 303 | JMP | Unconditional jump to memory location START |
| 005 | 000 | START | LO address byte of START |
| 006 | 003 | — | HI address byte of START |

In this program, we have made the assumption that there exists no memory at location HI = 200 and LO = 000.

If you would execute this program in the single-step mode, you would observe the following bytes, in succession, on the bidirectional data bus:

| Data Bus Byte | Comments |
|---|---|
| 062 | FETCH machine cycle for STA instruction code |
| 000 | FETCH machine cycle for byte <B2> of STA instruction |
| 200 | FETCH machine cycle for byte <B3> of STA instruction |
| accumulator contents | MEMORY WRITE machine cycle, during which the accumulator contents are made available on the bidirectional data bus, the memory address <B2> and <B3> appears on the address bus, and an $\overline{\text{MEMW}}$ control pulse is generated. |
| 074 | FETCH machine cycle for INR A instruction code |
| 303 | FETCH machine cycle for JMP instruction code |
| 000 | FETCH machine cycle for LO address byte of START |
| 003 | FETCH machine cycle for HI address byte of START |

You observe such information on the data bus because (a) *All* instruction bytes move over the data bus from read/write memory or EPROM to the instruction register within the 8080A chip, and (b) The contents of the accumulator is output to the data bus during the fourth machine cycle of the STA instruction.

The program increments the contents of the accumulator during each loop. Also, it outputs the accumulator contents to the memory I/O device, HI = 200 and LO = 000. The decoder circuit shown in Fig. 21-1 would be used.

## SOME INPUT/OUTPUT CIRCUITS

Input-output circuits that employ memory I/O addressing are identical to those shown for accumulator I/O in Unit 20. The only difference is the type of select pulse used. We provide several examples here to demonstrate the similarity, and also refer the reader to the preceding unit.

A memory output circuit that is based upon the 74198 8-bit shift register is shown in Fig. 21-8. The address select pulse is an $\overline{\text{MEMW}}$ pulse coded for memory address HI = 200 and LO = 000, as would be generated by the circuit of Fig. 21-1. A related circuit is based upon a pair of 7475 D-type latches. It is shown in Fig. 21-9. This time, however, the memory address of the output latch is HI = 200 and LO = 001. Seven-segment displays are used for two different purposes in these two output circuits. For the 74198 shift register, we assume that the output is a pair of packed BCD digits, whereas for the 7475 chips, we assume that the output is in 8-bit binary, which we decode as three octal digits.

The third and final circuit is a microcomputer input circuit based upon the 8212 8-bit latch/buffer chip (Fig. 21-10). The address

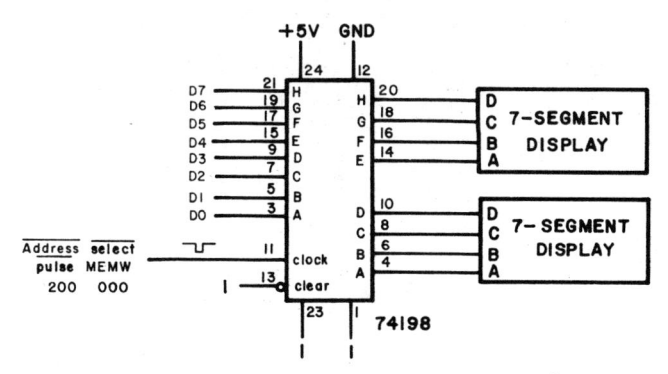

**Fig. 21-8.** Output circuit based upon the use of the memory I/O technique.

select pulse is generated from the $\overline{\text{MEMR}}$ control signal and the 16-bit address bus, and has a memory address of HI = 200 and LO = 002. Otherwise, the circuit is identical to the one shown in Fig. 20-19.

The resemblance between these figures and the corresponding ones for accumulator I/O in Unit 20 should be clear. In fact, Fig. 20-19 and 21-10 are identical except for the identification of the select pulse. We refer the reader to Figs. 20-7 through 20-18 for other

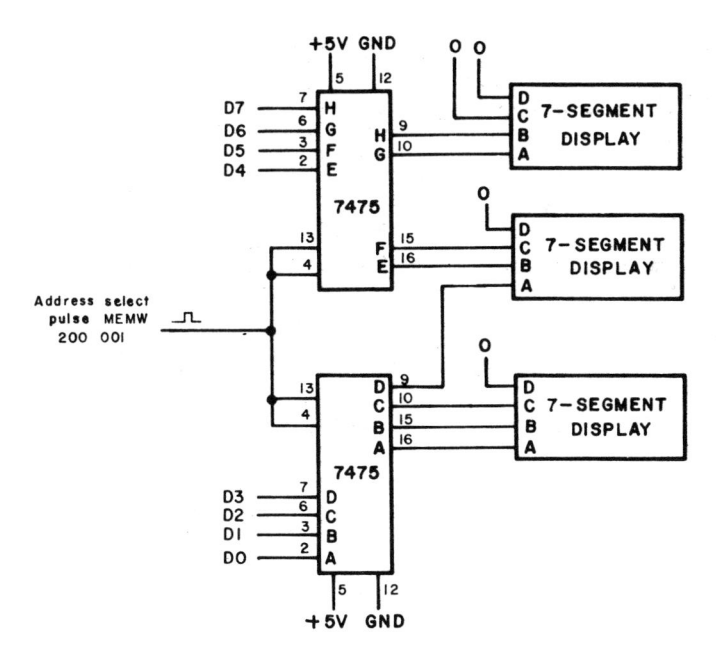

**Fig. 21-9.** Output circuit based upon a pair of 7475 D-type latches.

239

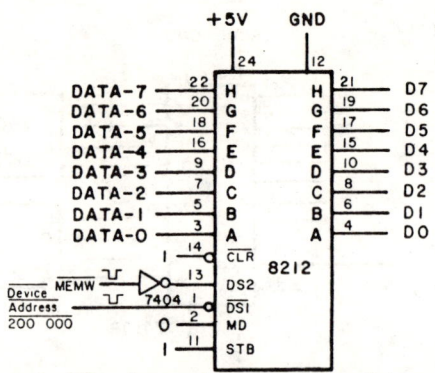

Fig. 21-10. Input circuit based upon the memory I/O techniques applied to an 8212 8-bit latch/buffer chip.

useful microcomputer input/output circuits that can be adapted to memory I/O.

## SECOND PROGRAM

A program that is, during execution, identical to the first program is as follows:

| LO Memory Address | Instruction Byte | Mnemonic | Description |
|---|---|---|---|
| 000 | 041 | START, LXI H | Load register pair H with the following two bytes |
| 001 | 000 | 000 | L register byte, the LO address byte of memory location M |
| 002 | 200 | 200 | H register byte, the HI address byte of memory location M |
| 003 | 167 | LOOP, MOV M,A | Write accumulator contents into memory location M |
| 004 | 074 | INR A | Increment accumulator |
| 005 | 303 | JMP | Unconditional jump to memory location LOOP |
| 006 | 003 | LOOP | LO address byte of LOOP |
| 007 | 003 | — | HI address byte of LOOP |

## THIRD PROGRAM

A third way to accomplish the desired result of the first and second programs is through the use of a STAX instruction.

| LO Memory Address | Instruction Byte | Mnemonic | Description |
|---|---|---|---|
| 000 | 001 | START, LXI B | Load register pair B with the following two bytes |
| 001 | 000 | 000 | C register byte, the LO address byte of memory location M |

| LO Memory Address | Instruction Byte | Mnemonic | Description |
|---|---|---|---|
| 002 | 200 | 200 | B register byte, the HI address byte of memory location M |
| 003 | 002 | LOOP, STAX B | Write the contents of the accumulator into memory location M identified by the contents of register pair B |
| 004 | 074 | INR A | Increment accumulator |
| 005 | 303 | JMP | Unconditional jump to memory location LOOP |
| 006 | 003 | LOOP | LO address byte of LOOP |
| 007 | 003 | — | HI address byte of LOOP |

Note that this time the identification of the output memory location M is contained within register pair B. Otherwise, the program execution is identical to that for the first and second programs.

## FOURTH PROGRAM

The D register pair can also be used to identify the memory location M. Thus:

| LO Memory Address | Instruction Byte | Mnemonic | Description |
|---|---|---|---|
| 000 | 021 | LXI D | Load register pair D with the following two bytes |
| 001 | 000 | 000 | E register byte, the LO address byte of memory location M |
| 002 | 200 | 200 | D register byte, the HI address byte of memory location M |
| 003 | 022 | LOOP, STAX D | Write the contents of the accumulator into memory location M identified by the contents of register pair D |
| 004 | 074 | INR A | Increment accumulator |
| 005 | 303 | JMP | Unconditional jump to memory location LOOP |
| 006 | 003 | LOOP | LO address byte of LOOP |
| 007 | 003 | — | HI address byte of LOOP |

This program is essentially the same as the third program.

## FIFTH PROGRAM

Memory I/O input programs are as simple as the output programs described above. Consider a system in which both the input and output ports have the same memory address, namely, $HI = 200$ and $LO = 000$. The following program will permit you to monitor the input data:

| LO Memory Address | Instruction Byte | Mnemonic | Description |
|---|---|---|---|
| 000 | 001 | LXI B | Load register pair B with the following two bytes |

| LO Memory Address | Instruction Byte | Mnemonic | Description |
|---|---|---|---|
| 001 | 000 | 000 | C register byte, the LO address byte of memory location M |
| 002 | 200 | 200 | B register byte, the HI address byte of memory location M |
| 003 | 012 | LOOP, LDAX B | Load the accumulator from the input port M identified by the contents of register pair B |
| 004 | 002 | STAX B | Write the contents of the accumulator into output port M identified by the contents of register pair B |
| 005 | 303 | JMP | Unconditional jump to memory location LOOP |
| 006 | 003 | LOOP | LO address byte of LOOP |
| 007 | 003 | — | HI address byte of LOOP |

If the memory address M is contained in register pair D rather than register pair B, you would substitute LDAX D and STAX D for the instruction bytes at LO = 003 and LO = 004.

## SIXTH PROGRAM

A program that, when executed, provides an identical result to that observed in the fifth program is as follows:

| LO Memory Address | Instruction Byte | Mnemonic | Description |
|---|---|---|---|
| 000 | 041 | LXI H | Load register pair H with the following two bytes |
| 001 | 000 | 000 | L register byte, the LO address byte of memory location M |
| 002 | 200 | 200 | H register byte, the HI address byte of memory location M |
| 003 | 176 | LOOP, MOV A,M | Load the accumulator with the contents of input port M |
| 004 | 167 | MOV M,A | Write the accumulator contents into output port M |
| 005 | 303 | JMP | Unconditional jump to memory location LOOP |
| 006 | 003 | LOOP | LO address byte of LOOP |
| 007 | 003 | — | HI address byte of LOOP |

As with the fifth program, the input and output ports have the same memory location, HI = 200 and LO = 000. What distinguishes data transfer between the two ports, shown in Fig. 21-11, is the way in which the internal data bus within the 8080A operates and also the existence of the different control signals, $\overline{\text{MEMR}}$ and $\overline{\text{MEMW}}$. We present Fig. 21-11 as a circuit that has educational value but not as one that you would wire in a microcomputer interface system. Why not? The answer is that *you need not wire an 8212 input port in order*

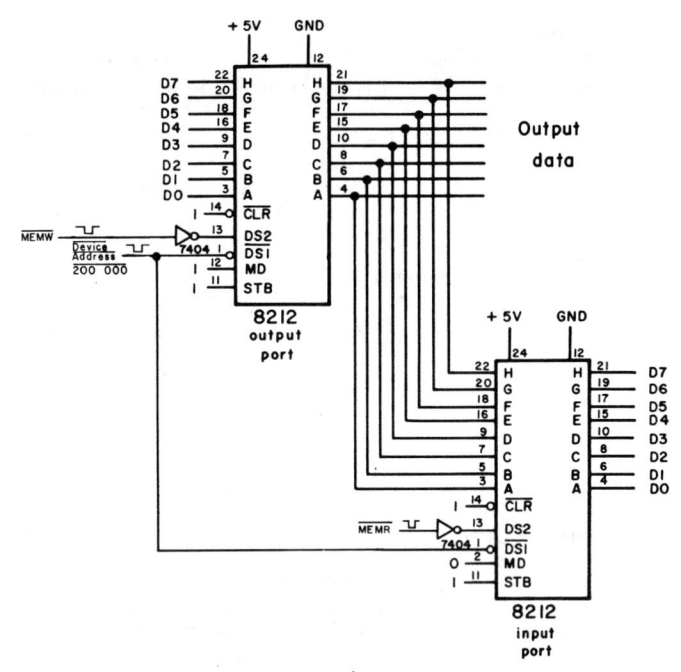

**Fig. 21-11. Memory I/O interface circuit.**

*to monitor the output from the 8212 output port* shown in Fig. 21-11. We recommend that you store the output contents in an internal register or a read/write memory location before or after you output the 8-bit word to the 8212 output port.

Remember that your objective in most cases is to substitute software for hardware. Use your read/write memory for the storage of control words and other types of temporary information. *Do not add* additional integrated-circuit chips to your interface circuit unless they are absolutely necessary.

## SEVENTH PROGRAM

It is not necessary to input and output data to and from the accumulator, as was done with all of the above programs. For example, in the sixth program, we could have exchanged data with register E. This is shown below.

| LO Memory Address | Instruction Byte | Mnemonic | Description |
|---|---|---|---|
| 000 | 041 | LXI H | Load register pair H with the following two bytes |
| 001 | 001 | 001 | L register byte, the LO address byte of memory location M |

243

| LO Memory Address | Instruction Byte | Mnemonic | Description |
|---|---|---|---|
| 002 | 200 | 200 | H register byte, the HI address byte of memory location M |
| 003 | 136 | MOV E,M | Load register E with the contents of input port M |
| 004 | 163 | MOV M,E | Write the contents of register E into output port M |
| 005 | 303 | JMP | Unconditional jump to memory location LOOP |
| 006 | 003 | LOOP | LO address byte of LOOP |
| 007 | 003 | — | HI address byte of LOOP |

Memory I/O input data can be exchanged with registers B, C, D, or E. We would recommend that register pair H not be used for such a purpose unless the SHLD and LHLD instructions are used.

### EIGHTH PROGRAM

Consider the following program:

| LO Memory Address | Instruction Byte | Mnemonic | Description |
|---|---|---|---|
| 000 | 041 | LXI H | Load register pair H with the following two bytes |
| 001 | 000 | 000 | L register byte, the LO address byte of memory location M |
| 002 | 200 | 200 | H register byte, the HI address byte of memory locations M through M+3 |
| 003 | 106 | MOV B,M | Load register B with contents of input port M |
| 004 | 054 | INR L | Increment register L |
| 005 | 116 | MOV C,M | Load register C with contents of input port M+1 |
| 006 | 054 | INR L | Increment register L |
| 007 | 126 | MOV D,M | Load register D with contents of input port M+2 |
| 010 | 054 | INR L | Increment register L |
| 011 | 136 | MOV E,M | Load register E with contents of input port M+3 |

This program illustrates two of the important advantages of memory I/O techniques:

1. The ease with which the I/O device code can be changed.
2. The speed with which four bytes of data can be input into an 8080A chip.

These advantages must be weighed against possible disadvantages of memory I/O techniques:

1. The additional circuitry required for absolute address decoding.
2. The loss of memory area when it is subdivided into memory and memory I/O blocks.
3. The additional software and register use.

Disadvantage number 2 is unimportant for small microcomputer systems. With large systems that require considerable amounts of memory, there is considerable incentive to add decoders so that only a very small section of memory is absolutely decoded into memory I/O address codes.

## INTRODUCTION TO THE EXPERIMENTS

The following simple experiments illustrate memory-mapped I/O techniques. More extensive memory-mapped I/O experiments are provided in Unit 22.

| Experiment No. | Comments |
| --- | --- |
| 1 | A simple memory-mapped input-output circuit consisting of a pair of 7475 latches and a pair of 8095 (74365) three-state buffers. Address select pulses are generated with the aid of a 74L20 4-input NAND gate chip. |
| 2 | Memory-mapped I/O to and from the accumulator. Demonstrates the use of different instructions that can transfer data between a memory-mapped I/O port and the accumulator, e.g., STA, LDAX B, STAX B, MOV A,M, and MOV M,A. |
| 3 | Use of the INR M, DCR M, and MVI M instructions. Demonstrates how a memory-mapped I/O port can be incremented or decremented. |
| 4 | Use of the ANA M instruction. Demonstrates how a memory-mapped input port can logically operate directly upon the contents of the accumulator. |

The memory-mapped I/O ports that you wire in Experiment No. 1 will be used in all of the experiments in this unit.

## EXPERIMENT NO. 1
## SIMPLE MEMORY-MAPPED INPUT-OUTPUT PORTS

### Purpose

The purpose of this experiment is to test the behavior of a simple memory-mapped input-output circuit based upon the 8095 three-state buffer and the 7475 latch.

## Pin Configurations of Integrated-Circuit Chips (Fig. 21-12)

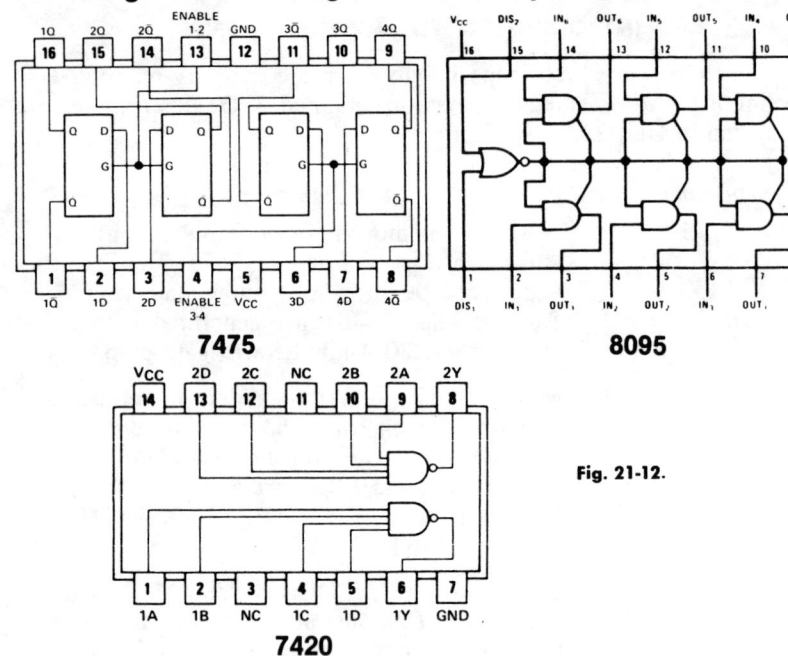

**7475**

**8095**

**7420**

Fig. 21-12.

## Schematic Diagrams of Circuits (Figs. 21-13 and 21-14)

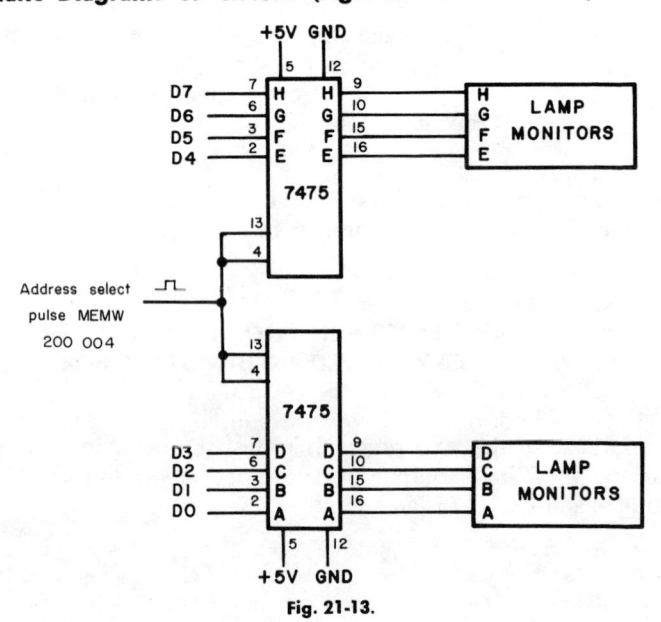

Fig. 21-13.

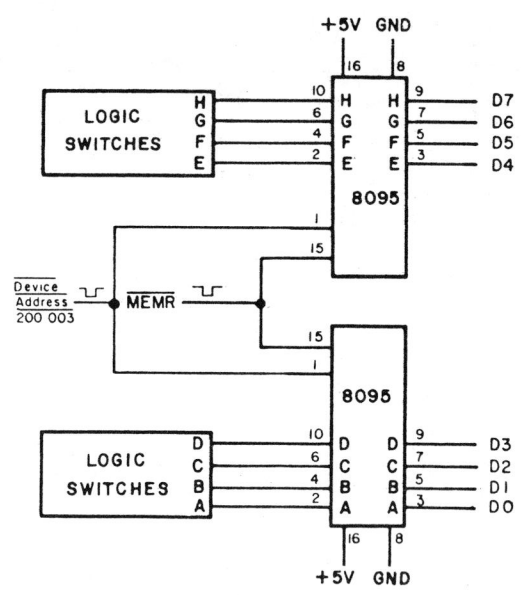

Fig. 21-14.

## Program

| LO Memory Address | Instruction Byte | Mnemonic | Description |
|---|---|---|---|
| 000 | 041 | LXI H | Load register pair H with the following two bytes |
| 001 | 003 | 003 | L register byte |
| 002 | 200 | 200 | H register byte |
| 003 | 106 | START, MOV B,M | Read into register B the contents of memory input port 200 003 |
| 004 | 043 | INX H | Increment register pair H |
| 005 | 160 | MOV M,B | Write register B contents into memory output port 200 004 |
| 006 | 053 | DCX H | Decrement register pair H |
| 007 | 303 | JMP | Jump to START and do it again |
| 010 | 003 | START | LO address byte of START |
| 011 | 003 | — | HI address byte of START |

## Step 1

Wire the circuits shown in Figs. 21-13 and 21-14. To generate the two memory address select pulses required, use either a decoder or use the 74L20 4-input NAND gate circuits shown in Fig. 21-15, which do not absolutely decode the 16-bit address bus and thus are only demonstration circuits. Don't forget the +5-Volt (pin 1) and GND (pin 7) power inputs to the 74L20 chip.

If you plan to acquire Address Bit A15 directly from the 8080A chip, we recommend that you use a 74L20 chip rather than a 7420

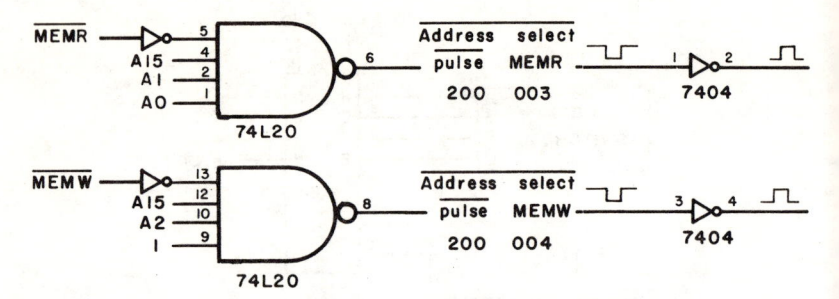

**Fig. 21-15.** NAND gate circuits.

in order to minimize fan-in. The control signals $\overline{\text{MEMR}}$ and $\overline{\text{MEMW}}$ must be inverted before being input into the 74L20. The write pulse, 200 004, must be inverted prior to being input into the pair of 7475 latches. Address Bit A15 is available as Pin 36 on the 8080A chip.

### Step 2

Load the program into memory. Execute the program. Change the logic switch settings and observe the output at Output Port 200 004. What happens?

We observed a one-to-one correspondence between the logic switch input to the 8095 chips and the lamp monitor output from the 7475 latches.

When the data is input into the microcomputer, where is it temporarily stored?

In register B.

### Step 3

What changes to the program are necessary if you desire to input and output data to and from register E?

All that is necessary is a change in the instruction bytes at LO memory address 003 and 005:

```
003   136   MOV E,M   Input into register E the contents of memory input port 200
                      003
005   163   MOV M,E   Output register E contents to memory output port 200 004
```

Make these changes and demonstrate the operation of the new program.

## Step 4

What would happen if you changed the instruction byte at LO memory address 005 to

```
005   160   MOV M,B   Output register B contents to memory output port 200 004
```

but left the instruction byte at LO memory address 003 unchanged? Would program execution change?

Yes. No longer would it be possible to input the logic switch data and then output it to the memory output port. The problem is that such data is input to register E where nothing further happens. Data output is from register B, which is not changed by the modified program. We observed an output of 000 when we tried this experiment.

## Step 5

Make the following program changes:

```
003   116   START, MOV C,M   Read into register C the contents of memory input
                             port 200 003
004   043   INX H            Increment register pair H
005   121   MOV D,C          Move contents of register C to register D
006   162   MOV M,D          Write register D contents into memory output port
                             200 004
007   053   DCX H            Decrement register pair H
010   303   JMP              Jump to START and do it again
011   003   START            LO address byte of START
012   003   —                HI address byte of START
```

Execute the program. What do you observe on the output port when you change the logic switch settings?

We observed a correspondence between the logic state of the output port and the logic switch setting.

### Step 6

Now change the instruction byte at LO memory address 005 to a NOP instruction:

    005   000    NOP   No operation

Execute the program once again. What happens? Why?

There no longer is a correspondence between the data at the memory input and output ports. The reason is that we have eliminated the MOV instruction that transfers the data byte from register C to register D, from which it is output.

### Step 7

What registers may be used when the memory-mapped I/O technique is used? Which registers are involved with accumulator I/O?

Memory-mapped I/O may use any of the general-purpose registers, including A, B, C, D, E, H, or L, as the source or destination of data. Accumulator I/O is restricted to register A as the source or destination of data.

Save the 7475 and 8095 I/O port circuits and continue to the following experiment.

## EXPERIMENT NO. 2
## MEMORY-MAPPED I/O TO AND FROM THE ACCUMULATOR

### Purpose

The purpose of this experiment is to test various memory-reference instructions that transfer data between input-output ports and the accumulator.

## Schematic Diagram of Circuit

Use the memory input and output ports described in Experiment No. 1.

### Program No. 1

| LO Memory Address | Instruction Byte | | Mnemonic | Description |
|---|---|---|---|---|
| 000 | 072 | LOOP, | LDA | Load the accumulator from the input port identified by the following memory address |
| 001 | 003 | | 003 | LO address byte of input port |
| 002 | 200 | | 200 | HI address byte of input port |
| 003 | 062 | | STA | Store accumulator contents in the output port identified by the following memory address |
| 004 | 004 | | 004 | LO address byte of output port |
| 005 | 200 | | 200 | HI address byte of output port |
| 006 | 303 | | JMP | Jump back to LOOP |
| 007 | 000 | | LOOP | LO address byte of LOOP |
| 010 | 003 | | — | HI address byte of LOOP |

### Program No. 2

| LO Memory Address | Instruction Byte | | Mnemonic | Description |
|---|---|---|---|---|
| 000 | 001 | | LXI B | Load register pair B with the following two bytes |
| 001 | 003 | | 003 | LO address byte of input port |
| 002 | 200 | | 200 | HI address byte of input port |
| 003 | 012 | LOOP, | LDAX B | Load accumulator from input port identified by the contents of register pair B |
| 004 | 003 | | INX B | Increment register pair B |
| 005 | 002 | | STAX B | Store accumulator contents in the output port identified by the current contents of register pair B |
| 006 | 013 | | DCX B | Decrement register pair B |
| 007 | 303 | | JMP | Jump back to LOOP |
| 010 | 003 | | LOOP | LO address byte of LOOP |
| 011 | 003 | | — | HI address byte of LOOP |

### Program No. 3

| LO Memory Address | Instruction Byte | | Mnemonic | Description |
|---|---|---|---|---|
| 000 | 041 | | LXI H | Load register pair H with the following two bytes |
| 001 | 003 | | 003 | LO address byte of input port |
| 002 | 200 | | 200 | HI address byte of input port |
| 003 | 176 | LOOP, | MOV A,M | Load accumulator from input port identified by the contents of register pair H |
| 004 | 043 | | INX H | Increment register pair H |
| 005 | 167 | | MOV M,A | Move accumulator contents to the output port identified by the current contents of register pair H |

| LO Memory Address | Instruction Byte | Mnemonic | Description |
|---|---|---|---|
| 006 | 053 | DCX H | Decrement register pair H |
| 007 | 303 | JMP | Jump back to LOOP |
| 010 | 003 | LOOP | LO address byte of LOOP |
| 011 | 003 | — | HI address byte of LOOP |

## Step 1

In this experiment, you are provided with three different types of memory-reference instructions that can be used to transfer data between the accumulator and external input-output devices. Even though the data transfer instructions have different mnemonics, Programs No. 2 and 3 are similar.

We assume that you have already wired the circuit described in Experiment No. 1. Memory address pulses can be generated using the simple 74L20 gate circuits shown in Step 1 of Experiment No. 1. Keep in mind, however, that the pair of 7475 latches require a positive select pulse; thus, the output from the 74L20 chip must be inverted.

## Step 2

Load and execute Program No. 1. Change the logic switch settings at the 8095 (74365) input port. What do you observe on the LED lamp monitors connected to the two 7475 latches that comprise the memory output port?

You should observe a one-to-one correspondence between memory input and memory output data. The response should be "instantaneous."

## Step 3

By changing the address bytes at LO memory addresses 004 and 005, would it be possible for you to output the memory input data to one of the three lamp monitor output ports on the MMD-1 microcomputer? Please explain your answer.

It would not be possible to convert any of the output ports on the MMD-1 microcomputer to memory output ports simply by modifying a pair of memory address bytes in Program No. 1. The three output ports on the MMD-1 are hardwired as *accumulator* output ports. If you wish to convert them to memory output ports, you would need to make a number of wiring changes on the printed-circuit boards. In addition, you would have to make a number of changes to the KEX monitor program. We do not suggest that you do this.

The point we wish to make here is that *both* hardware and software are required to determine the nature of an input-output port, i.e., whether it is a memory I/O port or an accumulator I/O port.

### Step 4

Load and execute Program No. 2. Change the logic switch settings to the memory input port and note the correspondence between such settings and the output from the pair of 7475 latches.

Why are the INX B and DCX B instructions needed in this program?

The memory input port has an address of 200 003, whereas, the output port has an address of 200 004. We use the INX B and DCX B instructions to change the address existing in register pair B prior to the LDAX B or STAX B instruction. In this way, we are able to address both ports. Memory I/O allows us to use instructions which can modify a memory address. This is difficult to implement with accumulator I/O.

### Step 5

Load and execute Program No. 3. Again change the logic switch settings to the memory input port and note the correspondence between such settings and the output appearing on the eight lamp monitors.

What differences do you observe in the execution of this program when compared to the execution of Programs No. 1 and 2?

You should observe no differences.

**Step 6**

Comment on the differences and similarities of the LDAX B and MOV A,M instructions.

Both instructions are similar in that the memory address is contained in a register pair and that it is this address that specifies the memory address of the memory input port. The only difference between the two instructions is the identity of the register pair. For LDAX B, register pair B contains the address, whereas, for MOV A,M, register pair H contains the address.

**Step 7**

Comment on the differences and similarities of the STAX B and MOV M,A instructions.

Both instructions are similar in that the memory address is contained in a register pair and that it is this address that specifies the memory address of the memory output port. The only difference between the two instructions is the identity of the register pair. For STAX B, register pair B contains the address, whereas, for MOV M,A, register pair H contains the address.

**Step 8**

Why might you prefer the use of an STA or LDA instruction in preference to an STAX, LDAX, MOV A,M, or MOV M,A instruction?

By specifying a memory address as a pair of immediate address bytes, you eliminate the need to use a register pair. There exist only three register pairs in the 8080A chip, so use them wisely.

Save your 7475 output and 8095 (74365) input circuits for the next two experiments.

## EXPERIMENT NO. 3
## USE OF THE INR M, DCR M, AND MVI M INSTRUCTIONS

### Purpose

The purpose of this experiment is to test the behavior of the INR M, DCR M, and MVI M instructions on a typical memory output port.

### Schematic Diagram of Circuit

Use the output port described in Experiment No. 1.

### Program No. 1

| LO Memory Address | Instruction Byte | Mnemonic | Description |
|---|---|---|---|
| 000 | 041 | LXI H | Load register pair H with the following two bytes |
| 001 | 004 | 004 | LO address byte of output port |
| 002 | 200 | 200 | HI address byte of output port |
| 003 | 066 | MVI M | Move following byte to output port identified by contents of register pair H |
| 004 | 111 | 111 | Immediate data byte to be output |
| 005 | 166 | HLT | Halt |

### Program No. 2

| LO Memory Address | Instruction Byte | Mnemonic | Description |
|---|---|---|---|
| 000 | 041 | LXI H | Load register pair H with the following two bytes |
| 001 | 004 | 004 | LO address byte of output port |
| 002 | 200 | 200 | HI address byte of output port |
| 003 | 066 | MVI M | Move following byte to output port identified by contents of register pair H |
| 004 | 111 | 111 | Immediate data byte to be output |
| 005 | 064 | INR M | Increment output port |
| 006 | 166 | HLT | Halt |

### Step 1

Wire the memory output port consisting of two 7475 latches, as described in Experiment No. 1.

### Step 2

Load and execute Program No. 1. What do you observe on the output port lamp monitors?

You should observe the output octal byte, 111.

### Step 3

Change the value of the data byte at LO = 004 and execute the program once again. For example, try the data byte 333. What do you observe on the output port now?

You should observe an output byte of 333.

If you would repeat this process with other data bytes, you should conclude that there is a one-to-one correspondence between the immediate data byte at LO = 004 and the output data on the output port once the program has been executed. When an MVI M, MOV r,M or MOV M,r instruction is used, the address of the memory location must be stored in registers H and L. They are loaded at the start of the program with an LXI H instruction.

### Step 4

Load and execute Program No. 2. What do you observe on the output port? What did you expect to observe, the byte at LO = 004?

We observed an output byte of 000. You probably should observe the same thing. We initially expected to observe the output byte 112. The reasons why we did not observe such a result are discussed in the next step.

### Step 5

What operations must the microprocessor execute in order to successfully perform an INR M instruction?

First, the current contents of memory location M must be input into the 8080A. Next, they must be incremented by one. Finally, the incremented value must be output back to memory location M. In other words, with a typical read/write memory location, the INR M performs both a read and a write.

**Step 6**

In memory mapped I/O, in order for you to successfully execute an INR M instruction, what conditions must exist at the I/O port?

The port must be similar to that shown in Fig. 21-11, i.e., you must be able to read from and write into the port. Such a condition exists naturally in read/write memory, but may not exist in memory-mapped I/O interface circuits.

**Step 7**

Why did we observe 000 as an output byte in Step 4 of this experiment?

We attempted to read from a nonexistent memory location, 200 004, and input the "default" data byte 377 (due to the data bus floating to all logic 1s), which was incremented and written into the memory output port as 000. The INR M and DCR M instructions are unusual in that both a read and write operation occur on a data byte.

**Step 8**

If the 8095 (74365) three-state input port is still connected, change the address decoding to the following:

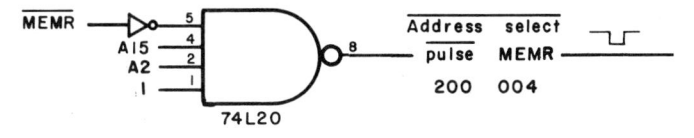

**Fig. 21-16.**

Now the input port and output port have the same address. Change the instruction bytes at LO = 003 and LO = 004 in Program No. 2 to NOP, 000. Set a value on the logic switches and execute Program No. 2. What output appears on the lamp monitors?

The lamp monitor output should be your 8-bit binary setting incremented by 1.

## EXPERIMENT NO. 4
## USE OF THE ANA M INSTRUCTION

### Purpose

The purpose of this experiment is to demonstrate the execution of an AND operation between a memory input port and the accumulator.

### Schematic Diagram of Circuit

Use the input port described in Experiment No. 1. Rewire it for address 200 003.

### Program

| LO Memory Address | Instruction Byte | Mnemonic | Description |
|---|---|---|---|
| 000 | 041 | LXI H | Load register pair H with the following two bytes |
| 001 | 003 | 003 | LO address byte of input port |
| 002 | 200 | 200 | HI address byte of input port |
| 003 | 076 | TEST, MVI A | Move immediate byte into accumulator |
| 004 | 001 | 001 | Mask byte |
| 005 | 246 | ANA M | AND contents of memory input port with contents of accumulator |
| 006 | 312 | JZ | If result is zero, jump back to TEST. Otherwise, continue to next instruction. |
| 007 | 003 | TEST | LO address byte of TEST |
| 010 | 003 | — | HI address byte of TEST |
| 011 | 323 | OUT | Flag bit is logic 1. Output it to following output port. |
| 012 | 000 | 000 | Output port 000 |
| 013 | 166 | HLT | Halt |

### Step 1

Wire the memory input circuit shown in Experiment No. 1 if it is not already wired on your breadboard.

### Step 2

Load and execute Program No. 1 with logic switch A at logic 0. Now set the logic switch to logic 1. What happens at Output Port 000?

Bit D0 becomes lighted.

## Step 3

Change the mask byte at LO = 004 to one of the following: 200, 100, 040, 020, 010, 004, or 002. Set all eight logic switches to logic 0. Execute the program once again and test each logic switch until you detect the one that is not masked. How do you know when you have found the right one?

The bit corresponding to the nonmasked bit becomes lighted at Output Port 000.

## Step 4

You can also test the other memory-reference instructions, including

| | |
|---|---|
| 206 | ADD M |
| 216 | ADC M |
| 226 | SUB M |
| 236 | SBB M |
| 256 | XRA M |
| 266 | ORA M |
| 276 | CMP M |

The CMP M instruction does not affect any data or the contents of the accumulator register. It only sets and clears flags.

## Step 5

Why would it be useful to be able to perform an arithmetic or logical operation between a memory-mapped input port and the contents of the accumulator?

It might be useful if you wish to externally set a mask byte for an ANA M operation using a set of eight logic switches.

## REVIEW QUESTIONS

The following questions will help you review memory-mapped I/O techniques.
1. What is meant by the term "memory-mapped input/output"?
2. List several differences between accumulator I/O and memory-mapped I/O. For example, what control signals are used, what instructions are used, and what registers are used in the two I/O techniques?
3. What is meant by the "absolute" decoding of the address bus in memory-mapped I/O?
4. Why is absolute decoding of the address bus in memory-mapped I/O important?
5. In this and preceding units, we have used the terms "device select pulse" and "address select pulse." What is the difference between these two terms?
6. We have heard it stated that the reason one uses memory-mapped I/O techniques is to be able to transfer data between more devices than with accumulator I/O. Do you agree? If so, why? If not, why not?

# 22

# Some Examples of Microcomputer Input/Output

### INTRODUCTION

One of the most important uses for microcomputers in the laboratory is as a data logger. This unit explores the principles of data logging and provides a number of experimental examples of data logging circuits.

### OBJECTIVES

At the completion of this unit, you will be able to do the following:

- Define data logging and discuss the most important considerations in the development of a data logging system.
- Describe various methods of generating time delays.
- Wire a simple data logging circuit.
- Interface an AD7522 digital-to-analog converter.

### DATA LOGGING WITH AN 8080 MICROCOMPUTER

A *data logger* can be defined as

*data logger*—An instrument that automatically scans data produced by another instrument or process and records readings of the data for future use.

It should be clear that a microcomputer can be a data logger. Data from an instrument can be input into the accumulator and then stored in memory. At a later time, this stored information can be read out in a variety of ways. Data logging will become a common application for microcomputers.

Perhaps the most important questions that you should ask when you plan to log data from an instrument are the following:

1. How many data points do you wish to log?
2. How much time will it take to log all of these points?
3. How much digital information is contained in a single data point?
4. What do you wish to do with the logged data once it has been acquired?
5. Do you need short-term or long-term data storage?

We shall now discuss these questions.

### How Many Data Points?

The number of data points that you wish to log and the time that you will need to store them will dictate the type of storage device required. If you need to data log one million four-BCD-digit data points, you will need a memory capacity of sixteen million bits and will, therefore, require some form of magnetic tape or magnetic disk. On the other hand, if you need to log one hundred data points, each containing only four BCD digits, and store the data for up to several hours, only 1600 bits of memory are required. A simple read/write memory board would do quite nicely for such an application. If you need to store more than two thousand bits of data, we would recommend the use of magnetic tape of some type, such as a cassette tape or a floppy disk. It provides a more secure data storage.

### Short-Term or Long-Term Storage?

Read/write memory is not, in general, suitable for the long-term storage of data. For one reason, read/write memory is volatile. If a power failure occurs, all of the data will be lost. Core memory is not volatile but, on the other hand, it is relatively expensive and generally not suited for long-term storage of data unless the amount of data stored is limited. The best data storage devices, as indicated above, include cassette tape and floppy disks. A high-quality tape cassette can store as much as 500,000 bits of information on a single cassette that costs no more than $10. Hardware costs for floppy disks are decreasing every year. The development of highly sophisticated LSI interface chips for floppy disks will reduce costs still further.

An inexpensive and long-term storage technique is the use of perforated paper tape. We should point out, however, that it takes considerable time to punch such tape as well as to read it back into a computer. At a teletype speed of 10 characters per second, almost seven minutes are required to read or punch 4096 bytes of program or data.

## How Much Information in a Single Data Point?

A typical data point is usually a three- or four-BCD digit number that also contains both a decimal point, or range, as well as a sign. Usually, the decimal point or range is fixed and the sign is positive, but such is not always the case. New digital devices are increasingly incorporating an *autoranging* capability, which means that the digital instrument decides where to place the decimal point.

Plan on a data point that contains at least sixteen bits of digital information. To obtain the total memory capacity required, multiply sixteen by the number of data points. Thus, for one hundred data points, 1600 bits of read/write memory would be required. Frequency meters typically have many more bits per data point. For example, a seven-digit frequency meter has at least 28 bits per data point.

## What Will You Do With the Logged Data?

Some logged data is only "raw" data that must be manipulated and interpreted in order to produce a useful final result. One example would be the conversion of a digital voltage to force. In such cases, the logged data will require mathematical computations that should be performed soon after the data is acquired. With data that requires additional mathematical treatment, we recommend that you keep the data in digital electronic form until it can be treated. Read/write memory, magnetic tape, and magnetic disk are all suitable for such a purpose. The printing of data is a form of long-term data storage. It certainly is the least expensive type of long-term storage around, but you pay a penalty in the time you must consume to convert it back to digital electronic signals.

## How Many Data Points Per Second?

This is a fundamental question for all data logging operations. The data can, for example (a) appear quite slowly and take considerable periods of time, such as a day, for its acquisition, or (b) appear extremely rapidly, and take only milliseconds for the acquisition of hundreds of data points. Both extremes in data-acquisition rates point to the need for automated data-acquisition techniques, such as the use of a microcomputer-based data logger. As the microcomputer decreases in price, more laboratory instruments will automatically log data via built-in microcomputers. Chart recorders will still be used, but they may not need to be of the quality previously required. A major use for chart recorders, in the future, will be to allow the eye to visually "integrate" a block of data for detection of curvature, linearity, etc.

## FIRST PROGRAM: LOGGING SIXTY-FOUR 8-BIT DATA POINTS

As a demonstration of the concept of data logging, we would like to provide a program that enables you to log sixty-four 8-bit data points as fast as the microcomputer can input and store them. As an example of an "instrument," assume that you are logging the data from the pair of 7490 counters shown in Fig. 22-1. The question

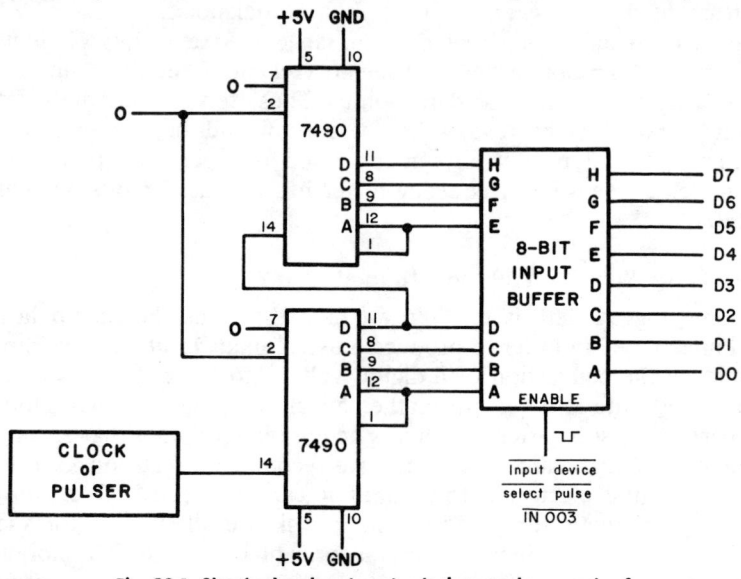

Fig. 22-1. Simple data logging circuit that employs a pair of cascaded 7490 decade counters.

that we wish to answer is, "What is the minimum amount of time required to log sixty-four 8-bit data points from the pair of counters?"

The program, which is an example of the use of accumulator I/O techniques, is as follows:

| LO Memory Address | Instruction Byte | Mnemonic | Clock Cycles | Description |
|---|---|---|---|---|
| 000 | 041 | LXI H | 10 | Load register pair H with the following two bytes |
| 001 | 100 | 100 | — | L register byte, the LO address byte of memory location M |
| 002 | 003 | 003 | — | H register byte, the HI address byte of memory location M |
| 003 | 006 | MVI B | 7 | Move the following byte to register B |

| LO Memory Address | Instruction Byte | Mnemonic | Clock Cycles | Description |
|---|---|---|---|---|
| 004 | 100 | 100 | — | Number of points that will be logged by the microcomputer, i.e., sixty-four points |
| 005 | 333 | LOOP, IN | 10 | Input data from pair of 7490 decade counter chips |
| 006 | 003 | 003 | — | Device code for input buffer in Fig. 22-1 |
| 007 | 167 | MOV M,A | 7 | Move contents of accumulator to memory location M addressed by contents of register pair H |
| 010 | 043 | INX H | 5 | Increment register pair H |
| 011 | 005 | DCR B | 5 | Decrement register B |
| 012 | 302 | JNZ | 10 | If register B is not equal to 000, jump to LOOP; otherwise, ignore this instruction and continue to the next instruction |
| 013 | 005 | LOOP | — | LO address byte of LOOP |
| 014 | 003 | — | — | HI address byte of LOOP |
| 015 | 166 | HLT | — | Halt |

The loop from LO = 005 to LO = 014 is executed sixty-four times before the microcomputer comes to a halt. During each LOOP pass, 37 clock cycles are required. Thus, the total time required to log sixty-four data points is 64 times 37 times the time per cycle for the 8080A microcomputer. For a microcomputer that operates at 2 MHz, the total time is 1.184 milliseconds. At 18.5 $\mu$s per 8-bit data point, a 2-MHz microcomputer can log approximately 54,000 bytes per second, which is an enormous amount of information.

If the clock in Fig. 22-1 operated at a frequency of 1 Hz, you would store one or two values in all sixty-four memory locations. The proper way to perform the above experiment is to use a clock input that has a frequency of at least 20 kHz. We obtained useful results using a clock that had a frequency of 90 kHz.

## SECOND PROGRAM: LOGGING SLOW DATA POINTS

It is not often that you will need to log data at a rate of 54,000 data points per second. A more common situation is a data logging rate of one byte per second. The only change required in the first program is the insertion of a time-delay loop that has a duration of approximately one second. Thus:

| LO Memory Address | Instruction Byte | Mnemonic | Description |
|---|---|---|---|
| 000 | 041 | LXI H | Load register pair H with the following two bytes |
| 001 | 100 | 100 | L register byte, the LO address byte of memory location M |

| LO Memory Address | Instruction Byte | Mnemonic | Description |
|---|---|---|---|
| 002 | 003 | 003 | H register byte, the HI address byte of memory location M |
| 003 | 006 | MVI B | Load register B with the following byte |
| 004 | 100 | 100 | Number of points that will be logged by the microcomputer, i.e., sixty-four data points |
| 005 | 333 | LOOP2, IN | Input data from pair of 7490 decade counter chips shown in Fig. 22-1 |
| 006 | 003 | 003 | Device code for input buffer shown in Fig. 22-1 |
| 007 | 167 | MOV M,A | Move contents of accumulator to memory location M addressed by contents of register pair H |
| 010 | 043 | INX H | Increment register pair H |
| 011 | 016 | MVI C | Load register C with the following timing byte, which determines the number of 10 ms time delay loop passes |
| 012 | 144 | 144 | Timing byte for register C, which corresponds to 100 loop passes |
| 013 | 315 | LOOP1, CALL | Call 10 ms time-delay routine DELAY |
| 014 | 277 | DELAY | LO address byte of DELAY |
| 015 | 000 | — | HI address byte of DELAY |
| 016 | 015 | DCR C | Decrement register C |
| 017 | 302 | JNZ | If register C is not equal to 000, jump to LOOP1; otherwise, continue to next instruction |
| 020 | 013 | LOOP1 | LO address byte of LOOP1 |
| 021 | 003 | — | HI address byte of LOOP1 |
| 022 | 005 | DCR B | Decrement register B |
| 023 | 302 | JNZ | If register B is not equal to 000, jump to LOOP2; otherwise, continue to next instruction |
| 024 | 005 | LOOP 2 | LO address byte of LOOP2 |
| 025 | 003 | — | HI address byte of LOOP2 |
| 026 | 166 | HLT | Halt |

It will require 1.0000345 seconds to log each data point, or a total of 64.0022 seconds to log all sixty-four data points. Clearly, the additional time required to perform the DCR, IN, CALL, INX, and JNX instructions is negligible when compared to the one-second time delay. A 10-ms time-delay subroutine will permit you to log data at rates between 23.4 data points/minute and 99.7 data points/second simply through a change in the value of the timing byte at LO = 012.

### THIRD PROGRAM: OUTPUT FROM A DATA LOGGER

Let us assume that you have stored sixty-four data points in read/write memory starting at HI = 003 and LO = 100 and now wish to output each point at the rate of one data point/second to an appropriate latch circuit, such as one of those shown in Figs. 20-7, 20-11, 20-13, and 20-14. What type of program is required? The answer to this question is that a program is needed that is almost identical to the second program. Only three instruction bytes in the second program need to be modified:

| | | | |
|---|---|---|---|
| 005 | 176 | MOV M,A | Move the contents of the memory location M to the accumulator |
| 006 | 323 | OUT | Output the accumulator contents to an output latch |
| 007 | 002 | 002 | Device code of output latch |

Otherwise, the second program can be used as written. For example, in the second program, you have already provided instruction bytes to (a) identify the memory location M, (b) establish the number of data points located in read/write memory, (c) initiate a one-second time delay between each data point, and (d) halt after all sixty-four data points have been output. As a dividend, you can repeatedly execute the modified second program, which we shall now call the third program, starting at $HI = 003$ and $LO = 000$. The third program is now an output program that does not modify the contents of read/write memory.

### FOURTH PROGRAM: DETECTING AN ASCII CHARACTER

While the concept of using input devices to input eight bits of data is straightforward, once you have input the data, you can perform

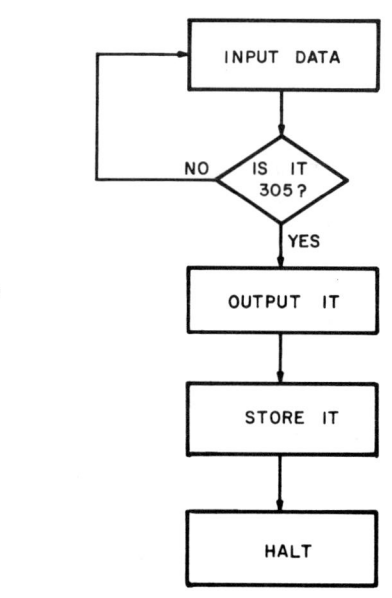

Fig. 22-2. Flow chart for the Fourth Program.

interesting programming tricks to take advantage of the power of the 8080A chip. For example, assume that the input data byte is the 8-bit ASCII code from a standard ASCII keyboard that has TTL output. Each time a new ASCII byte is input, it is tested to determine whether or not it is the ASCII equivalent of the letter "E", which has an ASCII code of 305. If it is an "E," the ASCII byte is output and also stored in read/write memory. If not, the program will immediately loop back to the IN instruction and input a new ASCII byte. The simple flow chart for this program is shown in Fig. 22-2.

The program is as follows:

| LO Memory Address | Instruction Byte | Mnemonic | | Description |
|---|---|---|---|---|
| 000 | 333 | START, | IN | Input ASCII character |
| 001 | 004 | | 004 | Device code for ASCII keyboard |
| 003 | 376 | | CPI | Compare the accumulator contents with the following data byte. If the two bytes are identical, set the zero flag. If not, clear (reset) the zero flag. |
| 004 | 305 | | 305 | ASCII code for the letter "E" |
| 005 | 302 | | JNZ | If the zero flag is reset, i.e., at logic 0, jump to START; otherwise, continue to the following instruction |
| 006 | 000 | | START | LO address byte of START |
| 007 | 003 | | — | HI address byte of START |
| 010 | 323 | | OUT | Output the ASCII code for the letter "E" |
| 011 | 002 | | 002 | Device code for output latch |
| 012 | 062 | | STA | Store the accumulator contents in memory location STORE |
| 013 | 200 | | STORE | LO address byte of STORE |
| 014 | 003 | | — | HI address byte of STORE |
| 015 | 166 | | HLT | Halt |

The compare immediate instruction, CPI, at LO = 003 and LO = 004 permit you to compare the ASCII byte 305 with the contents of the accumulator *without altering the accumulator contents*. Only the flags are changed. If the ASCII byte for the letter "E" *and* the accumulator contents are identical, the zero flag is set to logic 1; otherwise, the zero flag is reset (cleared) to logic 0. The condition of the zero flag is then tested by the JNZ instruction to determine whether or not to continue looping.

We have observed that the compare instruction is subtle and, on occasion, difficult to use properly. As indicated in Unit 18, the two flags that are tested after a compare instruction, such as CPI, are the Zero flag and the Carry flag. Four different conditional jump instructions can be inserted at LO = 005 in the above program. The questions that are implied by such instruction bytes can be summarized as follows:

| Instruction Byte at LO = 005 | Implied question |
|---|---|
| 302 | Is the input ASCII character the letter "E"? If not, continue looping until it is. |
| 312 | Is the input ASCII character any character other than the letter "E"? Continue looping until an ASCII character other than "E" is input. |
| 322 | Is the input ASCII character "A," "B," "C," or "D"? If not, continue looping until it is. |
| 332 | Is the input ASCII character "E" through "Z"? If not, continue looping until it is. |

We have tested the fourth program using each of the above four conditional branch instructions. With the ASCII code equivalents to the letters "D," "E," and "F," the program worked as expected.

## OTHER METHODS OF GENERATING TIME DELAYS

In the second program in this unit, a one-second time-delay loop was used to slow down the rate at which the microcomputer logged data from an input device. The use of such a loop represents a very inefficient application of a microcomputer, since the microcomputer could perform other useful functions during the one-second interval. Other methods of generating one-second time delays include the following:

- A *real time clock* based upon the 60-Hz line frequency.

  A 60-Hz square wave is produced by suitable analog circuitry and used to periodically *interrupt* program execution. As discussed in Unit 23, program control is directed to a small subroutine that acquires the data point and then returns control to the interrupted program.

- A *real time clock* based upon a high-frequency crystal-oscillator circuit.

  The interrupt approach is used here also.

- A *programmable interval timer*.

  A programmable interval timer such as the 8253 chip contains several 16-bit registers that can be counted down at the frequency of the microcomputer. An initial register word is loaded into the interval timer. Counting proceeds until the register contents is zero, at which time an interrupt pulse is sent to the microcomputer. A programmable interval timer represents a combined software-hardware approach to the problem of generating known time delays. Some hardware, i.e., an integrated-circuit chip, is needed, but the time delay is set with the aid of software.

If the microcomputer has no other functions to perform between data points, the use of a wait loop is quite acceptable.

## INTRODUCTION TO THE EXPERIMENTS

The following experiments provide examples of microcomputer input/output circuits, with an emphasis upon data logging.

| Experiment No. | Comments |
|---|---|
| 1 | Logging fast data points. Demonstrates a data logging circuit and program that can log 8-bit data points at a rate of 20,000 points/second. |
| 2 | Logging slow data points. The addition of a time-delay subroutine slows down the rate at which data points can be logged by the program of Experiment No. 1. |
| 3 | Detecting an ASCII character. Demonstrates a program that can detect the input of a specific ASCII character, such as ASCII "E." |
| 4 | Wiring a bus monitor. (See Experiment No. 1 in Unit 17 for a bus monitor circuit.) Describes and demonstrates the type of information that is latched when the following control signals are applied to the latch enable input ($\overline{STB}$) of the numeric indicator: $\overline{IN}$, $\overline{OUT}$, $\overline{MEMR}$, $\overline{MEMW}$, $\overline{INTA}$, and input and output device select pulses. |
| 5 | Bidirectional memory-mapped I/O using an 8216 chip. Demonstrates the use of the 8216 chip as a 4-bit bidirectional I/O port. |
| 6 | Accumulator I/O using the 8255 chip. The 8255 programmable peripheral interface chip is widely used in input/output circuits. This experiment demonstrates the mode 0 operation of this chip using accumulator I/O techniques. |
| 7 | Memory-mapped I/O using the 8255 chip. By changing the control signal inputs from $\overline{IN}$ and $\overline{OUT}$ to $\overline{MEMR}$ and $\overline{MEMW}$, it is possible to convert the circuit of Experiment No. 6 to memory-mapped I/O operation. |
| 8 | Interfacing a digital-to-analog converter. Demonstrates an interface circuit between an 8080A-based microcomputer and a 10-bit |

9    buffered, multiplying digital-to-analog (D/A) converter.

9. A staircase-ramp comparison analog-to-digital converter. With the aid of a suitable program and the addition of a comparator circuit based upon the LM311 comparator chip, you can convert the DAC circuit in Experiment No. 8 into an analog-to-digital converter.

Some of the above experiments employ expensive integrated-circuit chips. We encourage you to be careful when using such chips.

### EXPERIMENT NO. 1
### LOGGING FAST DATA POINTS

**Purpose**

The purpose of this experiment is to operate a simple 8080A-based data logger that can log data at high data rates.

**Pin Configurations of Integrated-Circuit Chips (Fig. 22-3)**

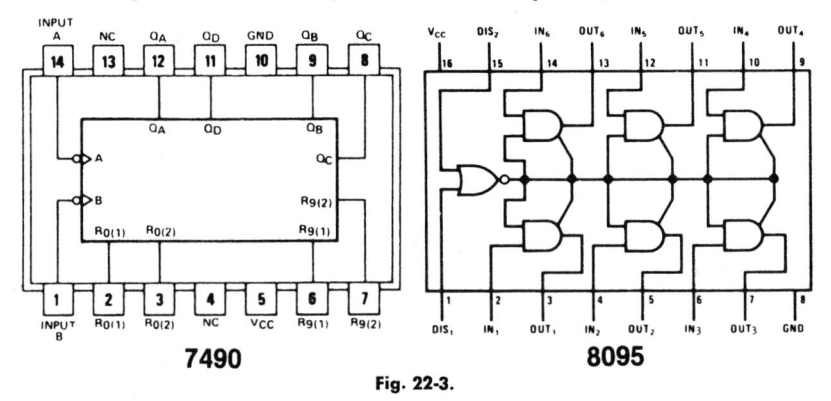

Fig. 22-3.

**Schematic Diagram of Circuit (Fig. 22-4)**

**Program**

| LO Memory Address | Instruction Byte | Mnemonic | Description |
|---|---|---|---|
| 000 | 041 | LXI H | Load register pair H with the following two bytes |
| 001 | 100 | 100 | L register byte, the LO address byte of memory location M |
| 002 | 003 | 003 | H register byte, the HI address byte of memory location M |

| LO Memory Address | Instruction Byte | Mnemonic | | Description |
|---|---|---|---|---|
| 003 | 006 | | MVI B | Move the following byte into register B |
| 004 | 100 | | 100 | Number of points that will be logged |
| 005 | 333 | LOOP, | IN | Input data from buffer |
| 006 | 004 | | 004 | Device code for input buffer |
| 007 | 167 | | MOV M,A | Move accumulator contents to memory location M addressed by contents of register pair H |
| 010 | 043 | | INX H | Increment register pair H |
| 011 | 005 | | DCR B | Decrement register B |
| 012 | 302 | | JNZ | If register B is not equal to 000, jump to LOOP and input another data point; otherwise, continue to the next instruction |
| 013 | 005 | | LOOP | LO address byte of LOOP |
| 014 | 003 | | — | HI address byte of LOOP |
| 015 | 166 | | HLT | Halt |

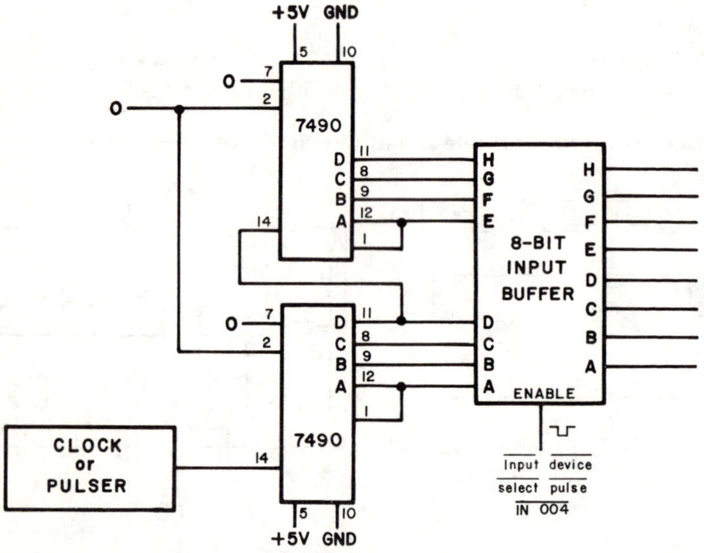

**Fig. 22-4.**

## Step 1

Study the schematic diagram in Fig. 22-4. Observe that a pair of cascaded 7490 decade counters are input into an input buffer. Two buffer circuits are given in Unit 20, Figs. 20-18 and 20-19. We would recommend the use of the pair of 8095 chips since they are less expensive and, in our experience, less subject to being damaged. Obtain the negative device select pulse from one of the decoder circuits described in Unit 17.

Wire the circuit required for this experiment. The clock frequency should initially be extremely slow, approximately 1 Hz.

### Step 2

Load the program into read/write memory starting at HI = 003 and LO = 000.

### Step 3

Execute the program at the full microcomputer speed. Now, go to memory location HI = 003 and LO = 100 and step through read/write memory, up to HI = 003 and LO = 200. What do you observe? Why?

We observed a reading of 120, in octal code, for all of the memory locations starting at LO = 100 and ending at LO = 177. Such a reading corresponds to an 8-bit binary word of 01010000, which is equivalent to the decimal number 50 in packed BCD. The microcomputer executed the program so quickly that the same output from the pair of 7490 decade counters was input into all memory locations.

### Step 4

What do we mean by the term "packed BCD"?

Binary coded decimal (BCD) is a 4-bit binary code for the decimal digits 0 through 9. Two BCD digits comprise a total of eight bits, which can be input as such into an 8-bit microcomputer such as the 8080A. By "packed BCD," we mean that an 8-bit data point contains two 4-bit BCD digits.

### Step 5

Execute the program at the full microcomputer speed several times. Observe what data you input starting at memory location HI = 003 and LO = 100. What do you conclude?

In each case, we input only a single pair of BCD digits into read/write memory.

**Step 6**

We concluded previously in this unit that it required 37 cycles to log a single 8-bit data point. If a microcomputer is operated at a clock rate of 750 kHz, how much time is required to log a single point? What is the data logging rate in bytes/second?

At 750 kHz, a cycle lasts for 1.333 microseconds. Thus, 37 cycles corresponds to 49.3 microseconds and a data rate of 20.27 kHz.

**Step 7**

If the clock input to the pair of 7490 counters has a frequency of 20 kHz, what can you conclude about the data stored in read/write memory starting at HI = 003 and LO = 100?

We would expect to see a series of increasing counts starting at HI = 003 and LO = 100, with an increment of one count between successive memory locations. The reason is that the input clock frequency to the 7490 chips is identical to the data logging rate of the 750-kHz microcomputer. For example, when we performed such an experiment on our microcomputer, we observed the following results:

| LO Address Byte of Read/Write Memory | Stored Data | |
|---|---|---|
| | Octal Code | Packed BCD |
| 100 | 106 | 46 |
| 101 | 107 | 47 |
| 102 | 110 | 48 |
| 103 | 111 | 49 |
| 104 | 120 | 50 |
| 105 | 121 | 51 |
| 106 | 122 | 52 |
| 107 | 123 | 53 |
| 110 | 124 | 54 |
| 111 | 125 | 55 |
| 112 | 126 | 56 |
| • | • | • |
| • | • | • |

## Step 8

Set the clock frequency of the clock input to the 7490 counters to 20 kHz or slightly less. Frequencies ranging between 5 kHz and 20 kHz would be quite acceptable. Execute the program once. Observe the data stored starting at HI = 003 and LO = 100. What do you conclude?

The 7490 counter data was stored sequentially in sixty-four read/write memory locations starting at HI = 003 and LO = 100. Our first data point was octal 070 and our last data point was octal 166.

## Step 9

If 49.33 microseconds are required to log a single data byte, and if the first data byte is 070 and the final data byte is 166 from a two-decade counter circuit, what is the frequency of the counter?

The total time required to log 64 data bytes is 49.33 $\mu$s $\times$ 64 points = 3157.12 $\mu$s. The number of counts between the first and last data byte is,

First count data = $070_8$ = $00111000_2$ = $38_{10}$ = $56_{10}$
Final count data = $166_8$ = $01110110_2$ = $76_{10}$ = $118_{10}$

In other words, a total of $76 - 38 = 38$ data bytes were logged during 3.15712 ms. The input clock frequency is therefore,

Clock frequency = 38/0.00315712 seconds
= 12.036 kHz

## Step 10

Calculate the input clock frequencies to your 7490 decade counter circuit using the calculation procedure described in the above step.

| Initial Count | Final Count | Calculated Frequency, |
|:---:|:---:|:---:|
| [LO = 100] | [LO = 177] | kHz |

We obtained the following results, where the counts are given in decimal:

| Initial Count (Decimal) | Final Count (Decimal) | Calculated Frequency, kHz |
|:---:|:---:|:---:|
| 09 | 305 | 93.8 |
| 20 | 147 | 40.2 |
| 11 | 105 | 30.0 |
| 36 | 74 | 12.0 |

## EXPERIMENT NO. 2
## LOGGING SLOW DATA POINTS

### Purpose

The purpose of this experiment is to operate a simple 8080A-based data logger that can log data at low data rates.

### Schematic Diagram of Circuit

See preceding experiment for details of the 7490 decade counter circuit.

### Program

| LO Memory Address | Instruction Byte | | Mnemonic | Description |
|:---:|:---:|:---:|:---|:---|
| 000 | 041 | | LXI H | Load register pair H with the following two bytes |
| 001 | 100 | | 100 | L register byte, the LO address byte of memory location M |
| 002 | 003 | | 003 | H register byte, the HI address byte of memory location M |
| 003 | 006 | | MVI B | Load register B with following byte |
| 004 | 100 | | 100 | Number of points that will be logged |
| 005 | 333 | LOOP, | IN | Input data from pair of counter chips |
| 006 | 004 | | 004 | Device code for counter buffer |

| | | | |
|---|---|---|---|
| 007 | 167 | | MOV M,A | Move accumulator contents to memory location M addressed by contents of register pair H |
| 010 | 043 | | INX H | Increment register pair H |
| 011 | 016 | | MVI C | Load register C with following timing byte |
| 012 | 144 | | 144 | Timing byte for register C |
| 013 | 315 | LOOP1, | CALL | Call 10 ms time-delay routine DELAY |
| 014 | 277 | | DELAY | LO address byte of DELAY |
| 015 | 000 | | — | HI address byte of DELAY |
| 016 | 015 | | DCR C | Decrement register C |
| 017 | 302 | | JNZ | If register C is not equal to 000, jump to LOOP1; otherwise, continue to next instruction |
| 020 | 013 | | LOOP1 | LO address byte of LOOP1 |
| 021 | 003 | | — | HI address byte of LOOP1 |
| 022 | 005 | | DCR B | Decrement register B |
| 023 | 302 | | JNZ | If register B is not equal to 000, jump to LOOP; otherwise, continue to next instruction |
| 024 | 005 | | LOOP | LO address byte of LOOP |
| 025 | 003 | | — | HI address byte of LOOP |
| 026 | 166 | | HLT | Halt |

## Step 1

The circuit is identical to that used in the preceding experiment. Load the new program into read/write memory starting at $HI = 003$ and $LO = 000$.

## Step 2

Execute the program. Remember that it will now require one second per data point, or a total of 64 seconds before all data points are logged and the program comes to a halt. Your clock input to the 7490 counters should be approximately 1 Hz. Go to memory location $HI = 003$ and $LO = 100$ and step through read/write memory. What do you observe?

We observed the data that we logged at one data point/second.

## Step 3

Now make the following changes to the above program and wire up an Output Port 002 if one is not already available on your microcomputer.

| 005 | 176 | MOV M,A | Move the contents of the memory location M to the accumulator |
| 006 | 323 | OUT | Output the accumulator contents to the output latch |
| 007 | 002 | 002 | Device code of output latch |

Execute the program and explain what you observe on the latch, which should be connected either to eight lamp monitors or to a pair of 7-segment displays.

We observed the data input that we stored in read/write memory before we made the modification to the program! In other words, by modifying three instruction bytes, we were able to convert our data input program into a data output program. Each data point was output at a rate of one data point/second, so it was very easy to study the data that we initially stored in memory.

### Step 4

Repeat Steps 2 and 3 as often as you desire. Each time you wish to input data into the memory, you must make certain that the proper instruction bytes are present at memory locations LO = 005 through LO = 007.

### Step 5

Rather than modifying the program each time, it probably would be more convenient to load a separate output program starting at HI = 003 and LO = 030. Assuming that you would do so, what would the addresses of LOOP and LOOP1 be?

LOOP would be at HI = 003 and LO = 035 and LOOP1 would start at HI = 003 and LO = 043.

## EXPERIMENT NO. 3
## DETECTING AN ASCII CHARACTER

### Purpose

The purpose of this experiment is to wire an interface and execute a program that demonstrates how to detect the ASCII character, "E."

## Pin Configuration of Integrated-Circuit Chip (Fig. 22-5)

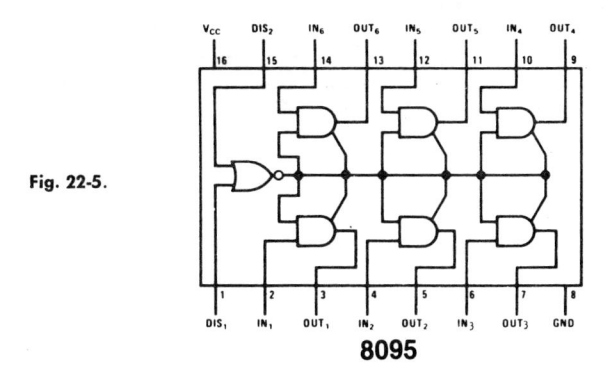

Fig. 22-5.

**8095**

## Schematic Diagram of Circuit (Fig. 22-6)

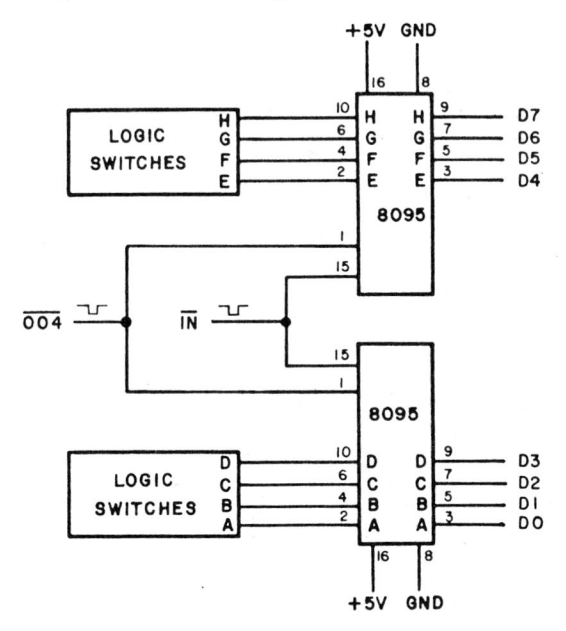

Fig. 22-6.

## Program

| LO Memory Address | Instruction Byte | Mnemonic | Description |
|---|---|---|---|
| 000 | 333 | START, IN | Input ASCII character from the switches |
| 001 | 004 | 004 | Device code for input port |
| 002 | 376 | CPI | Compare the accumulator contents with the following data byte. If the two bytes are identical, set the zero flag. If not, reset the zero flag. Set the carry flag if 305 is greater than the accumulator contents. |
| 003 | 305 | 305 | ASCII code for the letter "E" |
| 004 | 302 | JNZ | If the zero flag is reset, jump back to START; otherwise, continue to the following instruction |
| 005 | 000 | START | LO address byte of START |
| 006 | 003 | — | HI address byte of START |
| 007 | 323 | OUT | Output the ASCII code for the letter "E," which is contained in the accumulator |
| 010 | 002 | 002 | Device code for output port |
| 011 | 166 | HLT | Halt |

### Step 1

Wire the interface circuit shown in Fig. 22-6. Load the program in read/write memory starting at HI = 003 and LO = 000.

### Step 2

Set the logic switches to the 8095 three-state buffer chips to the ASCII equivalent to the letter "D," i.e., 304 in octal code or 1100-0100 in binary. Execute the program at the full microcomputer speed. What happens?

In our case, nothing happened. Outport Port 002 did not exhibit 305.

### Step 3

Set the logic switches to 306 while the microcomputer is running. Make certain that you do not set it to 305, even momentarily! Any change yet?

No. The reason is that so far a 305 input has not been detected. Such being the case, the program continues to loop back to START.

### Step 4
Now change the logic switches to 305. What happens?

The ASCII code for "E," 305, is output to Port 002 and the microcomputer comes to a halt. It does so in response to the query posed by the instruction byte at LO = 004:

    004   302   JNZ   Is the input ASCII character the letter "E"? If not, continue looping back to START until it is.

### Step 5
Change the instruction byte at LO = 004 to

    004   312   JZ   Is the input ASCII character any character other than the letter "E"? Continue looping until a character other than "E" is input.

Set the logic switches to 305 and execute the microcomputer at its full speed. What do you observe?

The program continues to loop. During each loop, it detects the character "E."

### Step 6
Now set the logic switches to 304 and execute the program once more. What happens?

The microcomputer immediately comes to a halt after 304 is output.

### Step 7
You may also wish to substitute either of the following instruction bytes at LO = 004:

    004   322   JNC   Is the input ASCII character greater than 304? If not, continue looping until it is.
                  or
    004   332   JC   Is the input ASCII character less than 305? If not, continue looping until it is.

See the text in this unit for a discussion of these two instruction bytes.

## EXPERIMENT NO. 4
## WIRING A BUS MONITOR

### Purpose

The purpose of this experiment is to wire a pair of circuits that you may find useful in subsequent experiments: (1) A three-octal-digit *bus monitor,* which permits you to monitor all information that passes over the bidirectional data bus, and (2) A latched 7490 counter, which permits you to detect and count different types of synchronization pulses.

### Pin Configurations of Integrated-Circuit Chips (Fig. 22-7)

**REAR VIEW**

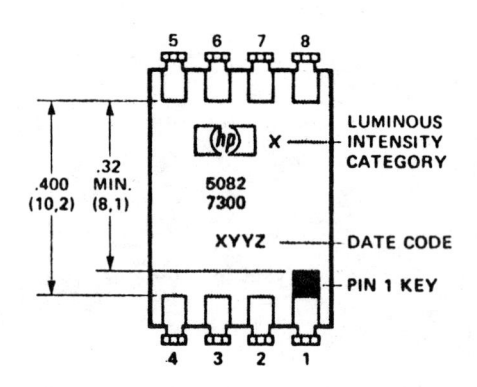

| PIN | FUNCTION | |
|---|---|---|
| | 5082-7300 and 7302 Numeric | 5082-7340 Hexadecimal |
| 1 | Input 2 | Input 2 |
| 2 | Input 4 | Input 4 |
| 3 | Input 8 | Input 8 |
| 4 | Decimal point | Blanking control |
| 5 | Latch enable | Latch enable |
| 6 | Ground | Ground |
| 7 | $V_{cc}$ | $V_{cc}$ |
| 8 | Input 1 | Input 1 |

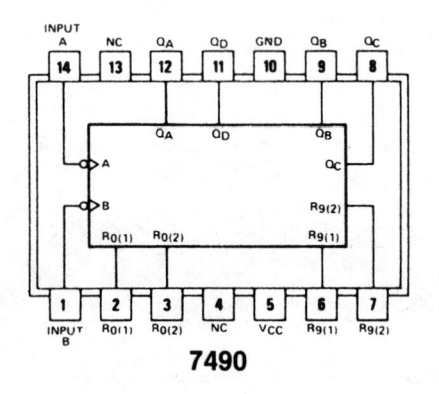

Fig. 22-7.

7490

## Schematic Diagrams of Circuits (Figs. 22-8 and 22-9)

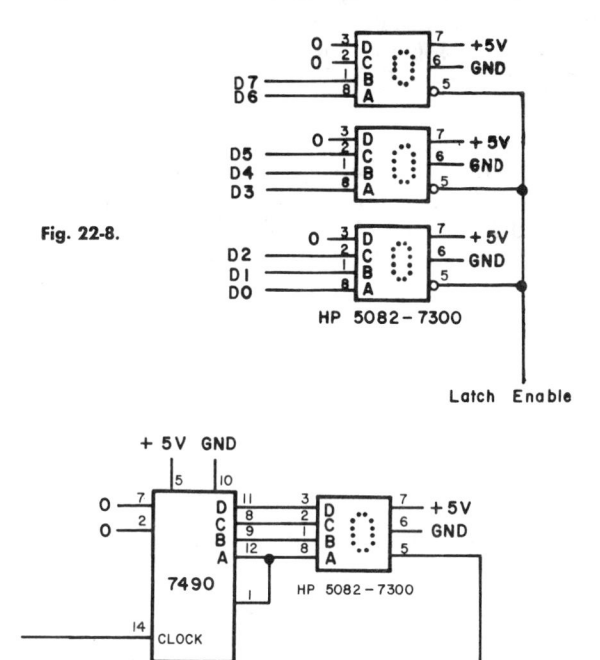

Fig. 22-8.

Fig. 22-9.

## Step 1

Wire the circuits shown in Figs. 22-8 and 22-9, preferably on a single breadboarding socket. You will find them useful as monitors of device select pulses and control signals as well as the data that appears on the bidirectional data bus.

## Step 2

The Hewlett-Packard 5082-7300 display contains a 4-bit latch of the 7475 type that is enabled by a logic 0 $\overline{\text{STROBE}}$ pulse. What does this mean?

The 7475 latch is a D-type latch that follows the input when the latch is enabled. Thus, a logic 0 applied to the HP 5082-7300 means that the output will be the same as the input as long as the $\overline{\text{STROBE}}$ input remains at the logic 0 state. Latching of input data occurs on the positive edge of the $\overline{\text{STROBE}}$ input pulse.

### Step 3

A variety of control and other signals can be used as the $\overline{\text{STROBE}}$ input to the HP 5082-7300 latch/displays that are connected to the bidirectional data bus, D0 through D7. List some of these signals and explain what information they permit you to latch from the data bus.

Some useful signals include:

| | |
|---|---|
| $\overline{\text{OUT}}$ | Latches all data output via an OUT instruction |
| $\overline{\text{IN}}$ | Latches all data input via an IN instruction |
| $\overline{\text{MEMR}}$ | Latches all data input via a memory-read type instruction |
| $\overline{\text{MEMW}}$ | Latches all data output via a memory-write type instruction |
| Output DS pulse | Latches all data output to a specific output device |
| Input DS pulse | Latches all data input from a specific input device |
| $\overline{\text{INTA}}$ | Latches data on the bus that appears during an interrupt acknowledge control signal |

We shall call the three-digit circuit that latches the bidirectional data bus, D0 through D7, a *bus monitor,* since it permits you to monitor all information that appears on the data bus. You will use this monitor in subsequent experiments, so save it.

### Step 4

The second circuit in this experiment consists of a Hewlett-Packard latch/display wired to the output of a 7490 counter. What is the function of this circuit?

The circuit permits you to detect individual control signal or device select pulses, provided that only a few are generated. Such a circuit is generally used when the program contains a Halt instruction or when there are long times between pulses.

## EXPERIMENT NO. 5
## BIDIRECTIONAL MEMORY-MAPPED I/O USING AN 8216 CHIP

### Purpose

The purpose of this experiment is to operate an 8216 chip as a bidirectional memory-mapped I/O port.

### Pin Configuration and Logic Diagram of Integrated-Circuit Chip (Figs. 22-10 and 22-11)

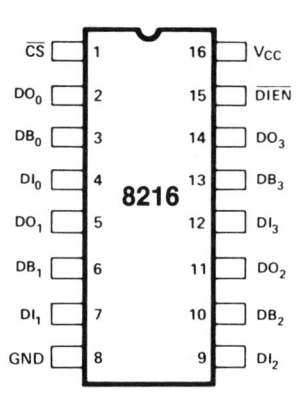

| Pins | Function |
|---|---|
| $DB_0\text{-}DB_3$ | DATA BUS BI DIRECTIONAL |
| $DI_0\text{-}DI_3$ | DATA INPUT |
| $DO_0\text{-}DO_3$ | DATA OUTPUT |
| $\overline{DIEN}$ | DATA IN ENABLE DIRECTION CONTROL |
| $\overline{CS}$ | CHIP SELECT |

**Fig. 22-10.**

### Schematic Diagram of Circuit (Fig. 22-12)

### Program

| LO Memory Address | Instruction Byte | Mnemonic | | Description |
|---|---|---|---|---|
| 000 | 041 | START, | LXI H | Load register pair H with the following two bytes |
| 001 | 000 | 000 | | L register byte, the LO address byte of memory location M |
| 002 | 100 | 100 | | H register byte, the HI address byte of memory location M |
| 003 | 176 | | MOV A,M | Move contents of memory location M to accumulator |
| 004 | 167 | | MOV M,A | Move contents of accumulator to memory location M |
| 005 | 323 | | OUT | Output contents of accumulator to output port 002 |
| 006 | 002 | 002 | | Device code for port 002 |
| 007 | 303 | | JMP | Unconditional jump to memory location START |
| 010 | 000 | START | | LO address byte of START |
| 011 | 003 | — | | HI address byte of START |

| DIEN | CS | |
|---|---|---|
| 0 | 0 | DI → DB |
| 1 | 0 | DB → DO |
| 0 | 1 | HIGH IMPEDANCE |
| 1 | 1 | |

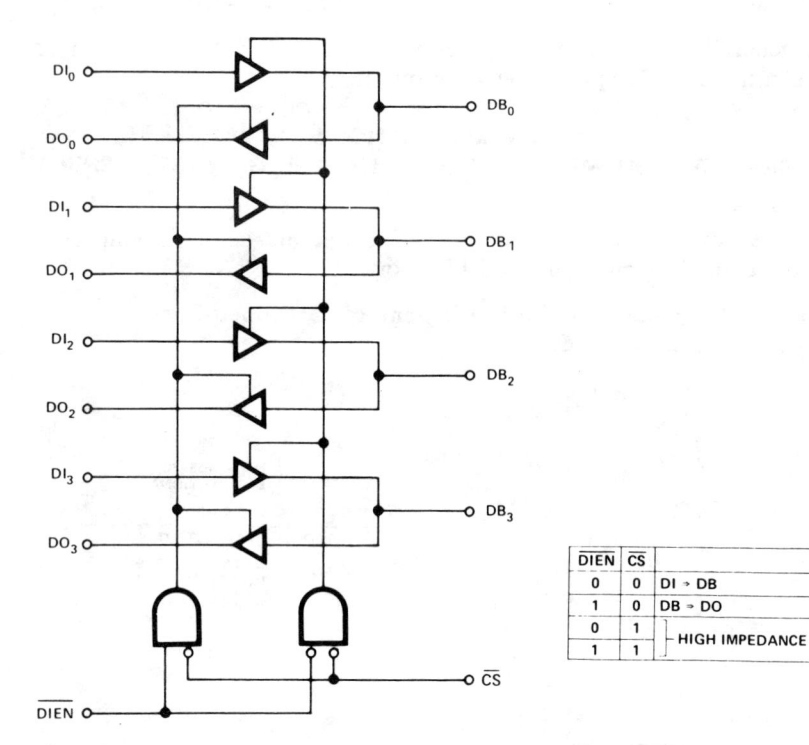

**Fig. 22-11.**

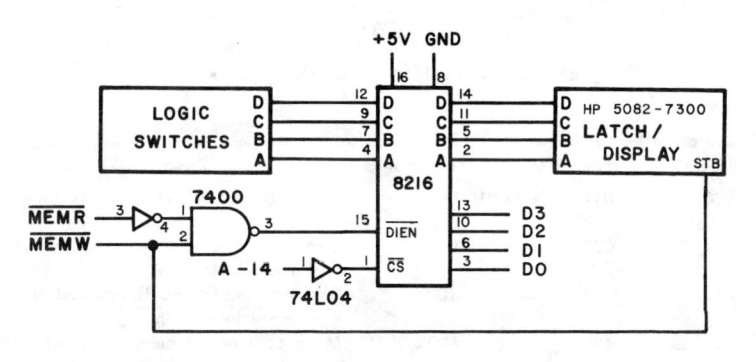

**Fig. 22-12.**

## Step 1

Wire the circuit shown. Load the program in read/write memory starting at $HI = 003$ and $LO = 000$.

## Step 2

Execute the program. What do you observe on Output Port 002 and also the latch/display as you change the logic switches from 0000 to 1001?




The least significant four bits in Output Port 002 and the latch/display exhibit the same reading as the logic switch input to the 8216 chip. The most significant four bits in Output Port 002 all remain at logic 1. The program reads the logic switch data using the MOV A,M instruction, and then outputs the accumulator contents to the 8216 latch/display using the MOV M,A instruction. Finally, the OUT 002 instruction outputs the accumulator contents to the output port.

Fill in the following truth tables for the $\overline{\text{DIEN}}$ and $\overline{\text{CS}}$ inputs to the 8216 chip.

| A-14 | $\overline{\text{CS}}$ |
|---|---|
| 1 | |
| 0 | |

| $\overline{\text{MEMR}}$ | $\overline{\text{MEMW}}$ | $\overline{\text{DIEN}}$ |
|---|---|---|
| 0 | 0 | |
| 0 | 1 | |
| 1 | 0 | |
| 1 | 1 | |

| $\overline{\text{DIEN}}$ | $\overline{\text{CS}}$ | Operation of the 8216 chip |
|---|---|---|
| 0 | 0 | |
| 0 | 1 | |
| 1 | 0 | |
| 1 | 1 | |

Explain the significance of these tables in the space below.

The first truth table indicates whether or not the chip is enabled.

| A-14 | $\overline{CS}$ | Operation of the 8216 chip |
|---|---|---|
| 1 | 0 | Chip enabled |
| 0 | 1 | Chip disabled (high impedance state) |

The second truth table indicates the direction of data transfer through the chip.

| $\overline{MEMR}$ | $\overline{MEMW}$ | DIEN | |
|---|---|---|---|
| 0 | 0 | 1 | Not allowed |
| 0 | 1 | 0 | Memory read |
| 1 | 0 | 1 | Memory write |
| 1 | 1 | 1 | Memory write |

The final truth table summarizes the operation of the 8216 chip.

| DIEN | $\overline{CS}$ | Operation of 8216 chip |
|---|---|---|
| 0 | 0 | Data input to 8080A chip |
| 0 | 1 | Chip disabled |
| 1 | 0 | Data output from 8080A chip |
| 1 | 1 | Chip disabled |

**Step 3**

In other words, when $\overline{MEMR}$ and $\overline{CS}$ are both at logic 0, the 8216 serves as an input port. When $\overline{MEMW}$ and $\overline{CS}$ are both at logic 0, the 8216 serves as an output port. $\overline{MEMR}$ and $\overline{MEMW}$ cannot both be at logic 0 since the 8080A microprocessor cannot read and write at the same time. The $\overline{MEMW} = \overline{MEMR} = 0$ state is never observed by external control logic.

**Step 4**

Is this chip useful as an I/O port?

Perhaps, but an additional latch is required if it is to be used as an output port. The succeeding experiment provides a better scheme. In general, the 8216 is used as a bidirectional bus driver/buffer that has a fan-in of 0.1 and a fan-out of 30.

### EXPERIMENT NO. 6
### ACCUMULATOR I/O USING THE 8255 CHIP

**Purpose**

The purpose of this experiment is to demonstrate the use of the 8255 programmable peripheral interface chip as an accumulator I/O port.

## Pin Configuration and Block Diagram of Integrated-Circuit Chip

The pin configuration and block diagram for the 8255 chip are given in Fig. 22-13.

## Schematic Diagram of Circuit (Fig. 22-14)

### Program

| LO Memory Address | Instruction Byte | Mnemonic | Description |
|---|---|---|---|
| 000 | 076 | MVI A | Move the following control word into the accumulator |
| 001 | 231 | 231 | Control word that establishes the mode 0 operation of the 8255 chip, with ports A and C being input ports and port B being an output port |
| 002 | 323 | OUT | Output accumulator contents to following output latch |
| 003 | 203 | 203 | Device code for control register within 8255 chip |
| 004 | 333 | LOOP, IN | Input logic switch data at port C |
| 005 | 202 | 202 | Device code for port C |
| 006 | 323 | OUT | Output accumulator contents to port B |
| 007 | 201 | 201 | Device code for port B |
| 010 | 303 | JMP | Unconditional jump to memory location LOOP |
| 011 | 004 | LOOP | LO address byte of LOOP |
| 012 | 003 | — | HI address byte of LOOP |

### 8255 Basic Operation

The truth table for the three I/O ports and the control register is as follows:

| A1 | A0 | $\overline{RD}$ | $\overline{WR}$ | $\overline{CS}$ | INPUT OPERATION (READ) |
|---|---|---|---|---|---|
| 0 | 0 | 0 | 1 | 0 | PORT A → DATA BUS |
| 0 | 1 | 0 | 1 | 0 | PORT B → DATA BUS |
| 1 | 0 | 0 | 1 | 0 | PORT C → DATA BUS |
| | | | | | **OUTPUT OPERATION (WRITE)** |
| 0 | 0 | 1 | 0 | 0 | DATA BUS → PORT A |
| 0 | 1 | 1 | 0 | 0 | DATA BUS → PORT B |
| 1 | 0 | 1 | 0 | 0 | DATA BUS → PORT C |
| 1 | 1 | 1 | 0 | 0 | DATA BUS → CONTROL |
| | | | | | **DISABLE FUNCTION** |
| X | X | X | X | 1 | DATA BUS → 3-STATE |
| 1 | 1 | 0 | 1 | 0 | ILLEGAL CONDITION |

(A) Pin configuration.

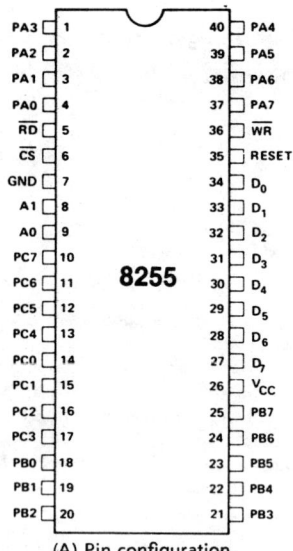

| $D_7$–$D_0$ | DATA BUS (BI DIRECTIONAL) |
|---|---|
| RESET | RESET INPUT |
| $\overline{CS}$ | CHIP SELECT |
| $\overline{RD}$ | READ INPUT |
| $\overline{WR}$ | WRITE INPUT |
| A0, A1 | PORT ADDRESS |
| PA7-PA0 | PORT A (BIT) |
| PB7-PB0 | PORT B (BIT) |
| PC7-PC0 | PORT C (BIT) |
| $V_{CC}$ | +5 VOLTS |
| GND | 0 VOLTS |

(B) Pin names.

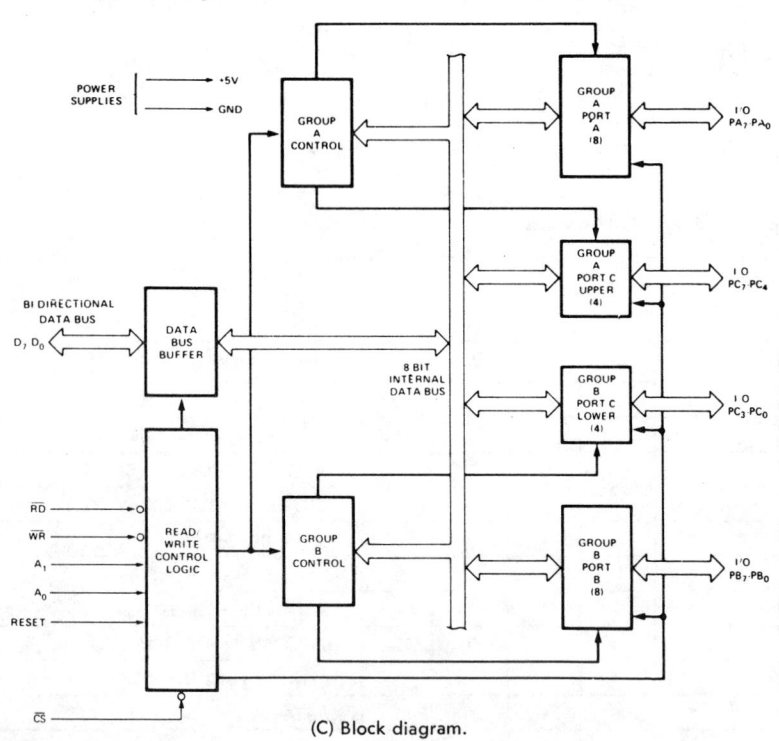

(C) Block diagram.

Fig. 22-13. The 8255 Programmable Peripheral Interface.

Fig. 22-14.

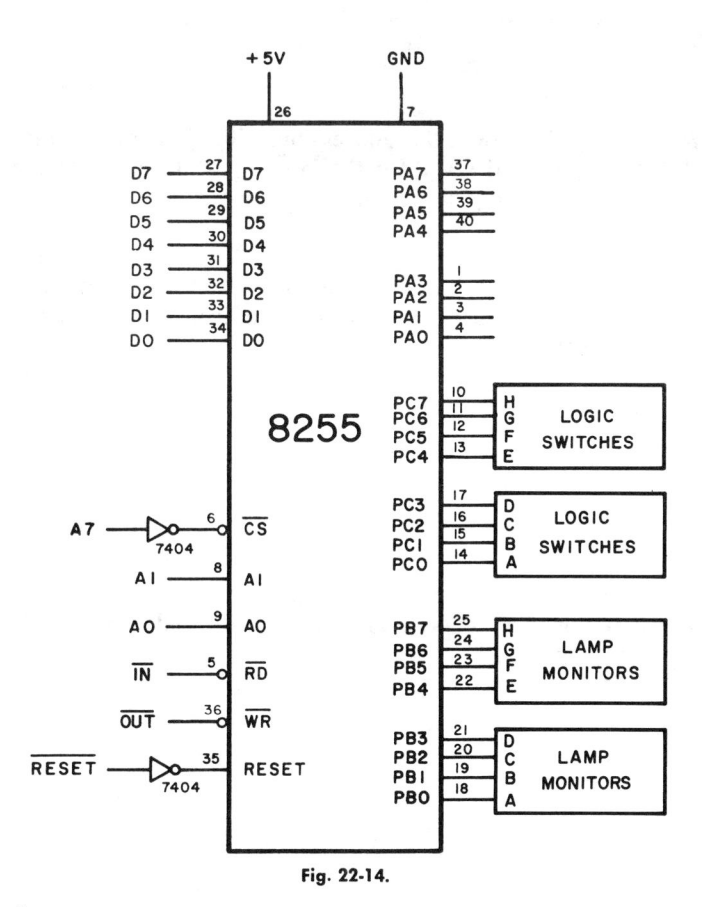

## Step 1

Study the truth table for the three I/O ports and the control register within the 8255 chip. Address bus bits A0, A1, and A7 are used to select the specific port or register desired. The control signals $\overline{\text{IN}}$ and $\overline{\text{OUT}}$ are connected to $\overline{\text{RD}}$ and $\overline{\text{WR}}$, respectively. Therefore, the 8-bit device code, A0 through A7, identifies the particular I/O device associated with an IN or OUT instruction. Since the address bus is not absolutely decoded, address bits A2 through A6 can be either 0 or 1. In defining the device codes, we have let these bits be logic 0.

| Device code | I/O port or register |
|---|---|
| 200 | Port A |
| 201 | Port B |
| 202 | Port C |
| 203 | Control register |

In the program, the instruction bytes at LO = 003, 005, and 007 are all device codes.

### Step 2

Wire the circuit shown. Use address bus bit A7 for the $\overline{CS}$ input, $\overline{IN}$ for the $\overline{RD}$ input, and $\overline{OUT}$ for the $\overline{WR}$ input. This use of the chip is an example of accumulator I/O.

### Step 3

Load the program in read/write memory starting at HI = 003 and LO = 000. What do you think the significance of the instruction byte at LO = 001 is?

As stated in the program, it is a control word that establishes the mode 0 operation of the 8255 chip and whether Port A, Port B, and Port C are input or output ports. In this case, the control word corresponds to Ports A and C being input ports and Port B being an output port (Fig. 22-15).

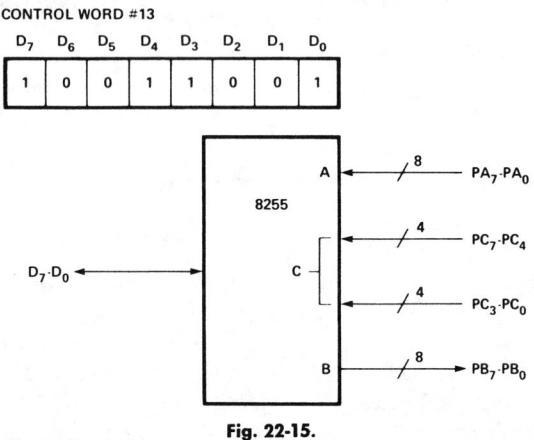

CONTROL WORD #13

| $D_7$ | $D_6$ | $D_5$ | $D_4$ | $D_3$ | $D_2$ | $D_1$ | $D_0$ |
|---|---|---|---|---|---|---|---|
| 1 | 0 | 0 | 1 | 1 | 0 | 0 | 1 |

**Fig. 22-15.**

### Step 4

Execute the program at the full microcomputer speed. While the microcomputer is running, change the logic switch settings and observe the output at Port B. What happens?

Port B displays the logic switch input to Port C. Any changes in the logic switch settings occur essentially instantaneously at Port B. This behavior demonstrates that we have input data into the accumulator and output it from the accumulator to Port B.

### Step 5

Change the control word at LO = 001 to 213. Execute the program at the full microcomputer speed and observe whether or not there is any output at Port B when you change the Port C logic switch settings.

We observed no change in output at Port B, which remained at 000. The reason is that the control word now assumes that Port B is an input port (Fig. 22-16).

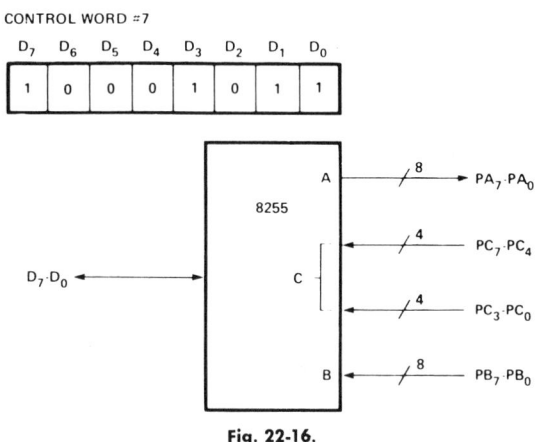

CONTROL WORD =7

| $D_7$ | $D_6$ | $D_5$ | $D_4$ | $D_3$ | $D_2$ | $D_1$ | $D_0$ |
|---|---|---|---|---|---|---|---|
| 1 | 0 | 0 | 0 | 1 | 0 | 1 | 1 |

Fig. 22-16.

### Step 6

Finally, change the control word at LO = 001 to 201, which corresponds to the situation shown in Fig. 22-17. Note that now, only four of the bits in Port C are input bits. Remove the logic switches from Bits PC4 through PC7. Change the control word at LO = 001 and execute the program at the full microcomputer speed. What do you observe?

**CONTROL WORD #1**

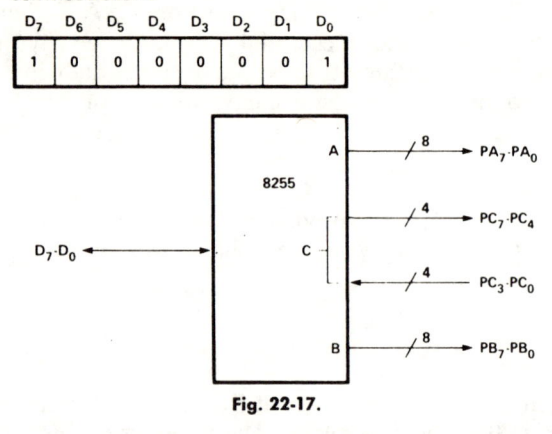

| $D_7$ | $D_6$ | $D_5$ | $D_4$ | $D_3$ | $D_2$ | $D_1$ | $D_0$ |
|---|---|---|---|---|---|---|---|
| 1 | 0 | 0 | 0 | 0 | 0 | 0 | 1 |

**Fig. 22-17.**

The output at Port B mirrors the input at Port C, Bits PC0 through PC3. As far as our circuit is concerned, Port B is once again an output port.

### Step 7

The 8255 chip is an interesting but somewhat complicated interface chip that is manufactured by the Intel Corporation. For further details, obtain a copy of Intel's *8080 Microcomputer Peripherals User's Manual* and a copy of their application note AP-15, *8255 Programmable Peripheral Interface Applications*.

Do not dismantle this circuit but save it for the following experiment, in which you will use the 8255 chip as a memory I/O device.

## EXPERIMENT NO. 7
## MEMORY-MAPPED I/O USING THE 8255 CHIP

### Purpose

The purpose of this experiment is to demonstrate the use of the 8255 programmable peripheral interface chip as a memory I/O port.

### Pin Configuration of the Integrated-Circuit Chip

The pin configuration drawing for the 8255 interface chip was given in Fig. 22-13A. The names for the various pins of the chip are given in Fig. 22-13B.

### Schematic Diagram of Circuit

The schematic diagram was given in the preceding experiment (Fig. 22-14). You were asked at the end of the preceding experiment not to dismantle the circuit but to save the circuit for use in this experiment.

## Program

| LO Memory Address | Instruction Byte | | Mnemonic | Description |
|---|---|---|---|---|
| 000 | 041 | | LXI H | Load register pair H with the following two bytes |
| 001 | 003 | | 003 | L register byte, the LO address byte of memory location M |
| 002 | 200 | | 200 | H register byte, the HI address byte of memory location M |
| 003 | 006 | | MVI B | Move the following control word to register B |
| 004 | 231 | | 231 | Control word for the 8255 chip |
| 005 | 160 | | MOV M,B | Move register B contents to memory location M, which is the control register within the 8255 chip |
| 006 | 055 | | DCR L | Decrement register L |
| 007 | 176 | LOOP, | MOV A,M | Input logic switch data through port C |
| 010 | 055 | | DCR L | Decrement register L |
| 011 | 167 | | MOV M,A | Output accumulator contents to port B |
| 012 | 054 | | INR L | Increment register L |
| 013 | 303 | | JMP | Unconditional jump to memory location LOOP |
| 014 | 007 | | LOOP | LO address byte of LOOP |
| 015 | 003 | | — | HI address byte of LOOP |

### Step 1

The input to $\overline{CS}$ should now be the inverted Bit A-15. The inputs to $\overline{RD}$ and $\overline{WR}$ should now be $\overline{MEMR}$ and $\overline{MEMW}$, respectively. Make these wiring changes to the circuit shown in Fig. 22-14.

### Step 2

Load the program into read/write memory starting at $HI = 003$ and $LO = 000$.

### Step 3

Execute the program at the full microcomputer speed. Vary the logic switch settings and observe what happens at Port B. What do you conclude?

The logic switch input at Port C appears as a lamp monitor output at Port B.

### Step 4

Change the control word at LO = 004 to 213 and then to 201. Execute the microcomputer in each case at the full microcomputer speed. Explain your observations at Port B in the space below.




With the control word of 213, Port B no longer functions as an output port. For control word 201, Port B is again an output port, but only Port C bits, PC0 through PC3, are input bits. Bits PC4 to PC7 are output bits with this second control word.

### Step 5

Change the LO address byte at LO = 014 to 013. Execute the program at the full microcomputer speed with 231 as a control word at LO = 004. Can you conclude that the output to Port B is latched?




Yes, the output to Port B is latched, since any logic 1 bits remain at logic 1 despite the fact that no additional input to the accumulator occurs from Port C. Only the initial logic switch settings are latched. All subsequent changes are ignored by the program.

## EXPERIMENT NO. 8
## INTERFACING A DIGITAL-TO-ANALOG CONVERTER

### Purpose

The purpose of this experiment is to test a simple parallel input program for the AD7522 10-bit, buffered, multiplying, digital-to-analog (D/A) converter.

## Pin Configuration and Functional Diagram of Integrated-Circuit Chip (Fig. 22-18)

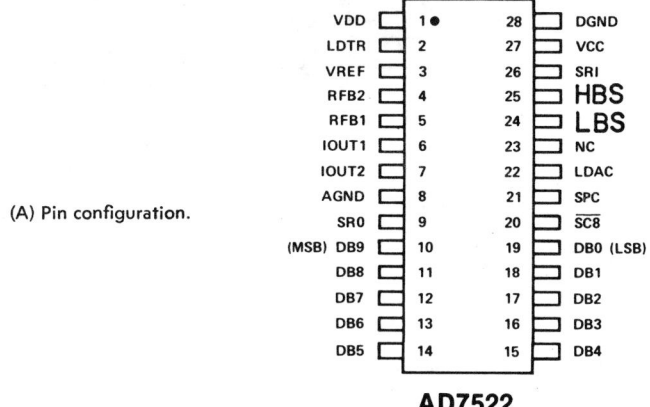

(A) Pin configuration.

```
          VDD  [ 1●   28 ]  DGND
          LDTR [ 2    27 ]  VCC
          VREF [ 3    26 ]  SRI
          RFB2 [ 4    25 ]  HBS
          RFB1 [ 5    24 ]  LBS
          IOUT1[ 6    23 ]  NC
          IOUT2[ 7    22 ]  LDAC
          AGND [ 8    21 ]  SPC
           SRO [ 9    20 ]  SC8
    (MSB) DB9  [ 10   19 ]  DB0 (LSB)
          DB8  [ 11   18 ]  DB1
          DB7  [ 12   17 ]  DB2
          DB6  [ 13   16 ]  DB3
          DB5  [ 14   15 ]  DB4
```

**AD7522**

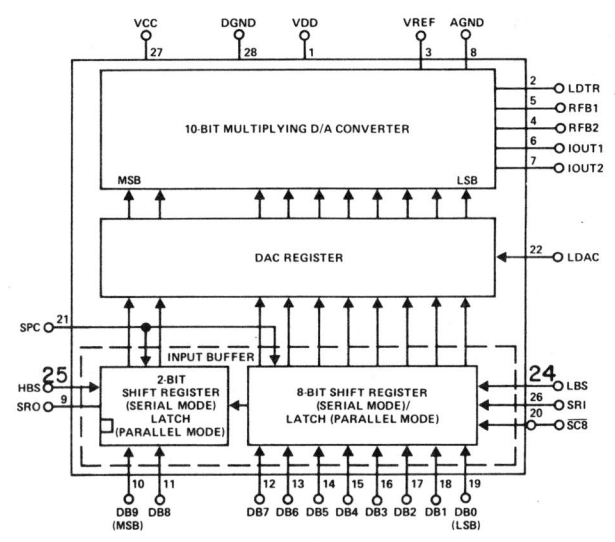

(B) Functional diagram.

**Fig. 22-18. The AD7522 D/A converter.**

## Schematic Diagram of Circuit (Fig. 22-19)

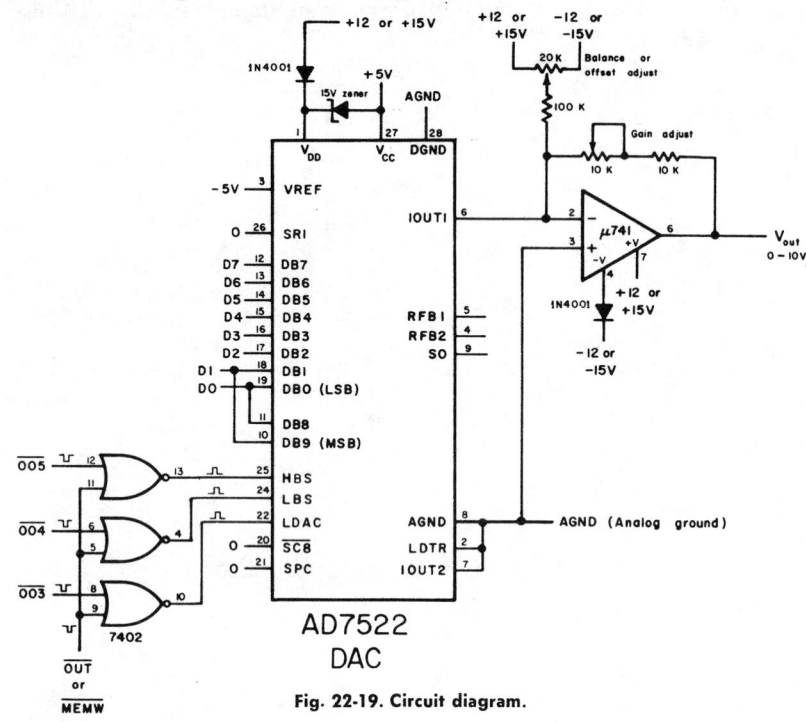

Fig. 22-19. Circuit diagram.

## Program

| LO Memory Address | Octal Instruction | Mnemonic | | Comments |
|---|---|---|---|---|
| 000 | 042 | START, | SHLD | Strobe ten bits of digital data into the AD7522 DAC registers |
| 001 | 004 | | 004 | HI = 000 and LO = 004 is the memory I/O device code for the LBS input to the DAC. HI = 000 and LO = 005 is the memory I/O device code for the HBS input to the DAC. [NOTE: We can use these device codes since memory block HI = 000 is EPROM.] |
| 002 | 000 | | 000 | |
| 003 | 062 | | STA | Send strobe pulse to the LDAC input of the AD7522 DAC. Ten bits of digital data are internally strobed within the DAC into the DAC register. |
| 004 | 003 | | 003 | HI = 000 and LO = 003 is the memory I/O device code for the LDAC input to the DAC |
| 055 | 000 | | 000 | |
| 006 | 043 | | INX H | Increment register pair H |
| 007 | 315 | | CALL | Call 10 ms time-delay routine located in the KEX EPROM (memory block HI = 000) |

| | | | |
|---|---|---|---|
| 010 | 277 | TIMEOUT | LO address byte of TIMEOUT |
| 011 | 000 | — | HI address byte of TIMEOUT |
| 012 | 303 | JMP | Jump back to START and repeat the execution of the program |
| 013 | 000 | START | LO address byte of START |
| 014 | 003 | — | HI address byte of START |

## Discussion

The above program causes the digital-to-analog converter to generate a slow linear ramp, which can be observed on a volt-ohm-milliammeter (VOM) or on an oscilloscope, as the voltage output from the DAC. The ramp output is subdivided into 1024 small steps, each step being approximately 5.0 to 5.5 mV in magnitude. The total time required to change from 0.0 volts to $+5.66$ volts output is 10.24 seconds.

The 10-bit digital-to-analog converter word consists of the Data Bits DB0 through DB7 strobed into the AD7522 8-bit shift register with the aid of strobe pulse LBS and, also, Data Bits DB8 and DB9 strobed into the AD7522 with the aid of strobe pulse HBS. Bits DB8 and DB9 appear on the microcomputer bidirectional data bus as Bits D0 and D1.

## Step 1

Study the schematic diagram of the circuit in Fig. 22-19. When we performed this experiment, $V_{out}$ was connected to a small volt-ohm-milliammeter (VOM). The digital ground, DGND, should be connected to the analog ground, AGND. Since you are using memory I/O techniques, $\overline{\text{MEMW}}$ should be used instead of $\overline{\text{OUT}}$.

To facilitate the wiring of this circuit, we have developed a small Outboard® module, a block diagram of which is shown in Fig. 22-20. The AD7522 DAC Outboard contains all of the necessary analog and digital circuitry needed to perform this experiment. If you have this Outboard, wire the circuit as shown. Inputs $\overline{003}$, $\overline{004}$, and $\overline{005}$ are the decoded output channels obtained from an appropriate LO address byte decoder circuit.

## Step 2

Wire the DAC circuit and load the program (shown at the start of this experiment) into memory starting at HI = 003 and LO = 000. If you are executing this program on the MMD-1 microcomputer, the 10-ms time-delay routine TIMEOUT is already loaded in the Keyboard *EX*ecutive EPROM. If you are using some other 8080A-based microcomputer, we have provided a TIMEOUT listing at the end of this experiment.

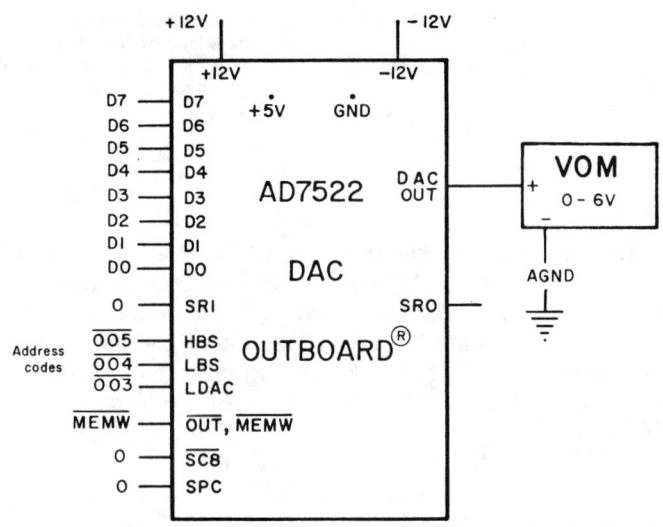

Fig. 22-20. Schematic diagram of the AD7522 DAC Outboard®.

## Step 3

Execute the program with the volt-ohm-milliammeter connected to the output of the DAC. What do you observe?

We observed a slow but steady increase in the VOM reading until +5.6 volts was reached, at which time the needle returned to 0 volts and repeated the process. The time required for the full range of readings was approximately 10 seconds.

## Step 4

Change the instruction byte at LO = 006 to the following:

    006    053    DCX H    Decrement register pair H

Execute the program once again. What change in behavior of the VOM do you observe? Why?

Now the VOM exhibits a slow and steady decrease from + 5.6 volts to 0 volts, at which time the needle returns to + 5.6 volts and repeats the process. We decrement the value instead of incrementing it as before.

### Step 5

Remove the time-delay subroutine by making the following program changes:

| | | | |
|---|---|---|---|
| 007 | 000 | NOP | No operation |
| 010 | 000 | NOP | No operation |
| 011 | 000 | NOP | No operation |

The DCX H instruction should still be present. Execute this modified program and explain what you observe on the VOM.

We observed that the VOM needle oscillated about the voltage reading of +2.75 volts. The magnitude of the oscillations was approximately ±0.02 volt. On a digital multimeter, the readings varied between +2.83 and +2.91 volts. In other words, the rather fast linear ramp could not be followed by either meter; only an average voltage reading was observed.

The negative linear ramp was easy to observe on an oscilloscope set to a sweep rate of 10 ms/division.

Save your interface circuit and continue to the following experiment. Additional information for this experiment is given on the following pages.

### ADDITIONAL INFORMATION

#### Time-Delay Routine TIMEOUT

A listing of the 10-ms time-delay routine TIMEOUT is as follows:

| LO Memory Address | Octal Instruction | | Mnemonic | Comments |
|---|---|---|---|---|
| 277 | 365 | TIMEOUT, | PUSH PSW | Push contents of accumulator and flags on stack |
| 300 | 325 | | PUSH D | Push contents of register pair D on stack |
| 301 | 021 | | LXI D | Load following two bytes into register pair D |
| 302 | 046 | | 046 | E register byte |
| 303 | 001 | | 001 | D register byte |

| LO Memory Address | Instruction Byte | | Mnemonic | Description |
|---|---|---|---|---|
| 304 | 033 | MORE, | DCX D | Decrement contents of register pair D by one |
| 305 | 172 | | MOV A,D | Move contents of register D to accumulator |
| 306 | 263 | | ORA E | OR contents of register E with contents of accumulator |
| 307 | 302 | | JNZ | Jump to LOOP if result of OR operation is not 000; otherwise, skip this instruction after testing the zero flag |
| 310 | 304 | | MORE | LO address byte of MORE |
| 311 | 000 | | — | HI address byte of MORE |
| 312 | 321 | | POP D | Pop stack into register pair D |
| 313 | 341 | | POP PSW | Pop stack into accumulator and flags |
| 314 | 311 | | RET | Return from subroutine |

## AD7522 Pin Functions

In Table 22-1, we provide a listing of the individual pin functions on the AD7522 digital-to-analog converter. We would like to acknowledge Analog Devices, Inc. for the use of this information.

### Table 22-1. Pin Function Description

| PIN | MNEMONIC | DESCRIPTION |
|---|---|---|
| 1 | VDD | +15V (nominal) Main Supply. |
| 2 | LDTR | R-2R Ladder Termination Resistor. Normally grounded for unipolar operation or terminated at IOUT2 for bipolar operation. |
| 3 | VREF | Reference Voltage Input. Since the AD7522 is a multiplying DAC, VREF may vary over the range of $\pm 10V$. |
| 4 | RFB2 | Rfeedback ÷ 2; gives full scale equal to VREF/2. |
| 5 | RFB1 | Rfeedback, used for normal unity gain (at full scale) D/A conversion. |
| 6 | IOUT1 | DAC Current OUT-1 Bus. Normally terminated at virtual ground of output amplifier. |
| 7 | IOUT2 | DAC Current OUT-2 Bus, terminated at ground for unipolar operation, or virtual ground of op amp for bipolar operation. |
| 8 | AGND | Analog Ground. Back gate of DAC N-channel SPDT current steering switches. |
| 9 | SRO | Serial Output. An auxiliary output for recovering data in the input buffer. |
| 10 | DB9 | Data Bit 9. Most significant parallel data input. |
| 11 | DB8 | Data Bit 8. |
| 12 | DB7 | Data Bit 7. |
| 13 | DB6 | Data Bit 6. |
| 14 | DB5 | Data Bit 5. |
| 15 | DB4 | Data Bit 4. |

(Note 1 applies to pins 10–15)

## Table 22-1 cont. Pin Function Description

| PIN | MNEMONIC | DESCRIPTION |
|---|---|---|
| 16 | DB3 | Data Bit 3. |
| 17 | DB2 (Note 1) | Data Bit 2. |
| 18 | DB1 (Note 1) | Data Bit 1. |
| 19 | DB0 | Data Bit 0. Least significant parallel data input. |
| 20 | $\overline{SC8}$ | 8-Bit Short Cycle Control. When in serial mode, if $\overline{SC8}$ is held to logic "0," the two least significant input latches in the input buffer are bypassed to provide proper serial loading of 8-bit serial words. If $\overline{SC8}$ is held to logic "1," the AD7522 will accept a 10-bit serial word. Data bits 0(LSB) and DB1 are in a parallel load mode when $\overline{SC8}$ = 0, and should be tied to a logic low state to prevent false data from being loaded. |
| 21 | SPC | Serial/Parallel Control. If SPC is a logic "0," the AD7522 will load parallel data appearing on DB0 through DB9 into the input buffer when the appropriate strobe inputs are exercised (see HBS and LBS). If SPC is a logic "1," the AD7522 will load serial data appearing on Pin 26 into the input buffers. Each serial data bit must be "strobed" into the buffer with the HBS and LBS. |
| 22 | LDAC | Load DAC: When LDAC is a logic "0," the AD7522 is in the "hold" mode, and digital activity in the input buffer is locked out. When LDAC is a logic "1," the AD7522 is in the "load" mode, and data in the input buffer loads the DAC register. |
| 23 | NC | No Connection. |
| 24 | LBS | Low Byte Strobe. When in "parallel load" mode (SPC = 0), parallel data appearing on the DB0 (LSB) through DB7 inputs will be "clocked" into the input buffer on the positive going edge of the LBS. When in "serial load" mode (SPC = 1), serial data bits appearing at the serial input terminal, Pin 26, will be "clocked" into the input buffer on the positive going edge of HBS and LBS. (HBS and LBS must be clocked simultaneously when in "serial load" mode.) |
| 25 | HBS | High Byte Strobe. When in "parallel load" mode (SPC = 0), parallel data appearing on the DB9 (MSB) and DB8 data inputs will be "clocked" into the input buffer on the positive going edge of HBS. When in "serial load" mode (SPC = 1), serial data bits appearing at the serial input terminal, Pin 26, will be "clocked" into the input buffer on the positive going edges of HBS and LBS. (HBS and LBS must be clocked simultaneously when in "serial load" mode.) |
| 26 | SRI | Serial Input. |
| 27 | VCC | Logic Supply. If +5V is applied, all digital inputs/outputs are TTL compatible. If +10V to +15V is applied, digital inputs/outputs are CMOS compatible. |
| 28 | DGND | Digital Ground. |

NOTE 1: Logic "1" applied to a data bit steers that bit's current to the IOUT1 terminal.

# EXPERIMENT NO. 9
## A STAIRCASE-RAMP COMPARISON ANALOG-TO-DIGITAL CONVERTER

## Purpose

The purpose of this experiment is to use the *staircase-ramp comparison* technique to convert an AD7522 10-bit buffered multiplying digital-to-analog converter into an analog-to-digital converter (ADC) with the aid of an LM311 comparator.

## Pin Configurations of Integrated-Circuit Chips (Fig. 22-21)

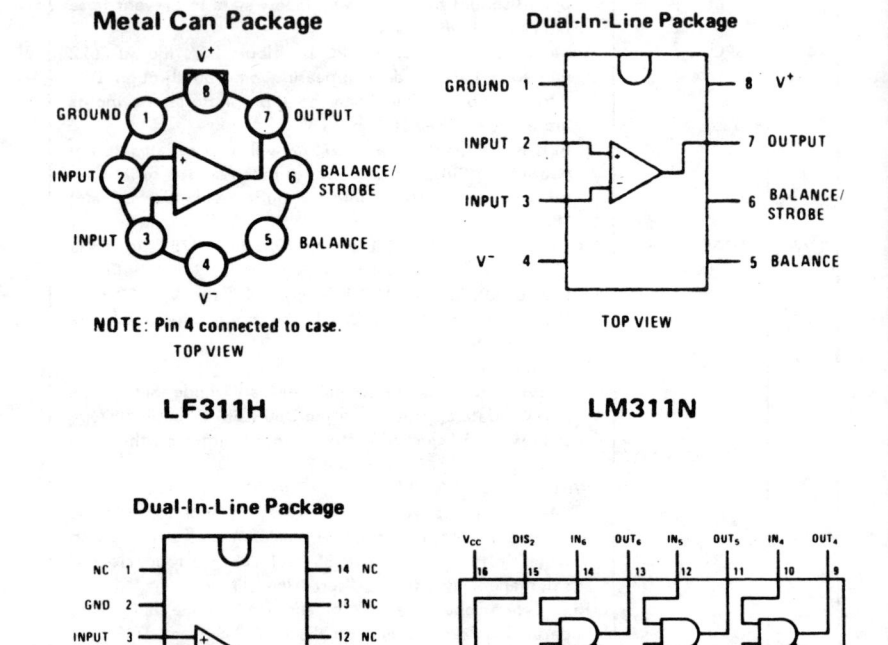

Fig. 22-21.

## Schematic Diagram of Circuit

The DAC circuit was shown in the preceding experiment (Fig. 22-19). The output from the DAC, $V_{out}$, should now be connected to input pin 2 of the LM311 comparator, as shown in Fig. 22-22.

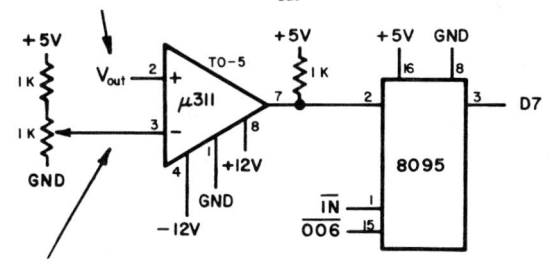

This is the output voltage, $V_{out}$, from the DAC

This is the unknown voltage, $E_x$

**Fig. 22-22.**

## Program

This program is written to be stored in EPROM, starting at HI = 001 and LO = 107 and terminating at HI = 001 and LO = 210. An asterisk, *, is used to indicate those absolute memory locations which must be changed to relocate the program elsewhere in memory. Since most EPROM programmers operate in hexadecimal code, the program is also listed in hex.

| LO Memory Address | Instruction Octal | Hex | Mnemonic | | Comments |
|---|---|---|---|---|---|
| • | | | | | |
| • | | | | | |
| 107 | 305 | C5 | CONVRT, | PUSH B | Push contents of register pair B on stack |
| 110 | 325 | D5 | | PUSH D | Push register pair D on stack |
| 111 | 345 | E5 | | PUSH H | Push register pair H on stack |
| 112 | 365 | F5 | | PUSH PSW | Push accumulator and flags on stack |
| 113 | 323 | D3 | | OUT | Generate synchronization pulse |
| 114 | 007 | 07 | | 007 | Device code of synchronization pulse |
| 115 | 041 | 21 | | LXI H | Initialize register pair H |
| 116 | 000 | 00 | | 000 | L register byte |
| 117 | 000 | 00 | | 000 | H register byte |
| 120 | 042 | 22 | AGAIN, | SHLD | Strobe ten bits of digital data into the AD7522 DAC shift registers |
| 121 | 004 | 04 | | 004 | HI = 000 and LO = 004 is memory I/O device code for LBS input; HI = 000 and LO = 005 is memory I/O device code for HBS input. |
| 122 | 000 | 00 | | 000 | |

**305**

| LO Memory Address | Instruction Byte | | Mnemonic | Description |
|---|---|---|---|---|
| 123 | 062 | 32 | STA | Send strobe pulse to LDAC input of AD7522 DAC; load data into DAC register |
| 124 | 003 | 03 | 003 | HI = 000 and LO = 003 is memory |
| 125 | 000 | 00 | 000 | I/O device code for LDAC input |
| 126 | 043 | 23 | INX H | Increment register pair H |
| 127 | 333 | DB | IN | Input into bit D7 the output from the AD311 comparator |
| 130 | 006 | 06 | 006 | Device code for comparator bit |
| 131 | 346 | E6 | ANI | Mask all bits in accumulator except bit D7 |
| 132 | 200 | 80 | 200 | Mask Byte |
| 133 | 312 | CA | JZ | If comparator, bit D7, is logic 1, continue to steps below; otherwise, jump to AGAIN and continue to increment the DAC output |
| 134 | 120* | 50* | AGAIN | LO address byte of AGAIN |
| 135 | 001* | 01* | — | HI address byte of AGAIN |
| 136 | 175 | 7D | MOV A,L | Move the low eight bits of the DAC word to the accumulator |
| 137 | 323 | D3 | OUT | Output low eight bits of DAC word to following output port |
| 140 | 002 | 02 | 002 | Device code for output port 002 |
| 141 | 174 | 7C | MOV A,H | Move the high two bits of the DAC word to the accumulator |
| 142 | 323 | D3 | OUT | Output high two bits of DAC word to following output port |
| 143 | 002 | 02 | 000 | Device code for output port 002 |
| 144 | 361 | F1 | POP PSW | Pop accumulator and flags off stack |
| 145 | 341 | E1 | POP H | Pop register pair H off stack |
| 146 | 321 | D1 | POP D | Pop register pair D off stack |
| 147 | 301 | C1 | POP B | Pop register pair B off stack |
| 150 | 311 | C9 | RET | Return from subroutine |

## Discussion

Observe that it is not difficult to relocate the preceding program. Only two address bytes need to be changed. The program generates a slow linear ramp output from the DAC. This DAC output voltage, $V_{out}$, is continuously tested by the LM311 comparator against an unknown voltage, $E_x$. In this case, one supplied by a 1-kilohm potentiometer circuit. If $V_{out} < E_x$, the output from the comparator is logic 0. This single bit is input into the D7 position of the accumulator, where it is masked and tested by the JZ instruction. The linear ramp continues to be generated until $V_{out} \geq E_x$, at which time the comparator output becomes logic 1. The JZ instruction is then skipped and a return occurs from the subroutine after the 10-bit DAC word is output to a pair of output ports.

## Step 1

Assume that the above program is already present in EPROM. If it is not, load it at an appropriate place in your read/write memory. The only two absolute address bytes present in the program are at LO = 134 and LO = 135.

## Step 2

If the program is in EPROM, you may have to write a short program to call the subroutine. At HI = 003 and LO = 000, load the following program into read/write memory:

| LO Memory Address | Instruction Byte | Mnemonic | Description |
|---|---|---|---|
| 000 | 315 | CALL | Call the staircase-ramp comparison subroutine CONVRT |
| 001 | 107 | CONVRT | LO address byte of CONVRT |
| 002 | 001 | — | HI address byte of CONVRT |
| 003 | 166 | HLT | Halt |

## Step 3

You will observe the results of the program execution with the aid of a VOM or a digital voltmeter. The program in read/write memory calls the program once, permits a single conversion to be made, and then halts. The measured analog voltage appears on the VOM. At the same time, the 10-bit DAC number also appears on the two output ports, 000 and 002.

With the 1-kilohm potentiometer set at 0 ohm, execute the program once. What do you observe?

We observed a 0.0-volt reading on our volt-ohm-milliammeter, a reading of +0.027 volt on our digital voltmeter (indicating that we had some offset error in our analog circuit), and a reading of HI = 000 and LO = 000 on Output Ports 000 and 002, respectively.

## Step 4

Now set the 1K potentiometer at its maximum value, approximately 1 kilohm. Execute the ADC conversion routine once, and observe both the output voltage on the VOM and the DAC word on the output ports. What readings do you observe? Is this what you would expect?

We observed a reading on the VOM of +2.30 volts and a reading on our digital multimeter of +2.529 volts. The resistance divider circuit should provide a reading of approximately +2.8 volts, based upon the results from our previous experiment. The tolerances of the resistor and potentiometer are such that our measured values are reasonable. The 10-bit output DAC word that appeared on Output Ports 000 and 002 was HI = 001 and LO = 371.

### Step 5

Vary the potentiometer setting and test the program for the measurement of voltages between 0.0 volt and +2.5 volts. Do you observe any difficulties?

We did not. Our measurements worked as expected.

### REVIEW QUESTIONS

The following questions will help you review data logging.

1. What is a data logger?
2. What are the important considerations in the design of a data logger?
3. How would you detect a specific ASCII character that is input into the microcomputer?
4. What methods are available to generate time delays?

# 23

# Flags and Interrupts

### INTRODUCTION

Flags and interrupts are useful interfacing techniques that find broad application in any type of computer interfacing. This unit explores their use with 8080A-based microcomputers and provides typical hardware and software examples. Interrupt timing problems are also discussed.

### OBJECTIVES

At the completion of this unit, you will be able to do the following:

- Define flag and give typical examples of its use.
- Design a simple flag circuit and explain its operation.
- Write flag servicing software for one or more flags and explain how such software is used.
- Describe three types of interrupts.
- Explain the use of the 8080A restart instructions, including the operation of the stack.
- Describe the operation of the 8080A microprocessor chip's interrupt capability and all of the signals involved.
- Design an interrupt instruction port and describe its use.
- Describe the software used in a typical interrupt service routine.
- Explain some of the timing problems associated with both flags and interrupts.

## WHAT IS A FLAG?

We have seen in previous units that it is fairly easy to transfer data in and out of a microcomputer using the IN and OUT instructions and some hardware. In many cases the computer will be ready for data to be input much faster than the data source can generate it. We can also have the case of an output device which may be much slower than the computer. For example, a teletypewriter can print only 10 to 30 characters per second, whereas a typical 8080A system can output a new character as fast as every 5 to 10 microseconds. Clearly, some method of synchronization is needed so that the computer responds only when an input device actually has data ready or when an output device needs more data. We need some sort of signal to indicate the state or status of our devices. This is called a *flag*.

> *flag*—Some sort of digital register or device used to indicate the state or status of a device. It can be cleared or set in response to an operation.

You have already used some flags that are internal to the 8080A microprocessor chip. These are the zero flag and the carry flag which, along with the sign, parity, and auxiliary carry flags make up the five internal 8080A flags that are useful to us in software. These flags, excluding the auxiliary carry flag, are the basis for the branching or transfer of control instructions. Note that the flags are cleared (logic 0) or set (logic 1) in response to various software instructions. This is consistent with our definition, since the software performs an *operation,* e.g., ADD, ROTATE, OR, etc. It is important to note that flags are used to detect conditions and to remember what condition has occurred.

External flags are used to indicate conditions of input/output devices and other digital systems or devices which the microcomputer must control. The following is a list of some of the types of conditions which flags are used to indicate:

- Data is available and ready to be input into the microcomputer.
- A device is ready for the next set of eight bits to be output to it.
- An external device is busy, or it is still performing an operation.
- An external device is ready for the next operation.
- A limit has been exceeded.
- A value is too low.

## FIRST EXAMPLE: INTERFACING A KEYBOARD

Let us consider a typical interfacing example and see how a flag can be used. We shall interface an 8-bit ASCII keyboard to our microcomputer by constructing an 8-bit input port. You should be able to do this based upon your experience with device decoding and three-state input ports. For additional details, see Units 17 and 20. Our interface circuit is shown in Fig. 23-1. A typical flow chart that illustrates how you would input characters and compare them to the letter "E" is shown in Fig. 23-2.

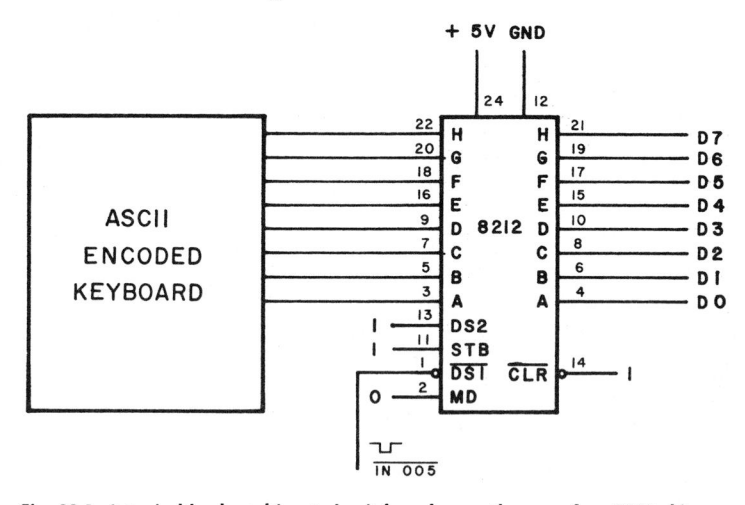

Fig. 23-1. A typical keyboard input circuit based upon the use of an 8212 chip as a three-state input port.

When the ASCII code for the letter "E," $305_8$, is finally input and detected by the microcomputer, the software will output all logic 1s, or $377_8$, to an output port—in this case, device 001. If there are

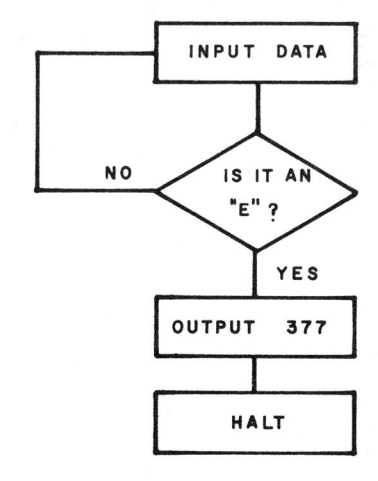

Fig. 23-2. Flow chart for detecting the ASCII letter "E."

LEDs at this output port, they will all be lit when an "E" has been detected. The software needed to accomplish this is as follows:

| LO Memory Address | Instruction Byte | | Mnemonic | Description |
|---|---|---|---|---|
| 000 | 333 | DETECT: | IN | Input keyboard data from input port 005 |
| 001 | 005 | | 005 | Device code 005 |
| 002 | 376 | | CPI | Compare accumulator contents with the ASCII byte for the letter "E" |
| 003 | 305 | | 305 | ASCII code for the letter "E" |
| 004 | 302 | | JNZ | If the keyboard input byte is not the same as the ASCII code for the letter "E," jump to memory location DETECT; otherwise, continue to the following instruction |
| 005 | 000 | | 000 | LO address byte of DETECT |
| 006 | 003 | | 003 | HI address byte of DETECT |
| 007 | 076 | | MVI A | The keyboard input byte is the ASCII letter "E." Input the following byte into the accumulator. |
| 010 | 377 | | 377 | Accumulator byte |
| 011 | 323 | | OUT | Output the contents of the accumulator to output port 001 |
| 012 | 001 | | 001 | Device code 001 |
| 013 | 166 | | HLT | Halt |

This software will continuously input data from the keyboard even when a key is not activated (and thus no real code is present). We would prefer to sense the key closure and have the computer only input the information when the code is valid. Most keyboards provide a pulse or level that indicates when data is ready or valid. This status signal is a flag. For example, in the case of our keyboard, it is a one-microsecond pulse, called VALID, which indicates that a valid code is present. We could input this pulse directly into the microcomputer and test to see if it were present, but in all likelihood the microcomputer would miss it since the pulse is so short. We need some means of stretching or holding the pulse until it can be sensed by the microcomputer. The solution is a flip-flop, which provides the means of holding the flag information. A typical flip-flop flag is shown in Fig. 23-3 for both the 7474- and 7476-type flip-flops.

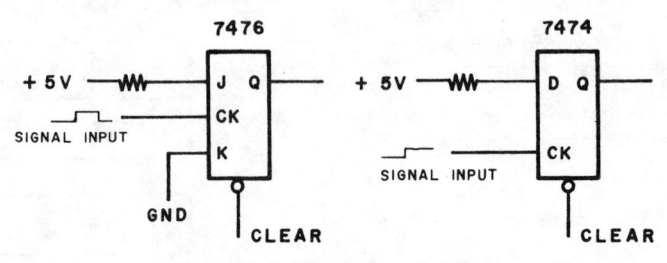

Fig. 23-3. Typical flip-flops used as flags.

The VALID pulse from the keyboard is used to set the flag, which may then be sensed by the microcomputer under software control. A three-state input port is used to input the flag information to the 8080A. This is exactly the same type of input port used to input data, except that it is now called a *sense register* since it is used with individual flags. Once a specific flag has been sensed, it must be cleared so that the next key closure will again set the flag. The hardware circuit is shown in Fig. 23-4, in which the two 8212 input ports have been simplified for clarity. This simplified circuit demonstrates how the VALID flag from the keyboard is tested by the 8080A microcomputer. The device decoders are not shown. (See Unit 17 for information on the generation of device select pulses.)

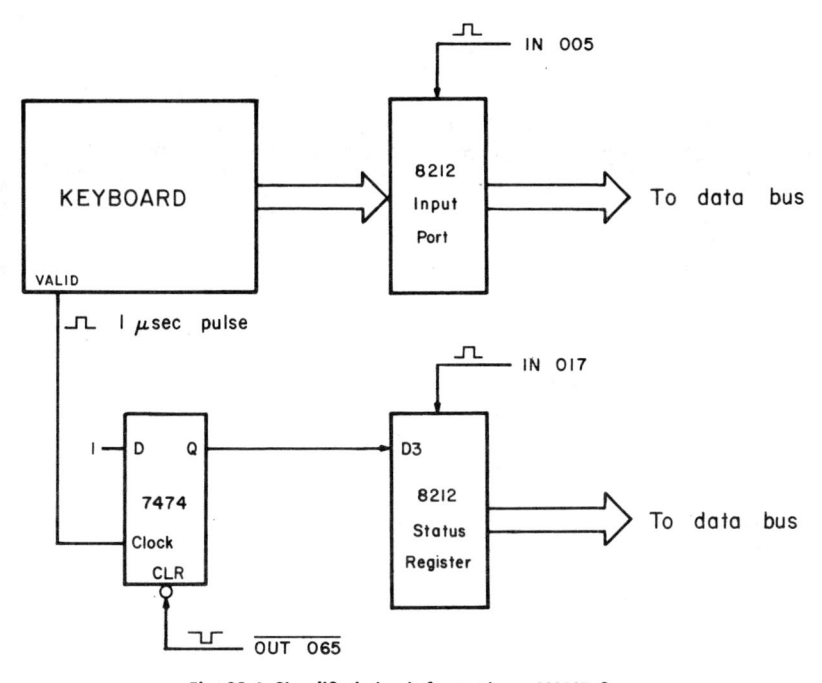

Fig. 23-4. Simplified circuit for testing a VALID flag.

## SECOND EXAMPLE: SOLVENT-LEVEL CONTROL

As a second example of the use of flags, we wish to control the level of a solvent in a storage tank. An empty/full switch and a digitally controlled valve are available. The system is represented in the diagram of Fig. 23-5. Two OUT device select pulses are used to control the valve through the use of a flip-flop, a buffer, and a solid-state relay, a technique that was previously shown in Unit 17. As a

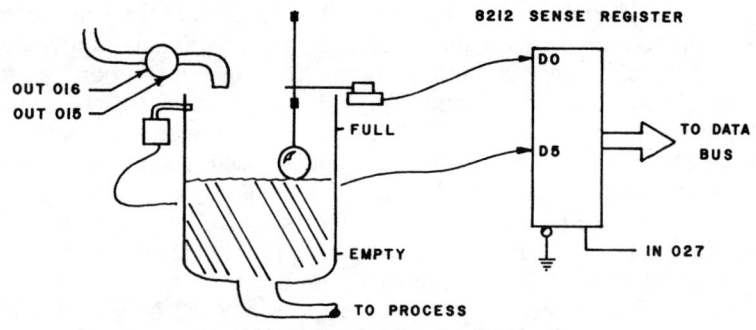

Fig. 23-5. Circuitry for detecting fluid levels.

precaution, an overflow indicator has been added that outputs a logic 1 when the solvent is about ready to overflow, and a logic 0 when there is no danger of overflow. The level switch is at a logic 1 when the solvent reaches the full point, and at a logic 0 when it reaches the empty point. Since the overflow and level indicators do not change rapidly, they can be used directly as flags without flip-flops. Flip-flop flag indicators might still be used in a real environment; you should be able to show how they could be added to this

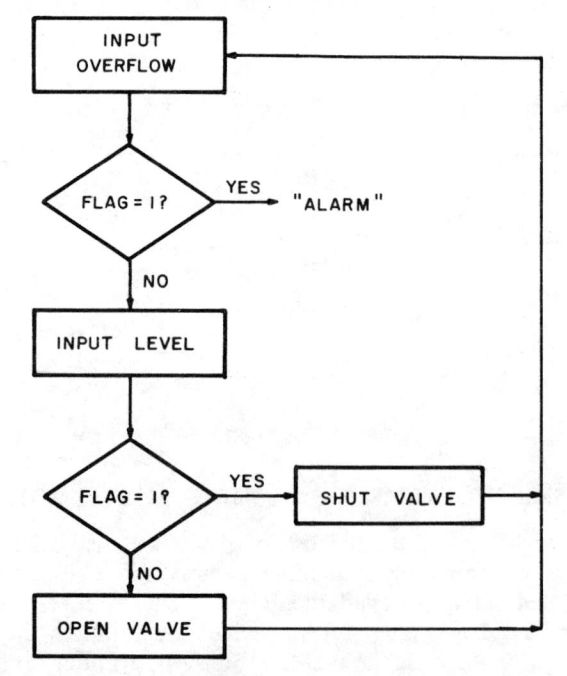

Fig. 23-6. Flow chart for controlling the level of solvent in a storage tank.

314

system. A three-state *sense register* is still used to input these signals into the microcomputer.

Your object is to write a software program to keep the liquid between the FULL and EMPTY limits and to sense an overflow condition, which might be caused by a poor switch or valve. Consider the flow chart shown in Fig. 23-6. The flags are sensed in software using two conditional jump instructions. Notice that in this flow chart example, symbolic addresses have been used. These address names or address symbols are used to simplify the programming task since actual address values do not have to be assigned *until* the program is finished or assembled.

The software for controlling the level of solvent in the tank can be written as follows:

| LO Memory Address | Instruction Byte | Mnemonic | | Description |
|---|---|---|---|---|
| 000 | 333 | START: | IN | Input flag data from device 027 |
| 001 | 027 | | 027 | Device code 027 |
| 002 | 346 | | ANI | Mask out bit D5, i.e., AND contents of accumulator with following mask byte |
| 003 | 040 | | 040 | Mask byte |
| 004 | 304 | | CNZ | If bit D5 is logic 1, call subroutine ALARM; otherwise, continue to next instruction |
| 005 | 100 | | ALARM | LO address byte of subroutine ALARM |
| 006 | 003 | | — | HI address byte of subroutine ALARM |
| 007 | 333 | | IN | Input flag data from device 027 once again |
| 010 | 027 | | 027 | Device code 027 |
| 011 | 017 | | RRC | Rotate bit D0 into the carry flag |
| 012 | 332 | | JC | If carry bit is logic 1, the tank is full; jump to the memory location of the FULL routine. Otherwise, continue to the next instruction. |
| 013 | 022 | | FULL | LO address byte of FULL routine |
| 014 | 003 | | — | HI address byte of FULL routine |
| 015 | 323 | | OUT | The tank is not full. Open the valve and let more solvent flow in. |
| 016 | 016 | | 016 | Device code 016, which generates a device select pulse that opens the valve |
| 017 | 303 | | JMP | Do it again, i.e., jump back to memory location START and execute the program once more |
| 020 | 000 | | START | LO address byte of START |
| 021 | 003 | | — | HI address byte of START |
| 022 | 323 | FULL: | OUT | The tank is full. Close the valve that lets solvent flow in. |
| 023 | 015 | | 015 | Device code 015, which generates a device select pulse that closes the valve |
| 024 | 303 | | JMP | Do it again, i.e., jump back to memory location START and execute the program once more |
| 025 | 000 | | START | LO address byte of START |
| 026 | 003 | | — | HI address byte of START |

We have assumed that the address of the ALARM subroutine is HI = 003 and LO = 100. The ALARM software might first contain a command to shut the valve, OUT 15, and then a routine to actually sound the alarm signal.

The above software is useful in understanding other aspects of flag bit manipulation. All of the bits except D5 have been masked out, or cleared, in the first AND operation, in which the overflow status is checked. This means that the flag information must be input again when the FULL and EMPTY limits are to be checked. Such a step could be eliminated if the status data were saved in a memory location—for example, in a register or on the stack. This might be important if the six other input bits are connected to other devices and must be checked as well.

In real projects such as this, there is the possibility of errors. For example, the FULL/EMPTY switch wires may be reversed, so that in the full position the valve will open and in the empty position it will shut. This could be disastrous, so the fluid-level control system must be checked or simulated before it goes into actual operation. You will do this in one of the experiments. In the software example, the microcomputer is dedicated to a small loop that continuously checks the solvent tank. In a real situation, other tanks and levels would also be checked. The time to fill the tank is extremely long compared to the time in which the microcomputer can check fifty or more tanks. If, however, the microcomputer is also performing other tasks, it may not be in a position to check each tank except once every several seconds. Depending upon the other tasks assigned to it, the microcomputer may actually miss a FULL level or even an OVERFLOW condition if it takes too long to do some of these other operations.

A final point to consider is the time required to turn the valve on or off. Depending upon the size of the valve, this time may range from one second to five or ten seconds. In properly written software, the valve will be given sufficient time to turn on or off before another decision on its state is made by the microcomputer. When the solvent tank is operating near its FULL position, software should be available to prevent the valve from opening or closing unnecessarily.

## POLLED OPERATION

The type of microcomputer operation, which we discussed above, that is used in both hardware and software for the keyboard and the solvent tank is called *polled operation*. *Polling* is defined in the following manner.

*polling*—A periodic checking of input-output or control devices to determine their condition or status, e.g., full/empty, on/off, busy/ready, done/not done, etc.

When devices are polled, they may require servicing or they may not. In polled operation, devices are checked one after the other in sequence. When a device needs to be serviced for input or output of data or for a control application, a *software driver* is used. The software driver is a series of steps in memory that are designed to serve that particular device. For example, the software for the keyboard input and solvent tank control could be called software drivers since they cause an action to be taken at the particular device. Each input/output device generally has a software driver routine, or perhaps even a set of software drivers for various types of operations. Polled operations are generally slow and are used with slow devices such as teletypewriters, paper-tape readers, paper-tape punches, games, coin-operated machines, etc. For faster response times, such as in a multitask problem, a faster way of servicing external devices is needed. This is discussed in the following sections.

### WHAT IS AN INTERRUPT?

If you were interrupted while reading this page, you would probably finish the sentence, mark your place (perhaps mentally), and then take care of the interrupt, i.e., a phone call, meal, child, etc. After finishing with the interrupt, you would continue reading where you left off. *Computers service interrupts in much the same way!* The term *interrupt* can be defined as follows:

> *interrupt*—In a computer, a break in the normal flow of a system or routine such that the flow can be resumed from that point at a later time.[2]

In a computer, interrupt operation is much more sophisticated than polled operation and has both advantages and disadvantages in comparison to polled operation. For example, in polled operation:

- The computer wastes time checking all possible I/O devices.
- Devices must wait their turn. All are treated as equal, in sequence. This establishes a sequential priority, but each device must still wait its turn before being serviced.
- Response times may be long.
- Software and programming are generally straightforward.

In interrupt-type systems:

- The computer may be doing other things not related to the I/O devices while waiting for them to require servicing.

- Priority can be established in software or hardware so that important devices are serviced first.
- Response times can be fast.
- Hardware and software can become very complex.

## TYPES OF INTERRUPTS

There are three basic interrupt modes, *single line, multilevel,* and *vectored.*

*single line*—An interrupt signal that is input to the computer on a single line and causes a well-defined action to take place. Multiple devices must be ORed onto this line and a polling routine must determine which device caused the interrupt. The PDP-8 family of minicomputers uses this method.

*multilevel*—Several independent interrupt lines are provided, each of which causes a specific action. Polling is not needed unless multiple devices are ORed to one of the inputs. The Motorola 6800 microprocessor chip uses this system with two interrupt input lines.

*vectored*—Each device points, or *vectors,* the computer's control to specific software drivers for the interrupting devices. The Intel 8080A and Digital Equipment Corporation PDP-11 family of minicomputers use this technique.

Each technique is shown in Fig. 23-7. In the single-line system, many devices may be added, but they must all be polled through a sense register using flag flip-flops. Servicing can be slow since a long time may be consumed in polling all of the devices in a large system. The multilevel interrupt is a mix between vector and single-line schemes. It has limitations and takes careful software management to use it effectively.

## RESTART: RST X

In this unit, we will be mainly concerned with the vectored interrupt techniques that are used on the 8080A microprocessor chip. This type of interrupt permits us to provide not only an interrupt pulse, but also an instruction to the microprocessor to tell it what to do. In simple 8080A systems, a single-byte instruction can be forced into the computer when it is interrupted. While rotate, increment, and other single-byte instructions might be useful in some applications, *restart* instructions, i.e., single-byte subroutine call instructions, are much more useful and flexible.

The usual 8080A call instructions, both conditional and unconditional, each specify a 16-bit address in the second (LO) and third

**(A) Single line.**

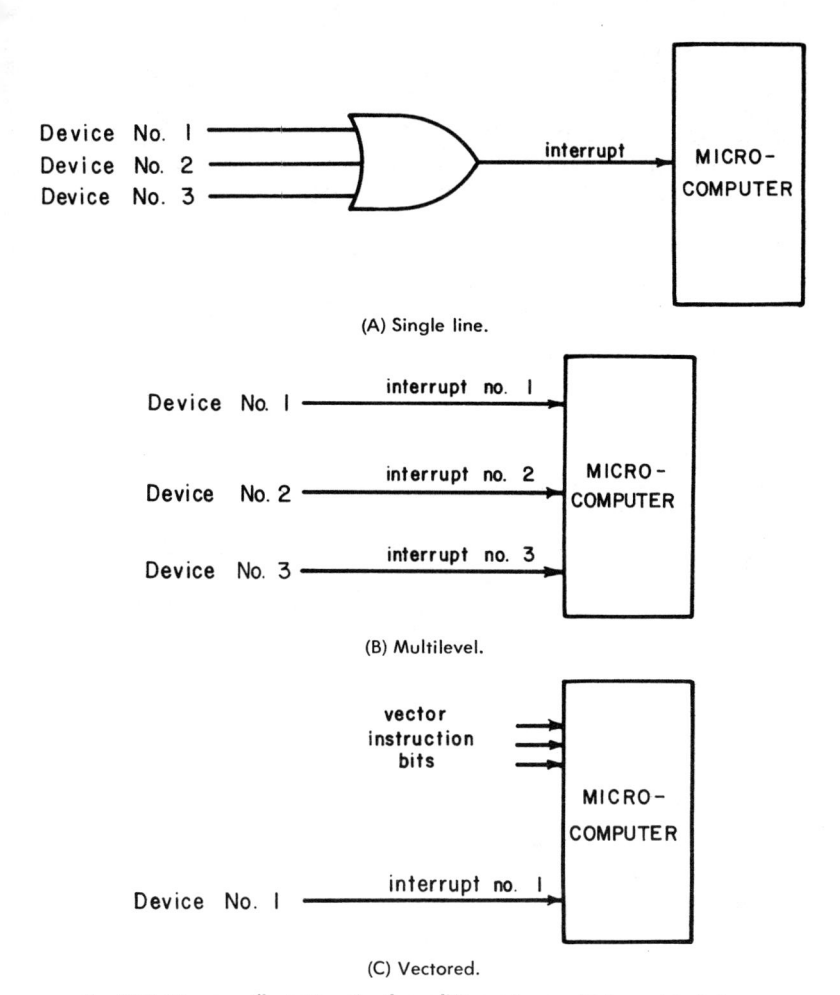

**(B) Multilevel.**

**(C) Vectored.**

**Fig. 23-7. Diagrams illustrating the three different types of interrupt techniques.**

(HI) instruction bytes. How can there be a single byte subroutine call? The answer is that *restart* instructions call subroutines at predefined addresses. These instructions are listed below,

| | | |
|---|---|---|
| 307 | RST 0 | Call subroutine at HI = $000_8$ and LO = $000_8$ |
| 317 | RST 1 | Call subroutine at HI = $000_8$ and LO = $010_8$ |
| 327 | RST 2 | Call subroutine at HI = $000_8$ and LO = $020_8$ |
| 337 | RST 3 | Call subroutine at HI = $000_8$ and LO = $030_8$ |
| 347 | RST 4 | Call subroutine at HI = $000_8$ and LO = $040_8$ |
| 357 | RST 5 | Call subroutine at HI = $000_8$ and LO = $050_8$ |
| 367 | RST 6 | Call subroutine at HI = $000_8$ and LO = $060_8$ |
| 377 | RST 7 | Call subroutine at HI = $000_8$ and LO = $070_8$ |

and can be summarized as follows:

3X7  RST X  Call subroutine at HI = $000_8$ and LO = $0X0_8$

The subroutine locations, HI = $000_8$ and LO = $0X0_8$, are preset in the 8080A chips and cannot be changed. There are other ways around this limitation if you wish to use other locations for interrupt software.

The *restart* instruction 3X7 is "jammed" into the 8080A chip only during an interrupt. As with other inputs such as memory read and accumulator I/O input, the data for the RST X instruction byte must be gated onto the 8080A bus at the proper time. An additional signal, *interrupt acknowledge* ($\overline{\text{INTA}}$ or $\overline{\text{IACK}}$), is provided to synchronize the input of the single-byte instruction. The interrupt acknowledge signal is used to strobe the instruction byte onto the data bus and into the 8080A chip. The instruction byte goes directly to the *instruction register* and not to any of the general-purpose registers. The interrupt signal flow is illustrated in Fig. 23-8. A standard three-state input port constructed from chips such as the 8212 buffer/latch or the DM8095 (74365) is used to input the interrupt instructions, using $\overline{\text{INTA}}$ rather than $\overline{\text{IN 006}}$ or some other input device select pulse as the strobe or enable pulse.

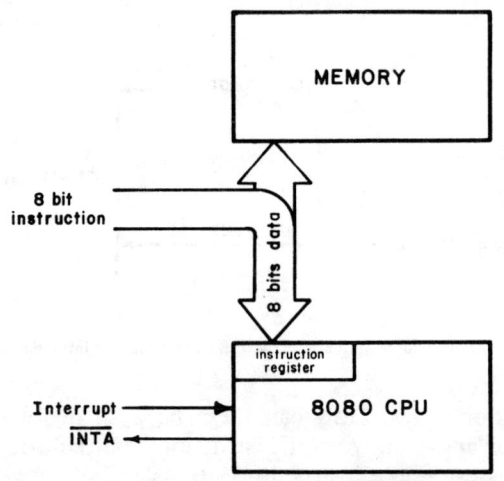

Fig. 23-8. Interrupt signal flow for a typical 8080A-based microcomputer.

## ENABLE AND DISABLE INTERRUPT: EI AND DI

The 8080A and other microprocessor chips have a very useful feature. The CPU has the ability to make itself immune to external interrupt requests. We may turn the interrupt on to allow them to be accepted, or we may turn it off and ignore them. There are times when we do not want the interrupt to be used at all. When the 8080A is started or reset, it turns the interrupt off. It is the responsibility of the programmer to enable the interrupt if the 8080A chip is to accept and service interrupts. This is done with a software instruction, *enable interrupt,* or EI. We can also perform the complementary operation, *disable interrupt,* or DI.

| 373 | EI | Enable the interrupt system and accept interrupts *after* execution of the *next* instruction. |
| 363 | DI | Disable the interrupt system and reject further interrupts. This takes place immediately. |

The interrupt capability can be enabled only under software control. Actually, we can think of the enable/disable process as an internal 8080A flag process. If the flag is enabled, interrupts are gated through to the 8080A chip's control section. When the flag is disabled, interrupts are blocked.

The interrupt input goes to another internal 8080A flag that can remember one interrupt event, or that can be triggered even if the interrupt is disabled. A third control output, the *interrupt enable* (INTE), which is Pin 16 on the 8080A chip, may be used to indicate to external devices and interfaces that the interrupt is enabled (logic 1) or disabled (logic 0). *The interrupt is always disabled after accepting an interrupt from an external device.*

## THIRD EXAMPLE: INTERRUPT-DRIVEN KEYBOARD INTERFACE

Let us take another look at the keyboard interface to see how an interrupt can be used in place of a flag. In some applications, where the keyboard and a large number of other I/O devices are connected to a computer, it may take a long time for the computer to get back and poll the keyboard. Characters may be missed or ignored if the software is not carefully written. The solution to this problem is the interrupt, which provides almost immediate servicing for external devices. To successfully use the interrupt, we need to connect the keyboard's VALID output pulse to the 8080A chip's interrupt input at Pin 14. The VALID output is a positive pulse, as required by the 8080A chip, so no inversion or buffering is required. We also need to provide the restart instruction byte, in this case, RST 5, which has an instruction code of 357. This instruction byte is sent directly to the 8080A chip's instruction register when the interrupt is acknowledged. With the aid of an 8212 buffer/latch chip, we can hardwire this instruction byte at an *interrupt instruction register* or *interrupt instruction port,* as shown in Fig. 23-9.

Let us now quickly review the operation of an interrupt. First, the interrupt flag must be enabled within the 8080A chip using the soft-

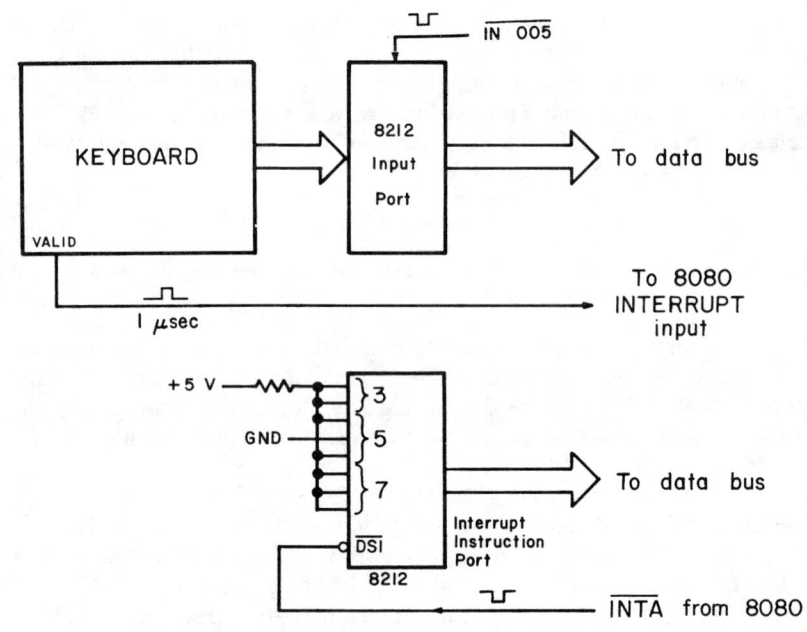

**Fig. 23-9.** Simplified vector interrupt circuit for the ASCII character keyboard. If there are other interrupting devices in the system, a NAND gate must be used prior to the INT input on the 8080A chip.

ware instruction, EI. Next, an external signal causes an interrupt and the 8080A acknowledges it by generating the interrupt acknowledge signal, $\overline{\text{INTA}}$, which is used to gate a single-byte restart instruction into the 8080A's instruction register. We use a restart instruction, specifically RST 5, to call the service subroutine at HI = 000 and LO = 050, the memory location where software for the keyboard input starts.

The keyboard interrupt acts to insert the keyboard input driver routine into the normal software flow. A typical example of how this occurs is shown below.

| LO Memory Address | Instruction Byte | Mnemonic | Description |
|---|---|---|---|
| 000 | 061 | LXI SP | Load stack pointer with following two bytes |
| 001 | 000 | 000 | LO stack pointer byte |
| 002 | 004 | 004 | HI stack pointer byte |
| 003 | 373 | EI | Enable interrupt |
| 004 | — | MAIN | These steps comprise the MAIN TASK of the |
| 005 | — | TASK | program |
| · | · | · · · | |
| · | · | · · · | |
| · | · | · · · | |

322

| | | | |
|---|---|---|---|
| ⋮ | ⋮ | ⋮ | |
| 050 | 333 | IN | Input keyboard data from input port 005 |
| 051 | 005 | 005 | Device code 005 |
| 052 | — | | |
| 053 | — | OTHER INTERRUPT SERVICE SOFTWARE | |
| ⋮ | ⋮ | | |
| — | 311 | RET | Return from subroutine |

Why have the LXI SP and RET instructions been included? Remember that the restart instructions are single-byte call instructions that call subroutines at specific addresses. Thus, when such instructions are used, *return addresses* or *linking addresses* will still be stored on the stack upon their execution.

While the restart instructions are very useful for interrupts, they are still valid 8080A instructions for normal program use. If you wish to employ a subroutine at one of the vector addresses, HI = 000 and LO = 0X0, use a restart instruction to call it. The above program will work if you try it, but it will not respond to more than the first key closure or interrupt. Why only a single key closure? The reason is that whenever the 8080A's interrupt flag is enabled and it accepts an interrupt, the interrupt flag becomes immediately disabled from accepting further interrupts. This protects the interrupting device's software task from being re-interrupted immediately. The 8080A chip will not accept further interrupts until the interrupt flag is re-enabled with an interrupt enable, or EI, instruction. In the keyboard example, there is no enable interrupt instruction in MAIN TASK or in the keyboard subroutine, so the flag cannot be re-enabled.

To re-enable the interrupt flag, an enable interrupt instruction should be placed immediately before the RET instruction byte. Further interrupts are not accepted until the next software instruction after the EI instruction, i.e., the RET instruction, is executed. Thus, program control can at least return to MAIN TASK before the 8080A accepts another interrupt. Why is this important? If an interrupting device could interrupt immediately after the EI instruction, the 8080A chip would accept the interrupt and the return instruction would *not* be executed. If the 8080A chip allowed this to happen many times, it is possible that the stack would fill with return addresses (since they would not be popped off the stack and used by the return instructions). This is why it is important that interrupts be accepted only after execution of the next software instruction after EI. The execution of the return allows us to "clean out" the stack after an interrupt subroutine is finished.

```
┌──────────────┐
│ PUSH         │
│ Instructions │
└──────────────┘
┌──────────────┐
│ Interrupt    │
│ Service      │
│ Software     │
└──────────────┘
┌──────────────┐
│ POP          │
│ Instructions │
└──────────────┘
┌──────────────┐
│ EI           │
└──────────────┘
┌──────────────┐
│ RET          │
└──────────────┘
```

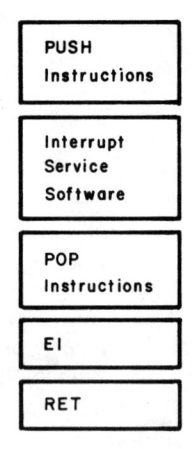

Fig. 23-10. A typical interrupt service subroutine.

We treat our vector subroutines as if they were normal subroutines, i.e., PUSH and POP instructions may be used to store and retrieve register data. A typical subroutine would appear as shown in Fig. 23-10. The first instructions, the PUSH instructions, save the microcomputer status. Near the end of the subroutine, the microcomputer status is popped back into the internal registers. Since there are only eight locations between the keyboard vector address, 050, and the next vector address, 060, how can all this software be used? If 060 is used as a vector address for another device, we certainly have a problem! We can circumvent the problem simply by placing a three-byte JMP instruction in locations 050, 051, and 052 that

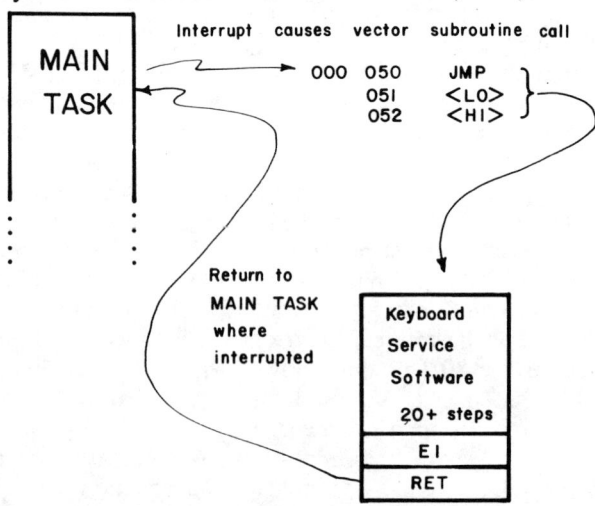

Fig. 23-11. Relationship between MAIN TASK, the vector subroutine jump, and the keyboard service software, which is located elsewhere in memory.

transfers program control to an area in memory where there is more room for the software. The penalty that we pay for doing this is a time delay of 10 clock states, i.e., 5 microseconds for a 2-MHz microcomputer and 13.33 microseconds for a 750-kHz microcomputer. The RET instruction at the end of the service routine still returns program control to the point where the MAIN TASK was interrupted and the RST 5 instruction is executed, as illustrated graphically in Fig. 23-11.

We could have made things considerably more complicated by including deferred interrupts and priority interrupts, but these become complex subjects that are beyond the scope of our simple keyboard example.

## PRIORITY INTERRUPTS

*Priority interrupts* are interrupts that are ordered in importance so that some interrupting devices take precedence over others. When a number of interrupts occur at the same time, or when this possibility exists, we need some method to determine which device should be serviced first. A priority must be established. The easiest way to do so is to poll the interrupting devices and, with the aid of software,

| LO Memory Address | Instruction Byte | Mnemonic | Description |
|---|---|---|---|
| 050 | 333 | POLL: IN | Input status bits |
| 051 | 057 | 057 | Device code 057, for the sense register |
| 052 | 057 | CMA | Complement the status bits in the accumulator (1 → 0 and 0 → 1) |
| 053 | 346 | ANI | Mask out all bits except bits D0, D1, and D2 |
| 054 | 007 | 007 | Mask byte |
| 055 | 037 | RAR | Rotate bit D0 into the carry flag |
| 056 | 332 | JC | If carry flag is at logic 1, jump to the cassette service routine CASSVC |
| 057 | 100 | CASSVC | LO address byte of CASSVC |
| 060 | 003 | — | HI address byte of CASSVC |
| 061 | 037 | RAR | Rotate original input bit D1 into the carry flag |
| 062 | 332 | JC | If carry flag is at logic 1, jump to the keyboard service routine KBRD |
| 063 | 200 | KBRD | LO address byte of KBRD |
| 064 | 003 | — | HI address byte of KBRD |
| 065 | 037 | RAR | Rotate original input bit D2 into the carry flag |
| 066 | 332 | JC | If carry flag is at logic 1, jump to the one-hour clock service routine CLOCK |
| 067 | 300 | CLOCK | LO address byte of CLOCK |
| 070 | 003 | — | HI address byte of CLOCK |
| 071 | 166 | HLT | Halt. If you got to this point, the program was interrupted, but it was not by one of the above three devices. |

determine which devices should be serviced and in what order. In the circuit shown in Fig. 23-12, only three interrupts are shown for clarity, but others could easily be added. An interrupt occurs whenever one of the flag flip-flops is set by a pulse applied at its clock

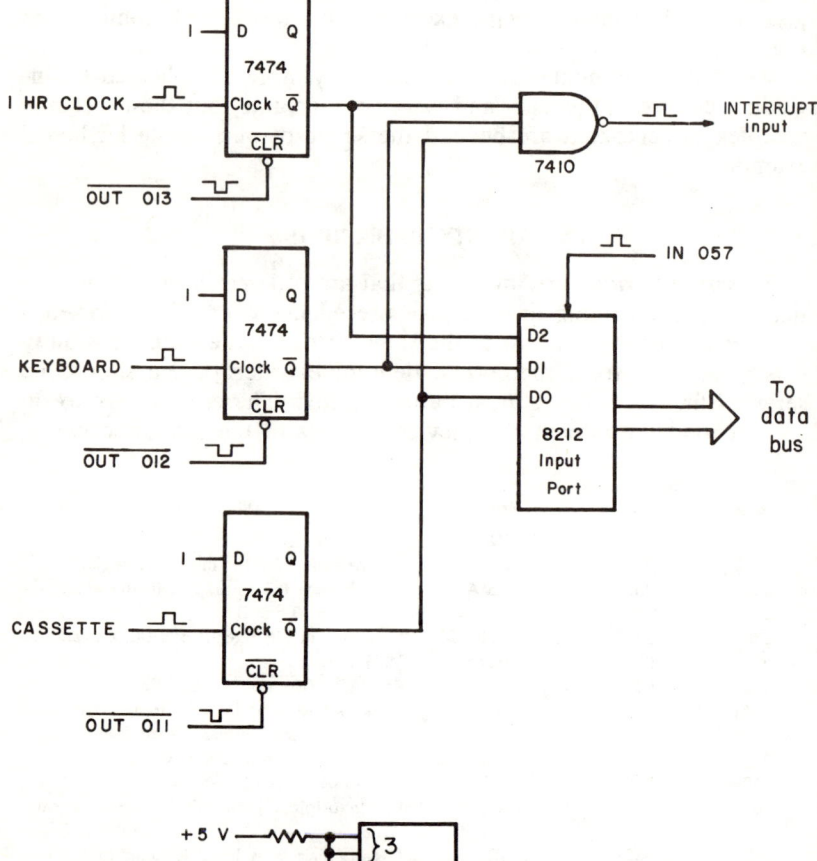

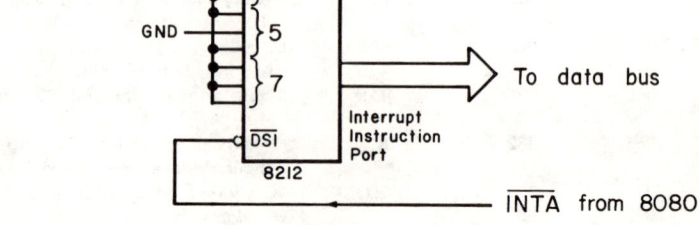

Fig. 23-12. Polled interrupt circuit that consists of three interrupt devices and a vector RST input.

input. When the interrupt occurs, the 8080A chip generates an $\overline{\text{INTA}}$ pulse and inputs the restart instruction code, 357, that is prewired at the 8212 interrupt instruction port. The 357 instruction causes a vector to the memory address HI = 000 and LO = 050, where the software for polling the interrupting devices starts. The vectoring and restart instructions should be well understood at this point. We will now discuss the polling routine.

Each flag bit is input as a logic 1 if service is *not* needed and as a logic 0 if service *is* needed. The three-input NAND gate provides a logic 1 to the 8080A microcomputer when any device generates an interrupt; this logic state is input to Pin 14 on the 8080A chip. Our priority is set up so that the cassette is highest (high-speed device), the keyboard next (low-speed device), and the one-hour clock last (extremely slow device). As input data to the accumulator, we would rather have a logic 1 if service is needed and a logic 0 if service is not needed. Our first program step would, therefore, be to invert, or complement, the input flag data with the use of a CMA instruction at LO = 052. This simple program step illustrates how easy it is to invert accumulator data and how easy it is to eliminate three 7404 inverters or else eliminate the need to rewire the hardware so that the Q output, rather than the $\overline{Q}$ output, is input to the 7410 gate.

After the CMA instruction, we mask out all other device bits except Bits D0, D1, and D2. We then proceed to rotate these bits into the Carry and to test them for a logic 1 state, which indicates that a specific device has generated an interrupt. Each service routine—CASSVC, KBRD, and CLOCK—is very similar to the interrupt service routine shown in Fig. 23-10, and ends with enable interrupt (EI) and return (RET) instructions.

Some additional comments about the polling routine are in order. Although the polling routine runs through vector addresses 060 and 070, in this case, it is not an error since we have no other interrupts that use them. We would probably start our polling routine with a PUSH PSW instruction, since *we do not know for what purpose MAIN TASK used the accumulator and flags when it was interrupted.* If we used PUSH PSW, each service routine would require a POP PSW immediately before the enable interrupt instruction. The first thing that we would do in each *service routine* is to clear the flag associated with the interrupting device. OUT instructions work well to generate pulses that clear the flip-flops, as shown in Fig. 23-12. Thus, an OUT 011 instruction clears the cassette flag, an OUT 012 clears the keyboard flag, and an OUT 013 clears the one-hour clock flag. Other polling software schemes and other bit testing methods work equally well, but the one given above is simple and effective.

The one-hour clock raises an important question. Why would you build an external one-hour hardware clock when software can do it under 50 bytes? The answer depends on how you use your microcomputer. If the microcomputer can just sit and perform the one-hour software loop, or if you are using interrupts and can tolerate error, you employ software. If you need an exact time and are using interrupts, you employ hardware. How do you reach this decision? When you use interrupts, you interject additional software into the MAIN TASK program flow. This all takes time since not only must the software check the interrupting device, but it must also service it. If you interrupt the one-hour software routine five times with a device service routine that takes two minutes to execute, you have really taken a total of one hour and ten minutes to reach your goal, i.e., the one-hour software operations are suspended when you interrupt and perform another task. Real time marches on while the software time is suspended. The one-hour external clock should be called a *real-time clock,* since it keeps real time, *not* computer software time!

## HARDWARE PRIORITY INTERRUPTS

Besides polled interrupts, interrupts may be also assigned a priority using hardware. This type of priority interrupt is important whenever a number of interrupting devices, all requiring fast service, are connected to a microcomputer. Each device generates its own restart instruction, RST X, which when input causes an immediate vector to memory location HI = 000 and LO = 0X0. Priority is assigned through the use of a 74148 priority encoder chip, which accepts up to eight flag inputs, each at logic 0 if an interrupt condition exists for each device, and *outputs the three-bit binary code for the highest numbered input that is at logic 0.* A truth table and chip diagram are shown in Fig. 23-13. The 74148 chip is used in conjunction with a regular interrupt instruction port in priority interrupt hardware, as is shown in Fig. 23-14.

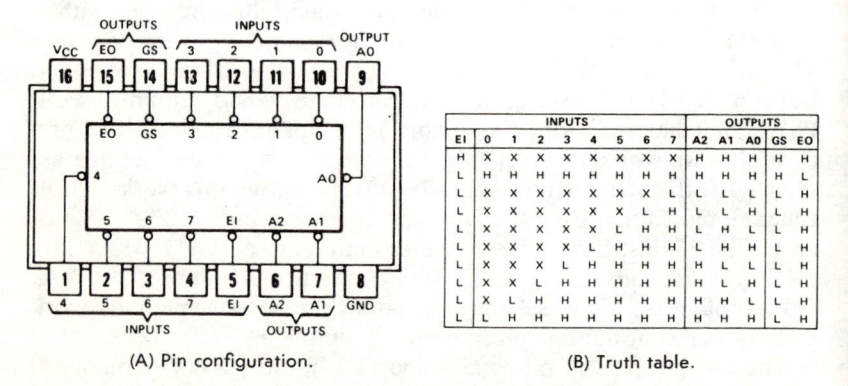

(A) Pin configuration.

| | INPUTS | | | | | | | | OUTPUTS | | | | |
|---|---|---|---|---|---|---|---|---|---|---|---|---|---|
| EI | 0 | 1 | 2 | 3 | 4 | 5 | 6 | 7 | A2 | A1 | A0 | GS | EO |
| H | X | X | X | X | X | X | X | X | H | H | H | H | H |
| L | H | H | H | H | H | H | H | H | H | H | H | H | L |
| L | X | X | X | X | X | X | X | L | L | L | L | L | H |
| L | X | X | X | X | X | X | L | H | L | L | H | L | H |
| L | X | X | X | X | X | L | H | H | L | H | L | L | H |
| L | X | X | X | X | L | H | H | H | L | H | H | L | H |
| L | X | X | X | L | H | H | H | H | H | L | L | L | H |
| L | X | X | L | H | H | H | H | H | H | L | H | L | H |
| L | X | L | H | H | H | H | H | H | H | H | L | L | H |
| L | L | H | H | H | H | H | H | H | H | H | H | L | H |

(B) Truth table.

**Fig. 23-13. The 74148 8-line-to-3-line priority encoder chip.**

The circuit of Fig. 23-14 generates eight different vector restart instructions, RST X, that have the priority $7 > 6 > 5 > 4 > 3 > 2 > 1 > 0$. If simultaneous interrupt requests are generated by Device 5 and Device 7, Device 7 has the highest priority and the 74148 chip and inverters in Fig. 23-14 supply a "7" for the 3X7 instruction. This vectors the microcomputer to memory location HI = 000 and LO = 070. While we have an RST 0 vector available, we do not often use it since its only effect is to reset the program counter and start the MAIN TASK program again.

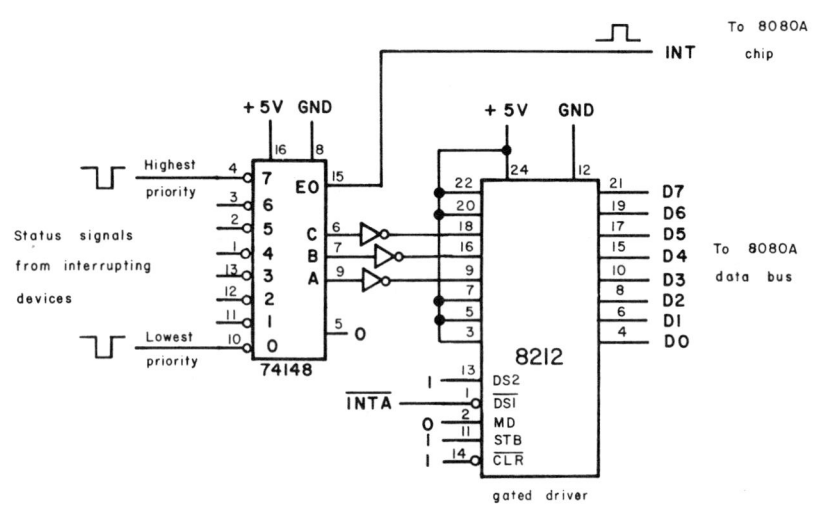

**Fig. 23-14. Hardware priority interrupt circuit.**

The necessary flags and flag setting or clearing lines are not shown in Fig. 23-14 for clarity. Keep in mind, however, that *a flag should be used for each interrupting device.* Additional hardware refinements could be added to the circuit to make it more efficient and effective. These would include an additional decoder to generate the flag clearing pulse without the need for an OUT instruction, and a mask register so that various devices could be masked on or off in external hardware. Such additions are shown in Fig. 23-15, which is a very sophisticated priority interrupt scheme that allows great flexibility in the use of vectored interrupts with an 8080A-based microcomputer.

In writing software, you must decide which devices are to be allowed interrupts and which are not. A mask bit pattern is developed in which devices that are allowed to interrupt are assigned a logic 1 and devices that are not allowed to interrupt are assigned a logic 0. The 8-bit mask pattern is placed in the accumulator and output to the two 7475 latches in Fig. 23-15. Bit position D7 corresponds to interrupt device 7, which has the highest priority and causes a vector to memory address $HI = 000$ and $LO = 070$. Active devices that are masked off use a sense register to request service. The mask can be changed under software control to achieve great flexibility in the use of interrupts.

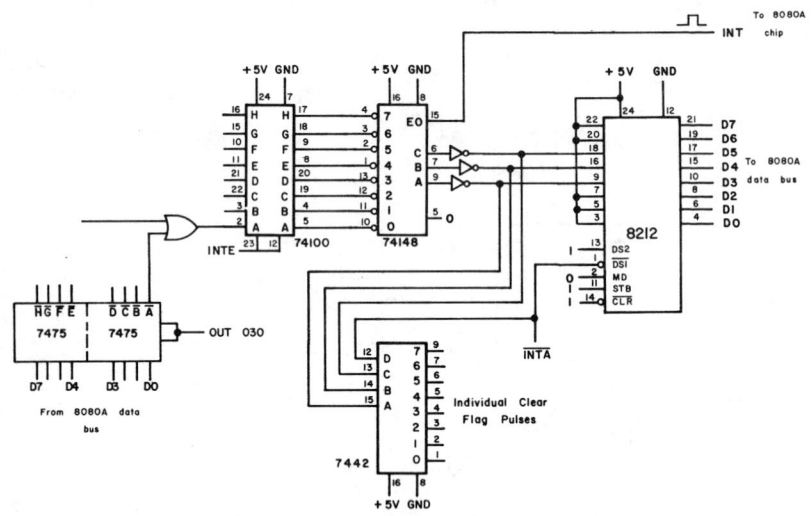

Fig. 23-15. A sophisticated priority interrupt circuit.

In Fig. 23-15, interrupt requests are gated with OR gates (one is shown) and nonmasked interrupt requests are passed through to the 74100 latch. Whenever the interrupt enabled output, INTE, from the 8080A chip indicates that interrupts will be accepted, the 74100 is "open" and passes interrupt requests through to the 74148 priority encoder. The priority encoder and interrupt instruction port have been discussed previously. When the interrupt is received, the 8080A chip disables its internal interrupt enable flip-flop and the INTE output goes to logic 0, thus "closing" the 74100 latch and latching any interrupt requests present at the inputs. The $\overline{\text{INTA}}$ control signal not only inputs the RST X instruction, it also pulses the 7442 decoder in the circuit of Fig. 23-15 to generate an interrupt flag clear pulse, which is routed back to the individual interrupt request flip-flop associated with the interrupting device. Many other interrupt schemes may be used, including the Intel 8214 interrupt controller chip and the 8259 programmable interrupt controller chip. Finally, keep in mind that interrupts, while permitting fast response to external events or demands for service, also can present problems.

## INTERRUPT SOFTWARE

Let us now consider the software necessary to serve some of our interrupt needs. Assume that we have only two devices, Device 7, which has the highest priority, and a low priority device, Device 2. Each has its own restart instruction that causes a vector to 000 070 or 000 020, respectively. We will further assume that the high priority device interrupts on a regular basis and that it is quickly serviced with its software service routine. Device 2, the low priority device, interrupts on an irregular schedule and takes some time to service. Perhaps Device 2 is another microcomputer that is dumping blocks of data. When not interrupted by these devices, the microcomputer will always be running the MAIN TASK software. Finally, we will assume that MAIN TASK initially assigns a stack pointer through the LXI SP <B2> <B3> instruction and also enables the interrupt flag.

Since our interrupts can occur at any time, we need both PUSH and POP instructions in the interrupt service routines, an example of which has been previously given in Fig. 23-10. These instructions will save and restore any registers that are altered in the service routines.

The execution of the software can be graphically represented by a *time line,* as shown in Fig. 23-16. Interrupts by the HIGH and LOW priority devices are denoted by the symbols, * and $\triangle$, respectively. Notice that the HIGH priority device has interrupted MAIN TASK four times and that the LOW priority device has interrupted only once. The HIGH priority interrupts on a regular basis, as shown by the spacing on the MAIN TASK time line. The heavy line indicates when the interrupt is enabled. The actual time line is deceptive since only the time spent in MAIN TASK is shown. It is more correct to show the real time spent in both MAIN TASK and in the subroutines, as we have done in Fig. 23-17.

In Fig. 23-17, the MAIN TASK starts operating and is then interrupted by the HIGH priority device. After executing the HIGH priority device service subroutine, control is returned to MAIN TASK, which is interrupted by the LOW priority device later on the time line. Control is eventually returned to MAIN TASK, which is then interrupted at repeated intervals by the HIGH priority device. *Note that it takes considerably longer to reach the point # in MAIN TASK when we keep interrupting it.* During a critical timing period, this could be disastrous if we are relying upon software timing loops.

Since the HIGH priority device interrupts on a regular basis, it probably tried to interrupt during the time that the microcomputer was working on the interrupt service software for the LOW priority device. If HIGH has higher priority, why couldn't it interrupt the

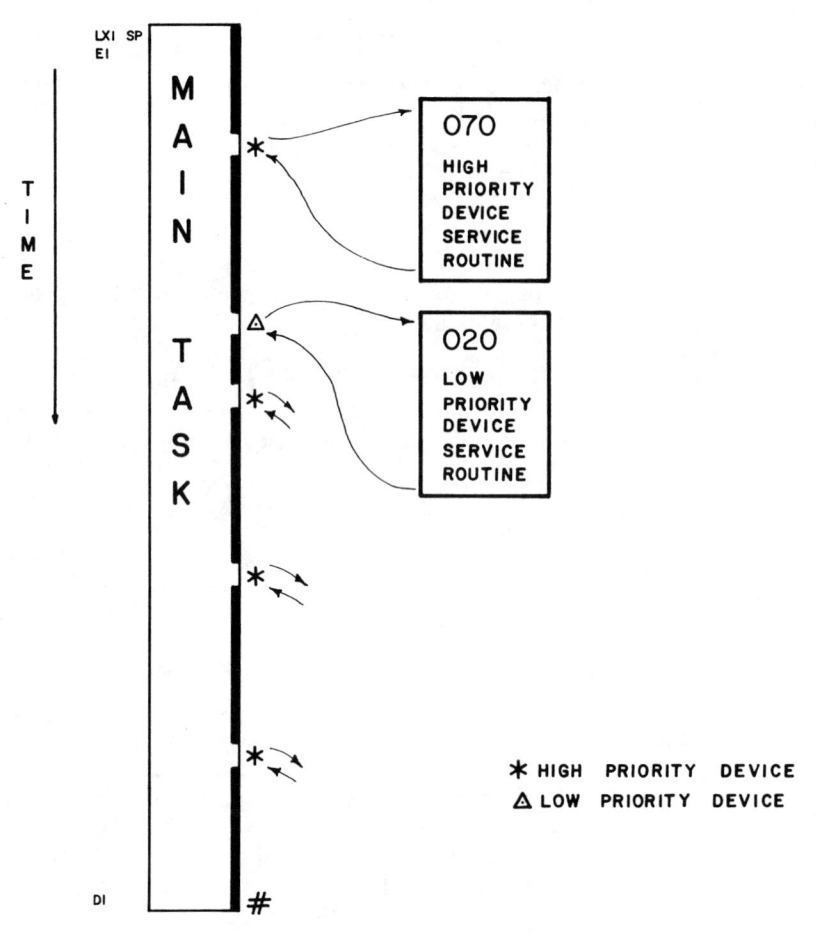

Fig. 23-16. Program execution time line.

LOW device software? The answer is obvious. *The interrupt flag was not enabled during the execution of the LOW priority device service subroutine.* Our first attempt at writing the interrupt service software did not take this possibility into account. Data or signals from the HIGH device were lost during this time. To solve this problem, we can correct our software by placing the enable interrupt instruction at the start of the LOW priority interrupt service subroutine rather than at the end. We can also design hardware to store data or signals associated with a missed interrupt.

By moving the enable interrupt instruction, EI, to the beginning of the LOW priority device service subroutine, we may encounter a new problem—a chopped-up LOW priority device software flow,

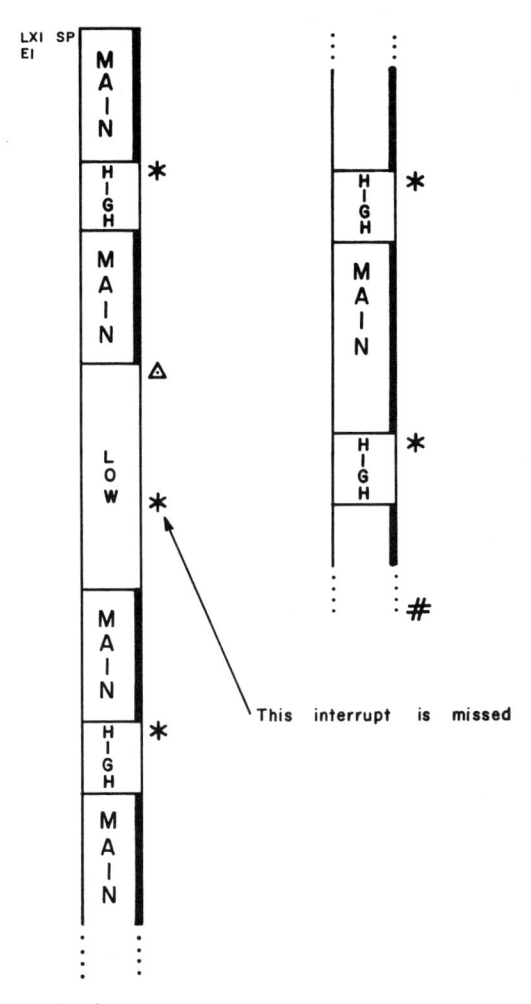

Fig. 23-17. Time line for MAIN TASK and both the LOW and HIGH priority device service subroutines.

as illustrated in Fig. 23-18. To emphasize the point, we have assumed that the HIGH priority device interrupts the LOW priority device service software twice, thus chopping the LOW software into three pieces. With the LOW priority device software so split up, we must inquire whether we are able to complete the LOW software *before the LOW priority device generates a new interrupt.* It is entirely possible for the LOW priority device to interrupt the microcomputer while it is still trying to service the last interrupt request from the LOW device. While the interrupt response is fast, the actual execution time may be much slower than the time required for a single pass through the interrupt service software. This is because we can interrupt our interrupts. Such considerations should give you a good idea of the care needed when using priority interrupts. It is very easy for a microcomputer to become *interrupt bound,* i.e., it spends all of its time checking and servicing interrupts and has no time left for its MAIN TASK software.

In our software, we may wish to prevent interrupts from taking place because of sensitive timing software or complex time-dependent tasks or calculations. The disable interrupt instruction allows the mi-

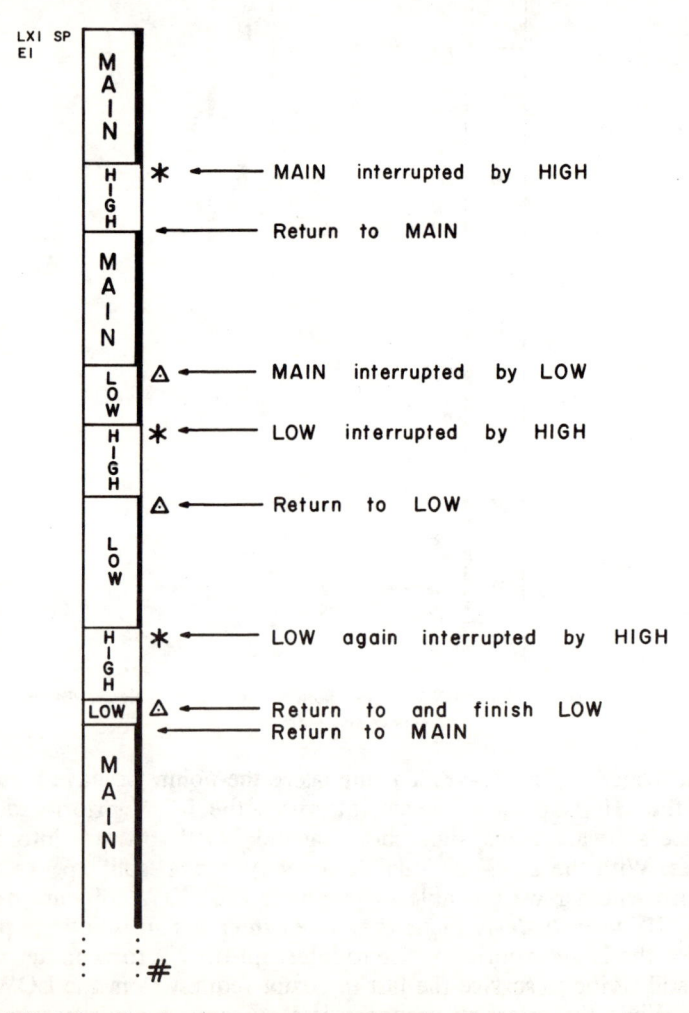

```
LXI SP
EI
```

```
MAIN
```
∗ ⟵ MAIN interrupted by HIGH
HIGH
⟵ Return to MAIN

MAIN

△ ⟵ MAIN interrupted by LOW
LOW
∗ ⟵ LOW interrupted by HIGH
HIGH
△ ⟵ Return to LOW

LOW

∗ ⟵ LOW again interrupted by HIGH
HIGH
△ ⟵ Return to and finish LOW
LOW
⟵ Return to MAIN

MAIN

⋮ #

**Fig. 23-18. A time line that demonstrates the interrupting of an interrupt service routine.**

crocomputer to operate under such conditions, insensitive to external interrupts. In our previous example, we could have included such a section in MAIN TASK when we needed to be immune from interrupts. We can always disable the interrupt flag and later re-enable it when we have completed a sensitive task. However, during the time that the interrupt flag is disabled, *we may lose signals or data that an interrupting device may need to input into the microcomputer.* Such a situation is represented in Fig. 23-19. Unless we provide some type of complex hardware back-up, such data is lost! We do not know exactly when an external device may interrupt MAIN TASK. Therefore, we cannot be sure that such an interrupt will not be during the period when the interrupt flag is disabled. How do we circumvent this problem? It is not easy to do, which is another reason why we must use a great deal of caution when we use interrupts.

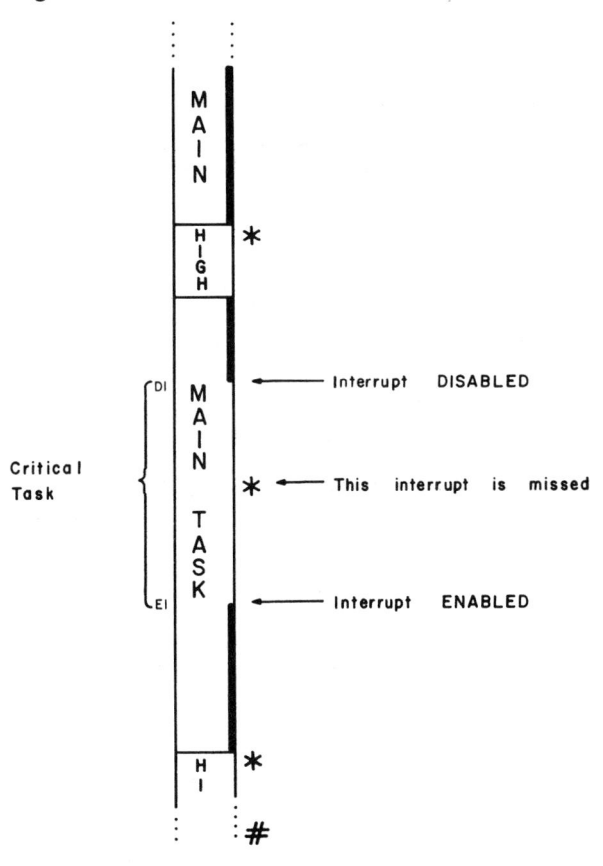

**Fig. 23-19.** Time line that uses DI and EI instructions to permit a critical task to be performed. However, an interrupt is missed while the critical task is being executed.

Another type of interrupt which may be of interest, although not generally used with an 8080A microcomputer, is a *time-oriented interrupt*. Only one interrupt is used, a clock. The clock interrupts every 10 milliseconds, or other reasonable period of time. When interrupted, the microcomputer uses a look-up table to determine which devices to *check* to see if they need service. Some devices are always checked, while other slower devices might be checked once every 1000 times the clock interrupt occurs. This is a good alternative interrupt technique, but it requires considerable amounts of software to work well.

The newer 8080A-type microprocessor chips allow multibyte instructions to be input during an interrupt, so that a complete three-instruction-byte call or jump could be inserted, thus doing away with the vector locations and providing much greater flexibility in both hardware and software. The key to multibyte "jammed" instructions is the 8228 controller chip, which has the capability to generate three $\overline{\text{INTA}}$ control signals in succession in response to an interrupt request. These three signals are used by hardware to successively jam the three instruction bytes of a call or jump instruction. The Intel 8259 programmable interrupt controller chip operates in conjunction with the 8228 chip to allow you to perform direct calls to interrupt service subroutines. If your 8080A-based microcomputer does not contain an 8228 chip, you will not be able to use the 8259, which is a complex device that is not for the beginner.

Some final notes of caution. Interrupts are difficult to debug. They can occur at almost any time, i.e., they occur asynchronously. Typical software debugging programs are not of much help. Special diagnostic software may need to be written to test interrupts in a specific application. When considering interrupts, try all other methods before settling on them. The time trying other methods will usually be well spent.

## INTRODUCTION TO THE EXPERIMENTS

The following experiments illustrate the use of flags and interrupts.

| *Experiment No.* | *Comments* |
| --- | --- |
| 1 | *A Simple Flag.* Demonstrates the operation of a simple external flag circuit constructed from a 7474 flip-flop and an 8095 (74365) three-state input buffer. |
| 2 | *Flag Response Time.* Illustrates the response of software to flags when the microcomputer has other tasks to perform. |

| | |
|---|---|
| 3 | *Nonideal Flags: Interfacing a Mechanical Switch.* Illustrates the operation of an external flag circuit that is connected to a single-pole single-throw (spst) switch, a nonideal mechanical device. |
| 4 | *Keyboard Characteristics of the MMD-1 Microcomputer.* Demonstrates how to use the keyboard flag, Bit D7, to signal that a key is pressed and data is ready to be input into the microcomputer. |
| 5 | *Simulation of Tank Liquid-Level Sensing.* Implements the hardware and software necessary to simulate the liquid-level-sensing example discussed in the text. |
| 6 | *Restart Instructions.* Illustrates the software characeristics of the 8080A restart instructions, RST X. |
| 7 | *A Simple Interrupt Instruction Register.* Illustrates the behavior of an instruction register constructed from an 8212 buffer/latch. |
| 8 | *Jamming a Restart Instruction.* Demonstrates the consequences of jamming a restart instruction into the 8212 instruction register wired in Experiment No. 7. |
| 9 | *Interrupt Response Time.* Illustrates the response of an 8080 system to interrupts when the microcomputer has other tasks to perform. |
| 10 | *Simple Priority Interrupts.* Illustrates the implementation of a simple priority interrupt scheme that includes both a low priority device and a high priority device. |
| 11 | *Priority Interrupt Timing.* Illustrates the timing relationships between HIGH and LOW priority devices and how the priority is assigned. |
| 12 | *Simultaneous Interrupts.* Illustrates the operation of simultaneous interrupts. |

## EXPERIMENT NO. 1
## A SIMPLE FLAG

**Purpose**

The purpose of this experiment is to demonstrate the operation of a simple external flag.

## Pin Configurations of Integrated-Circuit Chips (Fig. 23-20)

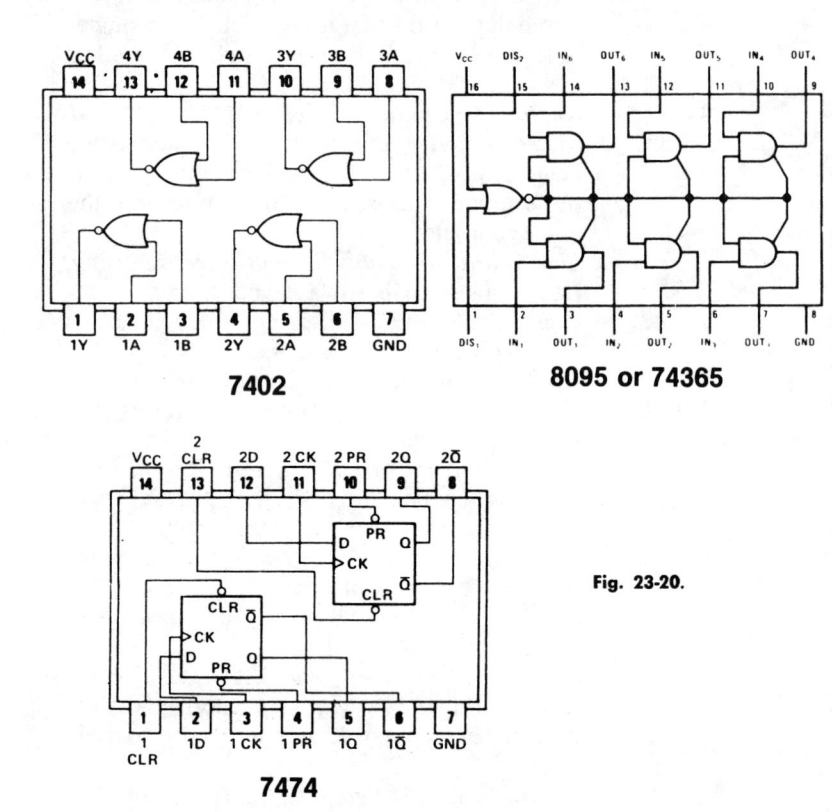

**7402**

**8095 or 74365**

**7474**

Fig. 23-20.

## Schematic Diagram of Circuit (Fig. 23-21)

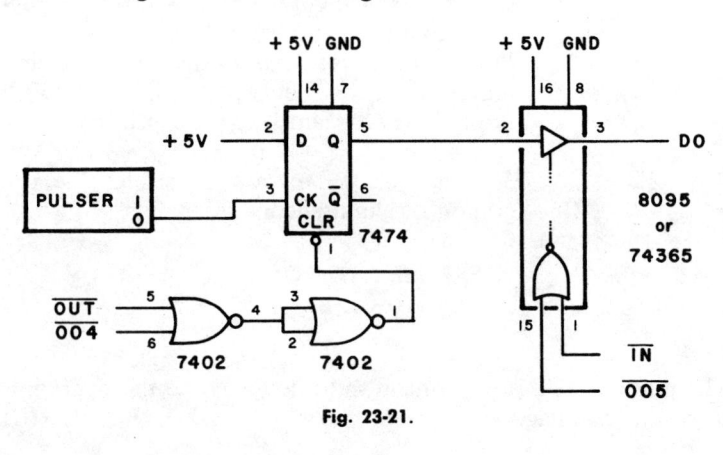

Fig. 23-21.

## Program

| Memory Address | Instruction Byte | | Mnemonic | Comments |
|---|---|---|---|---|
| 003 000 | 227 | | SUB A | /Clear A register (accumulator) |
| 003 001 | 107 | | MOV B,A | /Move A to B |
| 003 002 | 323 | | OUT | /Output contents of A to |
| 003 003 | 002 | | 002 | /output port 002 |
| 003 004 | 333 | INPUT, | IN | /Input data to A from |
| 003 005 | 005 | | 005 | /input port 005 |
| 003 006 | 037 | | RAR | /Rotate bit D0 to carry flag |
| 003 007 | 322 | | JNC | /Is CARRY = 0? If yes, jump to |
| 003 010 | 004 | | INPUT | /HI = 003 and LO = 004. If no, |
| 003 011 | 003 | | 0 | /continue to next instruction. |
| 003 012 | 323 | | OUT | /No, so output a device select pulse to |
| 003 013 | 004 | | 004 | /output port 004 |
| 003 014 | 170 | | MOV A,B | /Move B to A |
| 003 015 | 074 | | INR A | /Increment A |
| 003 016 | 107 | | MOV B,A | /Move A to B |
| 003 017 | 323 | | OUT | /Output contents of A to |
| 003 020 | 002 | | 002 | /output port 002 |
| 003 021 | 303 | | JMP | /Jump back to INPUT at |
| 003 022 | 004 | | INPUT | /LO = 004 and |
| 003 023 | 003 | | 0 | /HI = 003 |

## Step 1

Wire the digital circuit shown in Fig. 23-21. Make certain that +5 Volts and GROUND connections are made to all chips and to the power busses on the SK-10 breadboarding socket.

## Step 2

Enter the program into read/write memory starting at HI = 003 and LO = 000. Observe that the program format is different from that used in the text in this unit as well as in earlier units. This is the type of output that you would obtain from a commercial 8080A resident assembler such as the one available from Tychon, Inc. We will continue to use this output format in the following experiments so that you can get used to it. Note the use of the *delimiter*, /, which is a character that indicates the beginning of a comment.

The program will input the flag bit that we have wired, test the flag bit D0 to determine if it is a logic 1, and then increment the A register and output the data to Output Port 002.

### Step 3
Execute the program. What do you observe at Output Port 002?

The output port reading is 000, i.e., all eight LEDs are unlit.

### Step 4
Now depress and release the pulser. What changes do you observe at Output Port 002? Repeatedly press and release the pulser. What happens?

The clock pulse from the first pulser increments Output Port 002 by 1 so that the reading becomes 001. Additional clock pulses from the pulser continue to increment the output port.

### Step 5
If you make a mistake in wiring the circuit and wire the $\overline{Q}$ output (Pin 6) of the 7474 flip-flop to the 74365 chip, will the program operate correctly? Change the 7474 output connection to $\overline{Q}$ at Pin 6 and execute the program once more. What do you observe? Why?

With $\overline{Q}$ input into the microcomputer, we observed that all the lights on Output Port 002 appeared to be on. The reason is that the program detected that Bit D0 was continuously at logic 1. Owing to the nature of the program, the contents of register A were continuously incremented and output to Port 002.

## Step 6
Could you change the software to account for the error in wiring? If so, what changes would you make? Make these changes and execute the program once again. What do you observe?

To eliminate the effect of the error in wiring, the JNC instruction at 003 007 can be changed to a JC instruction, 332. When this change is made, the circuit and program now behave as described in Step 4.

## Step 7
Return the hardware and software to their original forms and continue to the next experiment.

## EXPERIMENT NO. 2
## FLAG RESPONSE TIME

### Purpose
The purpose of this experiment is to investigate the response of software to flags when the microcomputer has other tasks to perform.

### Schematic Diagram of Circuit
The circuit is identical to the circuit given in Experiment No. 1.

### Program

| Memory Address | Instruction Byte | | Mnemonic | Comments |
|---|---|---|---|---|
| 003 000 | | | | |
| • | | | | |
| • | | These program steps are the same as those given in Experiment No. 1. | | |
| • | | | | |
| 003 021 | 016 | | MVI C | /Load register C with the |
| 003 022 | 001 | | 001 | /data byte 001 |
| 003 023 | 315 | REPEAT, | CALL | /Call the KEX time delay subroutine, |
| 003 024 | 277 | | TIMEOUT | /TIMEOUT at LO = 277 and |
| 003 025 | 000 | | 0 | /HI = 000 |

| Memory Address | Instruction Byte | Mnemonic | Comments |
|---|---|---|---|
| 003 026 | 015 | DCR C | /Decrement register C |
| 003 027 | 302 | JNZ | /Is register C = 000? If not, jump |
| 003 030 | 023 | REPEAT | /to REPEAT at LO = 023 and |
| 003 031 | 003 | 0 | /HI = 003. Otherwise, continue. |
| 003 032 | 303 | JMP | /Yes, register C = 000. Jump back to |
| 003 033 | 004 | INPUT | /INPUT at LO = 004 and |
| 003 034 | 003 | 0 | /HI = 003 |

### Step 1

The hardware and software from the previous experiment will be used in this experiment. Some additional software steps have been added to keep the microcomputer busy for varying periods of time. The 10-millisecond KEX time-delay subroutine at 000 277 is used. A listing of TIMEOUT is provided at the end of this experiment.

In this experiment, we use the KEX stack area to save the return address for the CALL instruction. If you are not using KEX on an MMD-1 microcomputer, you will have to first establish a stack area. Use the LXI SP instruction to do so. Load the above program steps in read/write memory starting at 003 021.

### Step 2

Start execution of the program at 003 000. Press and release the pulser several times. Do you observe any difference between this experiment and the previous experiment where the count incremented for each pulser clock pulse?

We observed no difference.

### Step 3

The added software slows down the microcomputer by providing a 10-millisecond time-delay routine to execute. By setting register C to another value, you can cause the microcomputer to execute the time-delay routine many more times, thus slowing down the overall software loop even more.

Enter the following timing bytes, one at a time, into the program at memory location 003 022 and execute the program in each case. Test the influence of each timing byte by (a) applying several clock pulses from the pulser slowly, and (b) applying ten clock pulses from the pulser as fast as you can. Enter the number of counts that you observed in the chart below.

| Octal Timing Byte | Time Delay | Normal | Ten Fast Pulses |
|---|---|---|---|
| 012 | 100 ms | _____ | _____ |
| 024 | 200 ms | _____ | _____ |
| 062 | 500 ms | _____ | _____ |
| 144 | 1 s | _____ | _____ |
| 310 | 2 s | _____ | _____ |

When we applied clock pulses slowly to the flag, we observed normal behavior, i.e., the output port incremented once for each clock pulse. However, when we applied ten fast actuations of the pulser, we observed only five counted pulses with the 500-ms delay, three counted pulses with the 1-second delay, and only two counted pulses with the 2-second delay.

### Step 4

What do the results in Step 3 indicate about the use of flags when the microcomputer has other time-consuming tasks to perform?




Events or data may be lost since they are not sensed by the microcomputer when it is performing some other task.

### Step 5

Save your hardware and software for the following experiment.

### Listing of Subroutine TIMEOUT

With an 8080A-based microcomputer operating at 750 kHz, this time-delay routine will generate a delay of 10.0 milliseconds.

| Memory Address | Instruction Byte | | Mnemonic | Comments |
|---|---|---|---|---|
| 000 277 | 365 | TIMEOUT, | PUSH PSW | /Save accumulator and flags |
| 000 300 | 325 | | PUSH D | /Save register pair D |
| 000 301 | 021 | | LXI D | /Load D and E with value to be |
| 000 302 | 046 | | 046 | /decremented |
| 000 303 | 001 | | 001 | |
| 000 304 | 033 | MORE, | DCX D | /Decrement register pair D |
| 000 305 | 172 | | MOV A,D | /Move D to A |
| 000 306 | 263 | | ORA E | /OR E with A |

| Memory Address | Instruction Byte | Mnemonic | Comments |
|---|---|---|---|
| 000 307 | 302 | JNZ | /Is register A = 000? If not, jump |
| 000 310 | 304 | MORE | /to MORE at LO = 304 and |
| 000 311 | 000 | 0 | /HI = 000. Otherwise, continue to /next instruction. |
| 000 312 | 321 | POP D | /Restore register pair D |
| 000 313 | 361 | POP PSW | /Restore accumulator and flags |
| 000 314 | 311 | RET | /Return from subroutine TIMEOUT |

## EXPERIMENT NO. 3

## NONIDEAL FLAGS: INTERFACING A MECHANICAL SWITCH

### Purpose

The purpose of this experiment is to investigate the operation of an external flag circuit that is connected to a single-pole single-throw (spst) switch, a nonideal mechanical device.

### Schematic Diagram of Circuit

The circuit is identical to that given in Experiment No. 1 except for the pulser input to the 7474 flip-flop. In place of the pulser, use a single-pole single-throw (spst) mechanical switch wired as shown in Fig. 23-22.

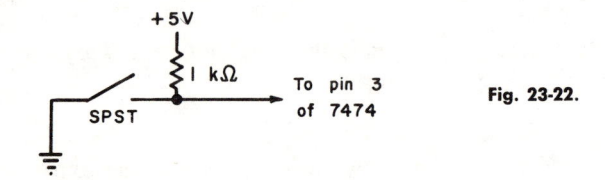

Fig. 23-22.

### Program

The program is identical to that given in Experiment No. 1. Make certain that the final jump instruction reads as follows:

| 003 021 | 303 | JMP | /Jump back to INPUT at |
|---|---|---|---|
| 003 022 | 004 | INPUT | /LO = 004 and |
| 003 023 | 003 | 0 | /HI = 003 |

### Step 1

The hardware and software from previous experiments will be used in this experiment as well. If they are not already set up, wire the circuit shown in Experiment No. 1. In place of the pulser, wire the spst switch circuit shown in Fig. 23-22 or else use a logic switch on the LR-2 or LR-25 Outboards.

### Step 2

The software used in this experiment is identical to that given in Experiment No. 1. Make certain that it is correctly loaded into read/write memory.

### Step 3

Execute the microcomputer program and actuate the single-pole single-throw (spst) nonideal mechanical switch (or equivalent wire circuit). Inspect the output at Output Port 000. Do you observe a single count, as in Experiment No. 1, or many counts? Why?

We observed many counts. This indicates that our spst switch is not an "ideal" switch, as is a "debounced" pulser. The difference between the two switches can be depicted as illustrated in Fig. 23-23. We observed many counts because the 7474 flag sensed each "bounce," or almost every one, as an individual switch closure.

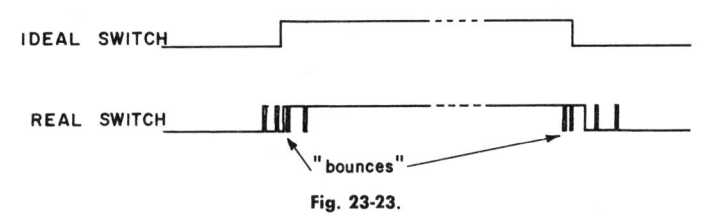

Fig. 23-23.

### Step 4

Actuate the spst switch ten times and observe the total number of counts. Repeat this process several times, being sure to restart the microcomputer before each trial. Summarize your results in the chart below, beside our results. Note that the number of counts is expressed in octal rather than decimal notation.

| Trial | Our results | Your results |
|---|---|---|
| 1 | 62 | _____ |
| 2 | 36 | _____ |
| 3 | 67 | _____ |
| 4 | 70 | _____ |
| 5 | 160 | _____ |

It is likely that the microcomputer loop was not fast enough to detect all the bounces of your imperfect mechanical switch, and the bounces are nonreproducible. While the bounces can be eliminated with hardware, as was done in the first experiment, they can also be eliminated using software. The following experiment investigates the keyboard interface and demonstrates how you use the KEX software to "filter out" the bounces.

### EXPERIMENT NO. 4

### KEYBOARD CHARACTERISTICS OF THE MMD-1 MICROCOMPUTER

**Purpose**

The purpose of this experiment is to demonstrate how to use the keyboard flag, Bit D7, to signal that a key is pressed and data is ready to be input into the 8080A microcomputer.

**Program No. 1**

| Memory Address | Instruction Byte | Mnemonic | | Comments |
|---|---|---|---|---|
| 003 100 | 333 | INPUT, | IN | /Input keyboard data from |
| 003 101 | 000 | | 000 | /input port 000 |
| 003 102 | 027 | | RAL | /Rotate bit D7 into carry flag |
| 003 103 | 322 | | JNC | /If CARRY is logic 0, jump to |
| 003 104 | 100 | | INPUT | /INPUT at LO = 100 and |
| 003 105 | 003 | | 0 | /HI = 003. Otherwise, continue to /next instruction. |
| 003 106 | 037 | | RAR | /If CARRY is logic 1, rotate data /back and |
| 003 107 | 323 | | OUT | /Output data to |
| 003 110 | 002 | | 002 | /output port 002 |
| 003 111 | 303 | | JMP | /Jump back to INPUT at |
| 003 112 | 100 | | INPUT | /LO = 100 and |
| 003 113 | 003 | | 0 | /HI = 003 and do it again. |

**Program No. 2**

| Memory Address | Instruction Byte | Mnemonic | | Comments |
|---|---|---|---|---|
| 003 200 | 315 | START, | CALL | /Call keyboard subroutine KBRD at |
| 003 201 | 315 | | KBRD | /LO = 315 and |
| 003 202 | 000 | | 0 | /HI = 000 |
| 003 203 | 323 | | OUT | /Output the keyboard data to |
| 003 204 | 002 | | 002 | /output port 002 |
| 003 205 | 303 | | JMP | /Jump back to START at |
| 003 206 | 200 | | START | /LO = 200 and |
| 003 207 | 003 | | 0 | /HI = 003 and do it again. |

**Step 1**

This experiment does not require an external interface circuit. You use the MMD-1 keyboard to generate the flag bit and keyboard data. Load program No. 1 into read/write memory starting at HI = 003 and LO = 100. Which bit in the accumulator will you use to signal the 8080A chip that a key is pressed?

You will use the keyboard flag, Bit D7, to signal the 8080A that a key is pressed and data is ready to be input.

## Step 2

Execute the program and note the four least significant bits at Output Port 002. Press each keyboard key in turn and list the 4-bit code that you observe under the "Step 2" heading. Do you observe a match between your code and the expected code?

| Key | Expected code | Your code Step 2 | Step 3 |
|---|---|---|---|
| 0 | 0000 | | |
| 1 | 0001 | | |
| 2 | 0010 | | |
| 3 | 0011 | | |
| 4 | 0100 | | |
| 5 | 0101 | | |
| 6 | 0110 | | |
| 7 | 0111 | | |
| H | 1100 | | |
| L | 1101 | | |
| G | 1011 | | |
| S | 1000 | | |
| A | 1110 | | |
| B | 1111 | | |
| C | 1010 | | |

We observed a match between our code for Step 2 and the expected code.

## Step 3

If your codes did not match our expected code, you may wish to use the KEX debounce and keyswitch filter subroutine, which is documented in the MMD-1 manual and also in the June, 1976 issue of *Radio-Electronics* magazine. Program No. 2 calls the keyboard input subroutine KBRD in KEX. Load Program No. 2 into read/write memory starting at HI = 003 and LO = 200 and execute it. Press each keyboard key in turn and note the least significant bits at Output Port 002. List these four bits under the Step 3 column in the table given in Step 2. Are the codes for Step 2 and Step 3 the same? If not, why not?

The codes for keys L, H, G, S, A, B, and C have been changed in a *look-up table* employed by the KEX software. Thus, they are not the same.

### Step 4

Can you suggest why this code translation of keys L through C might be useful?

It provides flexibility and allows us to redefine keys in software. KEX sets up a decimal keyboard if we want to use it that way. Note that the keys now go from 0 to 11 in sequence, skip 12, and finish up with 13, 14, and 15.

With the KEX EPROM in the system, the keyboard input routine at 000 315 may be called to input and encode the keyboard data. This keyboard routine uses a 10-millisecond delay subroutine at 000 277 that may also be called at any time. The delay routine is completely "transparent" and will not affect any flags or registers. It is listed at the end of Experiment No. 2 in this unit.

## EXPERIMENT NO. 5

### SIMULATION OF TANK LIQUID-LEVEL SENSING

#### Purpose

The purpose of this experiment is to implement the hardware and software necessary to simulate the liquid-level-sensing example discussed in the text.

#### Pin Configuration of Integrated-Circuit Chip (Fig. 23-24)

#### Schematic Diagram of Circuit (Fig. 23-25)

**Fig. 23-24.**

**8095 or 74365**

A = OVERFLOW
D = FULL / EMPTY

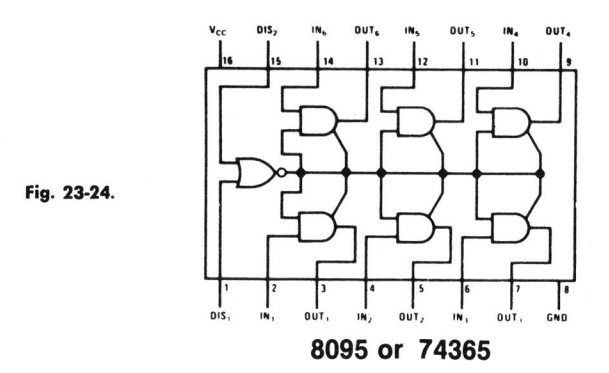

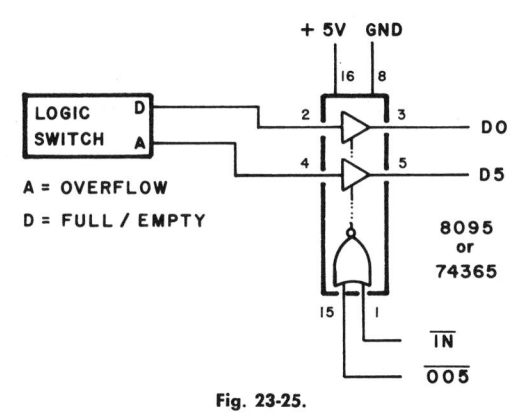

**Fig. 23-25.**

## Program

| Memory Address | Instruction Byte | Mnemonic | | Comments |
|---|---|---|---|---|
| 003 000 | 333 | START, | IN | /Input flag data from |
| 003 001 | 005 | | 005 | /input device 005 |
| 003 002 | 346 | | ANI | /Mask out all bits except bit D5 |
| 003 003 | 040 | | 040 | /Mask byte = 00100000 |
| 003 004 | 302 | | JNZ | /If result is 040, jump to ALARM at |
| 003 005 | 100 | | ALARM | /LO = 100 and |
| 003 006 | 003 | | 0 | /HI = 003. Otherwise, continue to |
| | | | | /next instruction. |
| 003 007 | 333 | | IN | /Overflow is OK. Input flag data from |
| 003 010 | 005 | | 005 | /input device 005 once again |
| 003 011 | 017 | | RRC | /Rotate bit D0 into carry flag |
| 003 012 | 332 | | JC | /If CARRY = 1, jump to FULL at |
| 003 013 | 024 | | FULL | /LO = 024 and |
| 003 014 | 003 | | 0 | /HI = 003. Otherwise, continue to |
| | | | | /next instruction. |
| 003 015 | 076 | | MVI A | /If CARRY = 0, the tank is not full. |
| 003 016 | 001 | | 001 | /Load register A with byte 001. |

| Memory Address | Instruction Byte | | Mnemonic | Comments |
|---|---|---|---|---|
| 003 017 | 323 | | OUT | /Output byte to |
| 003 020 | 000 | | 000 | /output device 000 |
| 003 021 | 303 | | JMP | /Check overflow and full/empty flags |
| 003 022 | 000 | | START | /again by jumping to LO = 000 and |
| 003 023 | 003 | | 0 | /HI = 003. |
| 003 024 | 227 | FULL, | SUB A | /The tank is full. Clear register A. |
| 003 025 | 323 | | OUT | /Output register A byte to |
| 003 026 | 000 | | 000 | /output device 000 |
| 003 027 | 303 | | JMP | /Check overflow and full/empty flags |
| 003 030 | 000 | | START | /again by jumping to START at LO = 000 |
| 003 031 | 003 | | 0 | /and HI = 003. |
| 003 100 | 166 | ALARM, | HLT | /Stop operation |

### Step 1

Wire the circuit shown in the schematic diagram of Fig. 23-25. Logic switch A is the overflow flag and logic switch D is the full/empty flag.

### Step 2

Load the program into read/write memory starting at HI = 003 and LO = 000.

### Step 3

With both logic switches set to logic 0, execute the program. You will simulate the valve with Bit D0 on Output Port 000. If Bit D0 is logic 1, the valve is open; if Bit D0 is logic 0, the valve is closed. What do you observe when the software is started? Why?




We found that Bit D0 was at logic 1, indicating that the valve was open. This is because the empty/full flag is at logic 0, indicating that the tank is not full (or perhaps empty).

### Step 4

Change logic switch D to logic 1 indicating that the tank is FULL. What happens to the valve bit D0? Why?

Valve bit D0 becomes logic 0, or off, indicating that the valve is now closed. The fluid level in the tank has tripped the full switch.

### Step 5

Switch logic switch A to a logic 1, indicating an overflow condition. What happens? Why?

Nothing happens. The microcomputer executes a halt instruction at memory location 003 100, the location of the ALARM routine.

### Step 6

Can you suggest a useful ALARM routine for your microcomputer? You may wish to test several different short software ALARM routines. Use the space below.

In developing an ALARM routine, we decided that the first thing we should do is to turn off the valve and then output an alarm condition at Port 001. The program is as follows:

| Memory Address | Instruction Byte | Mnemonic | Comments |
| --- | --- | --- | --- |
| 003 100 | 227 | ALARM, SUB A | /Clear register A |
| 003 101 | 323 | OUT | /Output register A byte to |
| 003 102 | 000 | 000 | /output device 000 |
| 003 103 | 057 | CMA | /Complement register A |
| 003 104 | 323 | OUT | /Output register A contents to |
| 003 105 | 001 | 001 | /output port 001 and |
| 003 106 | 166 | HLT | /Stop operation |

You should note in this experiment that you have not used flip-flops as flags. This is a valid procedure in this case since the overflow switch and the FULL/EMPTY switch will maintain their respective states until we take action. A flip-flop flag is generally used in those cases when the device requesting the attention of the microcomputer is generating a short pulse rather than a level.

## EXPERIMENT NO. 6

### RESTART INSTRUCTIONS

### Purpose

The purpose of this experiment is to investigate the software characteristics of the 8080A restart instructions, RST X.

### Program

| Memory Address | Instruction Byte | Mnemonic | Comments |
|---|---|---|---|
| 003 000 | 061 | LXI SP | /Load stack pointer with |
| 003 001 | 200 | 200 | /LO address byte and |
| 003 002 | 003 | 003 | /HI addressed byte |
| 003 003 | 357 | RST 5 | /Call subroutine at 000 050 |
| 003 004 | 166 | HLT | /Halt |
| 003 050 | 311 | RET | /Return from subroutine |

### Step 1

Enter the above program into read/write memory starting at 003 000. If you are executing this program on the MMD-1 microcomputer, please keep the following in mind:

> An RST X instruction calls the subroutine at 000 0X0, but the KEX monitor causes program control to jump to 003 0X0 for restart instructions RST 1 through RST 6. Restart instructions RST 0 and RST 7 are used by the KEX monitor.

You can confirm this fact by examining the contents of the EPROM memory locations at 000 010, 000 020, 000 030, 000 040, 000 050, and 000 060.

What instruction bytes do you find in KEX memory locations 000 050 through 000 052?

You will find the following instruction bytes:

```
000 050    303    JMP    /Jump to
000 051    050    050    /LO = 050 and
000 052    003    003    /HI = 003
```

## Step 2

Execute the above program. Observe that the stack pointer is set at address 003 200, so that the stack itself will start at one less than this memory location, or 003 177. After you execute the program, examine the contents of the two stack locations 003 176 and 003 177 and list the contents in the space below.

| Stack location | Contents |
|---|---|
| 003 176 | _____ |
| 003 177 | _____ |

Are these two bytes consistent with what you would expect for the execution of a CALL instruction?

We observed the following on the stack:

| Stack location | Contents |
|---|---|
| 003 176 | 004 |
| 003 177 | 003 |

The two bytes correspond to the memory address, 003 004, which is the address of the HALT instruction. This is exactly what we would expect for a call. You should note that the stack pointer will again point to memory address 003 200 after the program is executed. Why?

Not only did the program execute an RST 5 instruction, it also executed an RET instruction that popped the two bytes off the stack. Though they were popped off the stack, the original values remained in read/write memory. Remember that the stack pointer is an internal 8080A register and cannot be directly examined.

**Step 3**

The software routine at 003 050 can perform other tasks as well. Change the program steps to the following:

| Memory Address | Instruction Byte | Mnemonic | Comments |
|---|---|---|---|
| 003 050 | 170 | MOV A,B | /Move B to A |
| 003 051 | 074 | INR A | /Increment A |
| 003 052 | 107 | MOV B,A | /Move A back to B |
| 003 053 | 323 | OUT | /Output contents of A to |
| 003 054 | 000 | 000 | /output port 000 |
| 003 055 | 311 | RET | /Return from subroutine |

On your MMD-1 microcomputer, alternately press Keys RESET and G. What do you observe? Why?

We observed that Output Port 000 incremented its count for each RESET/G cycle. The increment software was called by the RST 5 subroutine.

**Step 4**

After several increments, examine the stack locations 003 176 and 003 177 once again. Are the stack byte values any different from those that you observed in Step 2? How do you explain this result in view of the fact that you have executed the restart subroutine several times?

You should observe no change in the stack bytes, which still contain the address of the HALT instruction, 003 004. Each time that you pushed an address on the stack as a result of the RST 5 instruction, you also popped the same two bytes when you executed the RET instruction. Clearly, for each RESET/G cycle and, thus, each calling of the subroutine, the return address was the same. Remember that the return address stored on the stack is for the instruction following the one- or three-byte CALL instruction. In this case, it happens to be a HLT instruction.

### EXPERIMENT NO. 7

### A SIMPLE INTERRUPT INSTRUCTION REGISTER

#### Purpose

The purpose of this experiment is to write and test a simple interrupt instruction register.

### Pin Configuration of Integrated-Circuit Chip (Fig. 23-26)

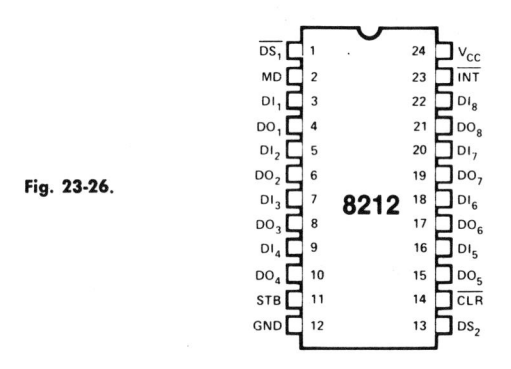

Fig. 23-26.

```
       ___
       DS₁ ┌─ 1      24 ─┐ V_CC
       MD  ┌─ 2      23 ─┐ INT
       DI₁ ┌─ 3      22 ─┐ DI₈
       DO₁ ┌─ 4      21 ─┐ DO₈
       DI₂ ┌─ 5      20 ─┐ DI₇
       DO₂ ┌─ 6  8212 19 ─┐ DO₇
       DI₃ ┌─ 7      18 ─┐ DI₆
       DO₃ ┌─ 8      17 ─┐ DO₆
       DI₄ ┌─ 9      16 ─┐ DI₅
       DO₄ ┌─ 10     15 ─┐ DO₅
       STB ┌─ 11     14 ─┐ CLR
       GND ┌─ 12     13 ─┐ DS₂
```

### Schematic Diagram of Circuit (Fig. 23-27)

### Program

| Memory Address | Instruction Byte | | Mnemonic | Comments |
|---|---|---|---|---|
| 003 000 | 061 | | LXI SP | /Load the stack pointer with |
| 003 001 | 200 | | 200 | /LO address byte and |
| 003 002 | 003 | | 003 | /HI address byte |
| 003 003 | 373 | | EI | /Enable interrupt |
| 003 004 | 000 | REPEAT, | NOP | /No operation |
| 003 005 | 000 | | NOP | /No operation |
| 003 006 | 000 | | NOP | /No operation |
| 003 007 | 170 | | MOV A,B | /Move B to A |
| 003 010 | 323 | | OUT | /Output register A to |
| 003 011 | 001 | | 001 | /output port 001 |
| 003 012 | 303 | | JMP | /Jump back to REPEAT at |
| 003 013 | 004 | | REPEAT | /LO = 004 and |
| 003 014 | 003 | | 0 | /HI = 003 and do it again. |

Fig. 23-27.

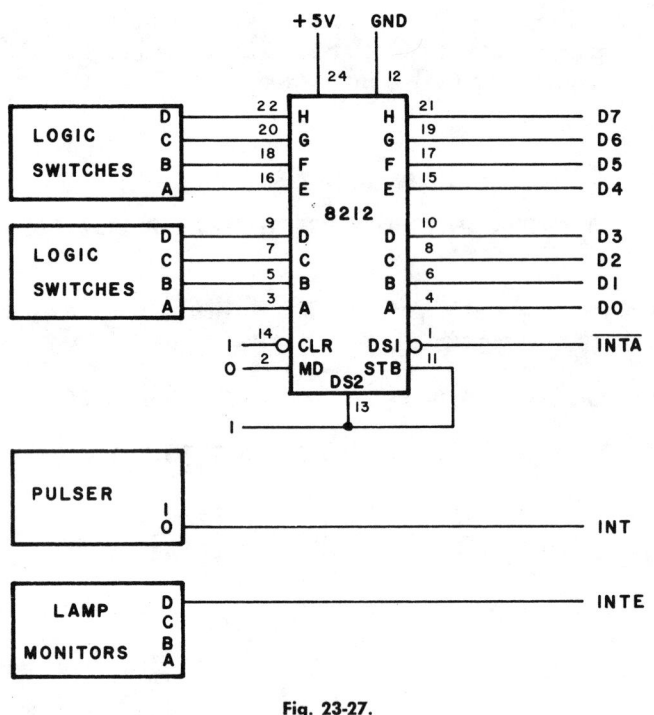

## Step 1

In the circuit shown for this experiment (Fig. 23-27), the 8212 buffer/latch is the interrupt instruction register. The pulser causes the actual interrupt and the lamp monitor indicates the status of the 8080A chip's interrupt enable flip-flop, which is located within the chip.

Turn the power to the microcomputer on and off several times. Note the condition of the interrupt enable (INTE) lamp monitor each time the microcomputer is on. Is the interrupt enabled when the computer is turned on?

Generally it is off, but this is not a rule with all 8080 microcomputer systems. With the MMD-1 microcomputer, there is no EI instruction in KEX, so it is not possible to enable the interrupt even if KEX was executed accidentally when the power was turned on.

## Step 2

Execute the program provided for this experiment. What is the state of the 8080A chip's interrupt enable flip-flop when you do?

The interrupt enable flip-flop should be at logic 1, i.e., it is enabled by the program.

## Step 3

After the software has been started (note that it loops continuously), observe the value of the byte present at Output Port 001. Write it in the space below.

Now set an 004 on the logic switches, HGFEDCBA = 00000100 = $004_8$, and depress the interrupt pulser. What byte now appears at Output Port 001? What is the condition of the INTE lamp monitor?

During an interrupt, the interrupt instruction register jams a single-byte instruction into the instruction register of the 8080A. In this case, the instruction was 004. What does this instruction accomplish? You may wish to refer to a listing of the 8080A instruction set.

After setting the logic switches to the instruction byte 004 and depressing the interrupt pulser, the value of the byte at Port 001 is incremented by one. The 004 instruction increments the contents of register B, INR B. The interrupt enable lamp monitor is logic 0 after the interrupt is serviced.

### Step 4

Reset the microcomputer and again execute the program. Note the values of the bytes at Port 001 before and after you actuate the interrupt pulser. Does the incrementing continue?

Yes it does. Keep in mind the fact that when the MMD-1 is first reset, KEX outputs a HI address byte of 003 to Output Port 001. It is this byte that is incremented each time you execute the program and press the interrupt pulser.

### Step 5

At memory address 003 007, replace the MOV A,B instruction byte with a NOP instruction byte (000). Set the instruction, 074, on the logic switches to the interrupt instruction port and execute the program. What happens when you cause an interrupt by activating the interrupt pulser?

The byte at Output Port 001 is incremented by one. In this case, we have executed an INR A instruction prior to outputting the accumulator contents to Port 001. This INR A instruction was jammed into the instruction register during the interrupt.

### Step 6

Substitute a 017 instruction on the logic switches, reset the microcomputer, execute the program, and press and release the interrupt pulser. What happens at Output Port 001?

The data byte is rotated one position to the right.

### Step 7

An interrupt causes the 8080A chip to accept a single-byte instruction from the interrupt instruction port. Does the nature of the

## Step 5

Make the following software changes to MAIN TASK:

```
  •
  •
003 003    373              EI            /Enable interrupt
003 004    000    LOOP,     NOP           /Do nothing
003 005    303              JMP           /Jump to LOOP
003 006    004              LOOP
003 007    003              0
```

## Step 6

Can you perform more than one interrupt with this software? Why or why not?

We could not. The interrupt enable instruction has been removed from the loop. Once it is used, we could never get back to it unless we reset the microcomputer.

## Step 7

Add the two enable interrupt (EI) instructions to the HIGH and LOW interrupt service routines as follows:

```
  •
  •
003 065    373    EI    /Enable interrupt
  •
  •
003 177    373    EI    /Enable interrupt
  •
  •
```

Note that the previous instruction bytes at these two locations were NOPs.

## Step 8

Repeat Steps 3 and 4 in this experiment. Are the results the same? Why?

Yes. We now re-enable the interrupt at the end of each service subroutine.

**Step 9**

Move the interrupt enable instruction in the LOW priority service software from 003 177 to 003 154. To do this, make the following changes:

```
   •
   •
003 154    373    EI    /Enable interrupt
   •
   •
   •
003 177    000    NOP   /No operation
   •
   •
```

**Step 10**

Repeat Step 3. Is the result the same?

Yes. No change was observed.

**Step 11**

Generate a LOW priority interrupt using the pulser. When the logic 1 bit has rotated toward the center, generate a HIGH priority interrupt. What happens? Why?

The HIGH priority interrupt interrupts the LOW priority interrupt software, performs the increment operation, and then returns control to the LOW priority software. The interrupt is now enabled at the start of the LOW priority interrupt software routine, thus allowing other devices of higher priority to interrupt it.

**Step 12**

Perform Step 11 again, but generate several HI priority interrupts during the LO priority service time. What happens?

The HIGH priority device is always serviced.

## EXPERIMENT NO. 12

## SIMULTANEOUS INTERRUPTS

### Purpose

The purpose of this experiment is to explore the operation of simultaneous interrupts.

### Schematic Diagram of Circuit

The circuit is identical to the circuit used in Experiments No. 10 and 11.

### Program

The software is the same as that used in Experiment No. 11.

### Step 1

The circuitry used in Experiment No. 10 to generate interrupts is shown in Fig. 23-37. Notice that one flip-flop has its Q output ap-

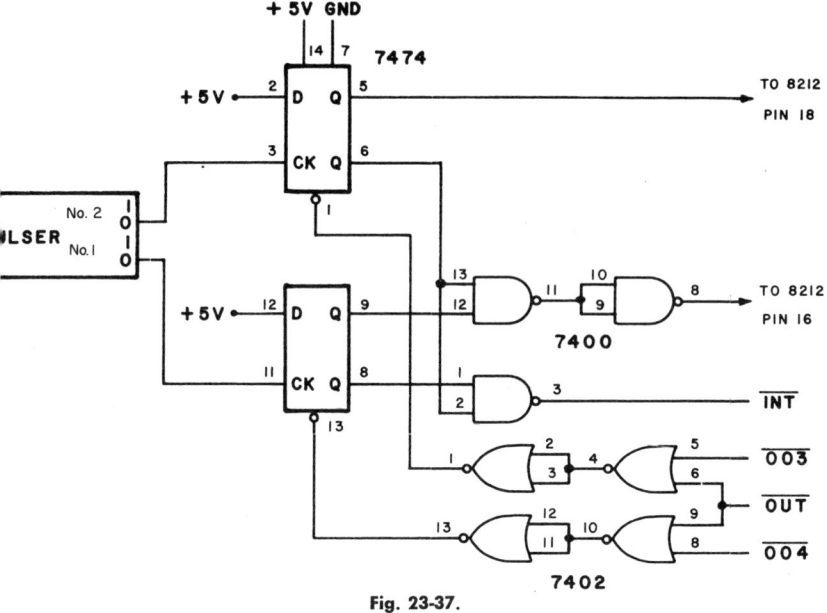

Fig. 23-37.

plied directly to the 8212 chip (Pin 18), while the Q output of the other flip-flop goes through some NAND gates and then to the 8212 (Pin 16). In this way, the restart instructions RST 2 (327) and RST 4 (347) are generated. Using the schematic diagram, fill in the following truth table:

| HIGH Priority Device Q Q̄ | LOW Priority Device Q Q̄ | 8212 Chip Pin 18 | 8212 Chip Pin 16 | Condition |
|---|---|---|---|---|
| 0 1 | 0 1 | _____ | _____ | _____ |
| 0 1 | 1 0 | _____ | _____ | _____ |
| 1 0 | 0 1 | _____ | _____ | _____ |
| 1 0 | 1 0 | _____ | _____ | _____ |

Our results, in the vertical order given, were as follows:

| 8212 Chip Pin 18 | Pin 16 | Condition |
|---|---|---|
| 0 | 0 | No interrupts |
| 0 | 1 | LOW priority device interrupts |
| 1 | 0 | HIGH priority device interrupts |
| 1 | 0 | Simultaneous interrupt |

### Step 2

Does the above truth table indicate what would happen if both the LOW and HIGH priority devices attempted to interrupt at the same time? What would actually happen?

Yes. It shows that the HIGH priority device would override the LOW priority device and cause the HIGH priority service subroutine to be executed.

### Step 3

Reset the microcomputer. *Do not* start the software. Remove the wire from the logic 0 output of Pulser No. 1 and place it in the logic 0 output of Pulser No. 2. Both interrupts will now be generated by the same pulser.

Now start the software. Interrupt software action may initially take place. Allow it to finish before you proceed with the experiment.

### Step 4

Observe the lamps at Ports 000 and 001 carefully. Press and release the interrupt pulser, Pulser No. 2. What happens?

We observed that the HIGH priority software (increment the LEDs) started. When it had finished, the LOW priority software (rotate a bit) started and completed its task.

### Step 5

Repeat Step 4 five more times. Does the HIGH priority device software always start the sequence?

Yes.

### Step 6

Reset the microcomputer. *Do not* start the software. Replace the wire, which you removed in Step 3, to the logic 0 output of Pulser No. 1. Cause a LOW priority interrupt and, in the middle of the service routine operation, again generate a LOW priority interrupt. What happens?

The second service routine goes to completion. The first routine proceeded halfway through and then appeared to start rotating again at the beginning.

### Step 7

Can you explain what you observed in Step 6? HINT: Register L is not saved on the stack in this program.

The second LOW priority interrupt actually interrupted the first LOW priority interrupt's service software. Since register L is not stored on the stack, it is left at the time of the second interrupt in some unknown state. When the first interrupt routine tries to use register L, it is not in the same state as it was before the second interrupt occurred.

What solutions could you suggest for the problem identified in Step 7?

We suggest:

1. Clear the low priority device's interrupt service request flag at the end of the service routine.
2. Push all of the registers to be used in the routine onto the stack.

You should *never* permit an interrupting device to interrupt its own software service routine. If it does, the external device is operating too fast for the microcomputer; the software must be simplified to speed it up. Why? Because by the time we get into the second interrupt's service software, a third interrupt will interrupt the second, etc. We will never be able to complete any of the service routines.

What conclusion can be drawn? *Exercise care in using interrupts!*

## REVIEW QUESTIONS

The following questions will help you review the use of flags and interrupts.

1. What types of devices can be used as flags and why are flags important?
2. What instructions in the 8080A instruction set can be used to detect internal flag conditions?
3. What advantage does an interrupt have over a polled flag?
4. What are the three types of interrupts, and which type is used in 8080A-based microcomputer systems?
5. What types of instructions are most useful with 8080A interrupts? How do they work?
6. What does a typical interrupt service subroutine look like?
7. What is a priority interrupt?
8. What are some of the potential problems with the use of interrupts?

# Appendixes

APPENDIX **A**

# References

1. *The Compact Edition of the Oxford English Dictionary,* Oxford University Press, 1971.
2. Rudolf F. Graf, *Modern Dictionary of Electronics,* Howard W. Sams & Co., Inc., Indianapolis, Indiana, 1972.
3. James Martin, *Telecommunications and the Computer,* Prentice-Hall, Inc., Englewood Cliffs, New Jersey, 1969.
4. Abraham Marcus and John D. Lenk, *Computers for Technicians,* Prentice-Hall, Inc., Englewood Cliffs, New Jersey, 1973.
5. Microdata Corporation, *Microprogramming Handbook,* Santa Ana, California, 1971.
6. J. Blukis and M. Baker, *Practical Digital Electronics,* Hewlett-Packard Company, Santa Clara, California, 1974.
7. Donald E. Lancaster, *TTL Cookbook,* Howard W. Sams & Co., Inc., Indianapolis, Indiana, 1974.
8. H. V. Malmstadt, C. G. Enke, and S. R. Crouch, *Instrumentation for Scientists Series, Module 3. Digital and Analog Data Conversions,* W. A. Benjamin, Inc., Menlo Park, California, 1973-4.
9. H. V. Malmstadt and C. G. Enke, *Digital Electronics for Scientists,* W. A. Benjamin, Inc., New York, 1969.
10. J. D. Lenk, *Handbook of Logic Circuits,* Reston Publishing Company, Inc., Reston, Virginia, 1972.
11. A. James Diefenderfer, *Principles of Electronic Instrumentation,* W. B. Saunders Company, Philadelphia, Pennsylvania, 1972.
12. P. R. Rony and D. G. Larsen, *Bugbook II. Logic & Memory Experiments Using TTL Integrated Circuits,* E&L Instruments, Inc., Derby, Connecticut, 1974.
13. Robert L. Morris and John R. Miller, Editors, *Designing with TTL Integrated Circuits,* McGraw-Hill Book Company, New York, 1971.
14. Charles J. Sippl, *Microcomputer Dictionary and Guide,* Matrix Publishers, Inc., Champaign, Illinois, 1976.

15. Donald Eadie, *Introduction to the Basic Computer,* Prentice-Hall, Inc., Englewood Cliffs, New Jersey, 1973.
16. Texas Instruments Incorporated, *Microprocessor Handbook,* Dallas, Texas, 1975.
17. Charles L. Garfinkel of Keithley Instruments, Inc. is the originator of this definition.

# APPENDIX B

# Definitions

In this appendix, we continue the summary of the definitions for important concepts of digital electronics and microcomputers that we started in Book 1. We acknowledge the following sources for the definitions used:

- Rudolf F. Graf, *Modern Dictionary of Electronics,* Howard W. Sams & Co., Inc., Indianapolis, Indiana, 1972.
- Microdata Corporation, *Microprogramming Handbook,* Santa Ana, California, 1972.
- Donald Eadie, *Introduction to the Basic Computer,* Prentice-Hall, Inc., Englewood Cliffs, New Jersey, 1973.
- Abraham Marcus and John D. Lenk, *Computers for Technicians,* Prentice-Hall, Inc., Englewood Cliffs, New Jersey, 1973.
- Peter R. Rony, David G. Larsen, and Jonathan A. Titus, The 8080A Bugbook®, Howard W. Sams & Co., Inc., Indianapolis, Indiana, 1977.

*absolute decoding*—The decoding of a binary number to produce a unique pulse, select a unique memory address, etc.

*accumulator*—The register and associated digital electronic circuitry in the arithmetic/logic unit (ALU) of a computer in which arithmetic and logical operations are performed.

*accumulator I/O*—A term associated with 8080-based microcomputer systems. The I/O instructions are IN and OUT and the data transfer occurs between the I/O device and the accumulator within the 8080 chip.

*addend*—A quantity which, when added to another quantity (called the augend), produces a result called the sum.

*address*—A group of bits that identify a specific memory location or I/O device. An 8080 microcomputer uses sixteen bits to identify a specific memory location and eight bits to identify an I/O device.

*address bus*—A unidirectional bus over which digital information appears to identify either a particular memory location or a particular I/O device. The 8080 address bus is a group of sixteen lines.

---

Bugbook is a registered trade mark of E&L Instruments, Inc., Darby, CT.

*address select pulse*—A software-generated clock pulse from a microcomputer that is used to strobe the operation of a memory-mapped I/O device.

*analog-to-digital converter*—A circuit that changes a continuously varying voltage or current into a digital output.

*augend*—In an arithmetic addition, the number increased by having another number, called the addend, added to it.

*bidirectional data bus*—A data bus in which digital information can be transferred in either direction. With reference to an 8080-based microcomputer, the bidirectional data path by which data is transferred between the CPU, memory, and input/output devices.

*bus*—A path over which digital information is transferred, from any of several sources to any of several destinations. Only one transfer of information can take place at any one time. While such transfer of information is taking place, all other sources that are tied to the bus must be disabled.

*bus monitor*—A binary, octal, or hexadecimal display that monitors and displays the data that appears on the bidirectional data bus.

*byte*—A group of eight contiguous bits that are operated on as a unit or occupy a single memory location.

*central processing unit (mainframe computer)*—Also called a central processor. That part of a computer system which contains the main storage, arithmetic unit, and special register groups. Performs arithmetic operations, controls instruction processing, and provides timing signals and other housekeeping operations.

*central processing unit (microprocessor)*—A single integrated-circuit chip that performs data transfer, control, input/output, arithmetic, and logical instructions by executing instructions obtained from memory.

*control*—Those parts of a computer which carry out instructions in proper sequence, interpret instructions, and apply proper signals.

*control bus*—A set of signals that regulate the operation of the microcomputer system, including I/O devices and memory. They function much like "traffic" signals or commands. They may also originate in the I/O devices, generally to transfer to or receive signals from the CPU. A unidirectional set of signals that indicate the type of activity—memory read, memory write, I/O read, I/O write, or interrupt acknowledge—in progress.

*controller*—An instrument that holds a process or condition at a desired level or status as determined by comparison of the actual value with the desired value.

*CPU*—Abbreviation for central processing unit.

*data logger*—An instrument that automatically scans data produced by another instrument or process, and records readings of the data for future use.

*device code*—In an 8080-based microcomputer, the 8-bit code for a specific input or output device.

*device select pulse*—A software-generated pulse from a microcomputer that is used to strobe the operation of an accumulator I/O device.

*digital-to-analog converter*—A circuit that changes a digital input into a continuously varying voltage or current.

*double-precision arithmetic*—1. Using two computer words to represent a number, usually to obtain greater accuracy than a single word of computer storage is capable of providing. 2. Arithmetic used when more accuracy is necessary than a single word of computer storage will provide.

*exclusive masking*—A masking technique in which one either clears or sets (seldom used) all bits not operated upon.

*fetch*—One of the two functional parts of an instruction cycle. The collective actions of acquiring a memory address and then an instruction or data byte from memory.

*flag*—Some sort of digital register or device used to indicate the state or status of a device. It can be cleared or set in response to an operation.

*general-purpose register*—In the 8080 microprocessor chip, 8-bit registers that can participate in arithmetic and logical operations with the contents of the accumulator.

*hardware*—The mechanical, magnetic, electronic, and electrical devices from which a computer is fabricated; the assembly of material forming a computer.

*inclusive masking*—A masking technique in which one leaves unaltered all bits not operated upon.

*input/output*—General term for the equipment used to communicate with a computer and the data involved in the communication.

*I/O device*—Input/output device. A card reader, magnetic tape unit, printer, or similar device that transmits to or receives data from a computer or secondary storage device. In a more general sense, any digital device, including a single integrated-circuit chip, that transmits data to or receives data or strobe pulses from a computer.

*instruction code*—A unique 8-bit binary number that encodes an operation that the 8080 microprocessor chip can perform.

*instruction decoder*—A decoder within the 8080 microprocessor chip that decodes the instruction code into a series of actions that the microprocessor performs.

*instruction register*—The 8-bit register in the 8080 microprocessor chip that stores the instruction code of the instruction being executed.

*interrupt*—In a digital computer, a break in the normal execution of a computer program such that the program can be resumed from that point at a later time.

*interrupt instruction register*—An external 8-bit register that permits an instruction to be jammed into the instruction register within an 8080 chip during an interrupt.

*machine cycle*—A subdivision of an instruction cycle during which time a related group of actions occur within the microprocessor chip. All instructions are combinations of one or more machine cycles.

*masking*—A logical technique in which certain bits of a word are blanked out or inhibited.

*memory*—Any device that can store logic 0 and logic 1 bits in such a manner that a single bit or group of bits can be accessed and retrieved.

*memory address*—See address.

*memory I/O*—See memory-mapped I/O.

*memory-mapped I/O*—A term associated with 6800, 8080, and other microcomputer systems. The I/O instructions are memory-reference instructions and the data transfer occurs, in the case of the 8080 chip, between the I/O device and any of the general-purpose registers within the chip.

*multilevel interrupt*—Several independent interrupt lines are provided, each of which causes a specific action. Polling is not needed unless multiple devices are ORed to one of the inputs.

*nibble*—A group of four contiguous bits that are operated on as a unit or occupy a single memory location.

*open-collector output*—An output from an integrated-circuit device in which the final pull-up resistor in the output transistor for the device is missing and must be provided by the user before the circuit is complete.

*polling*—A periodic checking of input/output or control devices to determine their condition or status, e.g., full/empty, on/off, busy/ready, done/not done, etc.

*priority interrupts*—Interrupts that are ordered in importance so that some interrupting devices take precedence over others.

*program counter*—The 16-bit register in the 8080 chip that contains the memory address of the next instruction byte that must be executed in a computer program.

*PSW*—Abbreviation for processor status word. The contents of the accumulator and the five status flags in the 8080 microprocessor chip.

*read*—To transmit data from a specific memory location to some other digital device. A synonym for retrieve.

*real-time clock*—Refers to a device that provides interrupts at regular time intervals, frequently twice the ac line frequency. It allows maintenance of an accurate time of day clock and the measurement of elapsed time.

*register*—A short-term digital electronic storage circuit, the capacity of which is usually one computer word.

*service routine*—A computer subroutine that services an interrupting device.

*single-line interrupt*—An interrupt signal that is input to the computer on a single line and causes a well-defined action to take place. Multiple devices must be ORed onto this line and a polling routine must determine which device caused the interrupt.

*software*—The totality of programs and routines used to extend the capabilities of computers, including compilers, assemblers, linker-loaders, narrators, translators, and subroutines. Contrasted with hardware.

*stack pointer*—The 16-bit register in the 8080 microprocessor chip that stores the memory address of the top of the stack, which is a region of read/write memory that stores temporary information.

*sync*—Short for synchronous, synchronization, synchronizing, etc.

*synchronization pulses*—Pulses originated by the transmitting equipment and introduced into the receiving equipment to keep the equipment at both locations operating in step.

*to synchronize*—To lock one element of a system into step with another.

*synchronous*—In step or in phase, as applied to two devices or machines. A term applied to a computer, in which the performance of a sequence of operations is controlled by clock signals or pulses. At the same time.

*synchronous computer*—A digital computer in which all ordinary operations are controlled by a master clock.

*synchronous inputs*—Those inputs of a flip-flop that do not control the output directly, as do those of a gate, but only when the clock permits and commands.

*synchronous logic*—The type of digital logic used in a system in which logical operations take place in synchronism with clock pulses.

*synchronous operation*—Operation of a system under the control of clock pulses.

*three-state device*—A semiconductor logic device in which there are three possible output states: (1) a logic 0 state, (2) a logic 1 state, and (3) a state in which the output is, in effect, disconnected from the rest of the circuit and has no influence upon it.

*timing loop*—A software loop that requires a precise period of time for its execution.

*TRI-STATE® device*—See three-state device.

*vectored interrupt*—Each device points, or vectors, the computer's control to specific software service routines for the interrupting devices.

*word*—A group of contiguous bits occupying one or more storage locations in a computer.

*write*—To transmit data from some other digital device into a specific memory location. A synonym for store.

APPENDIX **C**

# Octal/Hex Conversion Table

| OCTAL | HEX | OCTAL | HEX | OCTAL | HEX | OCTAL | HEX | OCTAL | HEX |
|---|---|---|---|---|---|---|---|---|---|
| 000 | 00 | 040 | 20 | 100 | 40 | 140 | 60 | 200 | 80 |
| 001 | 01 | 041 | 21 | 101 | 41 | 141 | 61 | 201 | 81 |
| 002 | 02 | 042 | 22 | 102 | 42 | 142 | 62 | 202 | 82 |
| 003 | 03 | 043 | 23 | 103 | 43 | 143 | 63 | 203 | 83 |
| 004 | 04 | 044 | 24 | 104 | 44 | 144 | 64 | 204 | 84 |
| 005 | 05 | 045 | 25 | 105 | 45 | 145 | 65 | 205 | 85 |
| 006 | 06 | 046 | 26 | 106 | 46 | 146 | 66 | 206 | 86 |
| 007 | 07 | 047 | 27 | 107 | 47 | 147 | 67 | 207 | 87 |
| 010 | 08 | 050 | 28 | 110 | 48 | 150 | 68 | 210 | 88 |
| 011 | 09 | 051 | 29 | 111 | 49 | 151 | 69 | 211 | 89 |
| 012 | 0A | 052 | 2A | 112 | 4A | 152 | 6A | 212 | 8A |
| 013 | 0B | 053 | 2B | 113 | 4B | 153 | 6B | 213 | 8B |
| 014 | 0C | 054 | 2C | 114 | 4C | 154 | 6C | 214 | 8C |
| 015 | 0D | 055 | 2D | 115 | 4D | 155 | 6D | 215 | 8D |
| 016 | 0E | 056 | 2E | 116 | 4E | 156 | 6E | 216 | 8E |
| 017 | 0F | 057 | 2F | 117 | 4F | 157 | 6F | 217 | 8F |
| 020 | 10 | 060 | 30 | 120 | 50 | 160 | 70 | 220 | 90 |
| 021 | 11 | 061 | 31 | 121 | 51 | 161 | 71 | 221 | 91 |
| 022 | 12 | 062 | 32 | 122 | 52 | 162 | 72 | 222 | 92 |
| 023 | 13 | 063 | 33 | 123 | 53 | 163 | 73 | 223 | 93 |
| 024 | 14 | 064 | 34 | 124 | 54 | 164 | 74 | 224 | 94 |
| 025 | 15 | 065 | 35 | 125 | 55 | 165 | 75 | 225 | 95 |
| 026 | 16 | 066 | 36 | 126 | 56 | 166 | 76 | 226 | 96 |
| 027 | 17 | 067 | 37 | 127 | 57 | 167 | 77 | 227 | 97 |
| 030 | 18 | 070 | 38 | 130 | 58 | 170 | 78 | 230 | 98 |
| 031 | 19 | 071 | 39 | 131 | 59 | 171 | 79 | 231 | 99 |
| 032 | 1A | 072 | 3A | 132 | 5A | 172 | 7A | 232 | 9A |
| 033 | 1B | 073 | 3B | 133 | 5B | 173 | 7B | 233 | 9B |
| 034 | 1C | 074 | 3C | 134 | 5C | 174 | 7C | 234 | 9C |
| 035 | 1D | 075 | 3D | 135 | 5D | 175 | 7D | 235 | 9D |
| 036 | 1E | 076 | 3E | 136 | 5E | 176 | 7E | 236 | 9E |
| 037 | 1F | 077 | 3F | 137 | 5F | 177 | 7F | 237 | 9F |

| OCTAL | HEX | OCTAL | HEX | OCTAL | HEX | OCTAL | HEX | OCTAL | HEX |
|-------|-----|-------|-----|-------|-----|-------|-----|-------|-----|
| 240 | A0 | 263 | B3 | 306 | C6 | 331 | D9 | 354 | EC |
| 241 | A1 | 264 | B4 | 307 | C7 | 332 | DA | 355 | ED |
| 242 | A2 | 265 | B5 | 310 | C8 | 333 | DB | 356 | EE |
| 243 | A3 | 266 | B6 | 311 | C9 | 334 | DC | 357 | EF |
| 244 | A4 | 267 | B7 | 312 | CA | 335 | DD | 360 | F0 |
| 245 | A5 | 270 | B8 | 313 | CB | 336 | DE | 361 | F1 |
| 246 | A6 | 271 | B9 | 314 | CC | 337 | DF | 362 | F2 |
| 247 | A7 | 272 | BA | 315 | CD | 340 | E0 | 363 | F3 |
| 250 | A8 | 273 | BB | 316 | CE | 341 | E1 | 364 | F4 |
| 251 | A9 | 274 | BC | 317 | CF | 342 | E2 | 365 | F5 |
| 252 | AA | 275 | BD | 320 | D0 | 343 | E3 | 366 | F6 |
| 253 | AB | 276 | BE | 321 | D1 | 344 | E4 | 367 | F7 |
| 254 | AC | 277 | BF | 322 | D2 | 345 | E5 | 370 | F8 |
| 255 | AD | 300 | C0 | 323 | D3 | 346 | E6 | 371 | F9 |
| 256 | AE | 301 | C1 | 324 | D4 | 347 | E7 | 372 | FA |
| 257 | AF | 302 | C2 | 325 | D5 | 350 | E8 | 373 | FB |
| 260 | B0 | 303 | C3 | 326 | D6 | 351 | E9 | 374 | FC |
| 261 | B1 | 304 | C4 | 327 | D7 | 352 | EA | 375 | FD |
| 262 | B2 | 305 | C5 | 330 | D8 | 353 | EB | 376 | FE |
|  |  |  |  |  |  |  |  | 377 | FF |

APPENDIX **D**

# Answers to Review Questions

### Unit 16

1.  a. Software
    b. Software
    c. Hardware
    d. Hardware
    e. Hardware
    f. Hardware
    g. Software
    h. Hardware, whatever it is.
    i. Hardware
2.  The bidirectional data bus, the control bus, and the address bus.
3.  For an 8-bit microprocessor chip, it reduces the number of pins required by eight. For a 16-bit microprocessor chip, it reduces the number of pins required by sixteen.
4.  $\overline{\text{OUT}}$, $\overline{\text{IN}}$, $\overline{\text{MEMR}}$, $\overline{\text{MEMW}}$, and $\overline{\text{I ACK}}$

### Unit 17

1.  a. Negative device select (DS) pulse at Pin 1
    b. Negative device select pulse at Pin 10
    c. Positive device select pulse at Pin 7, with Pin 6 at logic 0
    d. Positive device select pulse at Pin 2
    e. Negative device select pulse at Pin 18, with Pin 19 at logic 0
    f. Positive device select pulse at Pin 13
    g. Positive device select pulse at Pin 11
    h. Negative device select pulse at Pin 3

i. Negative device select pulse at Pin 1, with Pin 2 at logic 1
   j. Negative device select pulse at Pin 2
   k. Negative device select pulse at Pin 1, with Pins 2, 3, and 4 at logic 1

2. A device select pulse will be produced for the following output device code bytes (in octal code):

   | 005 | 105 | 205 | 305 |
   | 025 | 125 | 225 | 325 |
   | 045 | 145 | 245 | 345 |
   | 065 | 165 | 265 | 365 |

   You are not absolutely decoding the device code byte. If you were, the only device code that would provide an output on the 74154 decoder chip would be 005.

3. a. Three
   b. Five
   c. Three
   d. Three
   e. Two
   f. Three
   g. Two
   h. Three
   i. One
   j. One

## Unit 18

1. a. None
   b. None
   c. None (This is very important.)
   d. None
   e. None
   f. None
   g. None
   h. None
   i. None
   j. All flags affected
   k. None
   l. None
   m. None
   n. Carry flag only
   o. Carry flag only
   p. All flags affected

2. | LO Address Byte | Contents (Register) | |
|---|---|---|
| 377 | HI byte | (from program counter) |
| 376 | LO byte | (from program counter) |
| 375 | D | |
| 374 | E | |
| 373 | H | |
| 372 | L | |
| 371 | Accumulator | |
| 370 | Flags | |
| 367 | B | |
| 366 | C | |

3. LXI SP   (most useful)
   INX SP
   DCX SP
   SPHL

   In addition, all POP, PUSH, subroutine call, and subroutine return instructions influence the location of the stack.

4. a. In the 8080A chip, a register is one of the general-purpose registers (8 bits) or the accumulator. A register pair is a 16-bit register that is treated as a unit and consists of two general-purpose registers, such as B and C, or D and E.
   b. These days, one byte consists of eight bits. A bit is a single binary decision.
   c. A byte is a sequence of adjacent binary digits operated upon as a unit but usually shorter than a computer word, which may consist of two or more bytes.
   d. A memory address is a sequence of adjacent binary digits that define a single memory location. A word is a sequence of binary digits that are treated as a unit, and may represent data, instructions, or other binary quantities besides memory addresses.
   e. In an 8080A microcomputer, the HI address byte is the eight most-significant bits in the 16-bit memory address word; the LO address byte is the eight least-significant bits.
   f. Both are branch instructions. However, in a call instruction, the contents of the program counter are saved before the instruction is executed. In a jump instruction, the program counter is ignored.
   g. Both refer to branch instructions in the 8080A instruction set. In a conditional branch instruction, whether or not the branch occurs depends upon the logic state of the selected flag. An unconditional branch instruction ignores the logic states of all flags.

h. For an OR gate, when both inputs are at logic 1, the output is also logic 1. For an Exclusive-OR gate, when both inputs are at logic 1, the output is at logic 0. In other respects, the two gates are the same in their logic characteristics.
i. The zero flag is set only when the result of an arithmetic/logical instruction is zero. The sign flag refers to the logic state of the most-significant bit in the result, not to the total word or byte. The flags refer to different things.
j. The carry flag refers to a carry out of the most-significant bit in an 8-bit result in the 8080A microprocessor. The auxiliary carry flag refers to a carry out of Bit D3 (the fourth bit) in the result.
k. A PUSH instruction adds two bytes to the stack and decrements the stack pointer. A POP instruction removes two bytes from the stack and increments the stack pointer.
l. The accumulator is a single register in the arithmetic-logic unit (ALU), which contains other digital circuitry required for performing arithmetic and logic operations.
m. The data byte never gets loaded in the program counter. An address byte does.
n. Octal code is an eight-state code. Hexadecimal code is a sixteen-state binary code. The first eight states of the two codes are identical.
o. To increment means to increase by one. To decrement means to decrease by one.
p. To increment means to increase only by one. In an ADD operation, the addend is limited by the byte or word length; it is not limited to a single unit.
q. The IN instruction inputs data from an external device into the accumulator. The OUT instruction outputs data from the accumulator to an external device.
r. The ADC instruction is an ADD instruction in which you also add the contents of the Carry.
s. The byte being transferred in a MVI instruction is contained within the program as the second byte of the instruction. The byte being transferred in a MOV instruction is originally present in a register or in a specific memory location.
t. The MVI instruction transfers one program byte to one register. The LXI instruction transfers a pair of program bytes to a register pair.
u. The EI instruction enables the interrupt flag and permits the 8080A chip to be interrupted. The DI instruction disables the interrupt flag and prevents the 8080A chip from being interrupted.
v. Both instructions clear the accumulator and the carry flag.

w. Both refer to the logic state of the carry flag, but carry refers to the state of the flag after an addition operation, whereas, borrow refers to the flag after a subtraction operation.
x. Machine code is represented as binary, octal, or hexadecimal digits. Mnemonic code is represented as alphanumeric characters, usually alphabetic characters. Mnemonic code is easier to remember, but must be converted into machine code before it can be executed by a microcomputer.
y. The H register pair serves as a pointer address for all instructions that refer directly to memory location M. Though the B register pair can serve as a memory address, it has fewer instructions associated with it when it is used as a pointer address, ADD M, SUB M, INR M, DCR M, XRA M, ORA M, CMP M, etc., are some of the instructions that employ the H register pair as a pointer address.
z. The instruction register stores the 8-bit operation code in an 8080A chip. The instruction decoder decodes this 8-bit quantity into a series of actions.

## Unit 19

1. A digital bus is a set of common conducting paths over which digital information is transferred. Only one transfer of information can take place at any one time. While such a transfer is taking place, all other sources that are tied to the bus must be disabled. The fundamental purpose of a bus is to minimize the number of interconnections required to transfer information between digital devices.
2. Busses are present: (a) Within integrated-circuit chips such as the 8080A microprocessor chip and the 8251, 8253, 8255, 8257, and 8259 programmable interface chips; (b) On printed-circuit boards that contain collections of integrated-circuit chips between which information must be bussed; and (c) In digital instruments such as minicomputers, microcomputers, frequency meters, digital voltmeters, and the like.
3.

| Input Data | Gating Signal | Output Data |
|---|---|---|
| 0 | enable | 0 |
| 1 | enable | 1 |
| X | disable | High impedance |

X = irrelevant

4. Such a circuit can act in several different ways: (a) As a latch that stores input data but does not output it to a bus; (b) As a simple three-state buffer that does not latch input data; and (c) As a latch that stores input data and outputs it to a bus.

5. Buffer, latch, counter, inverter, multiplexer, read/write memory, read-only memory, line driver, transceiver, demultiplexer, counter/latch, flip-flop, and multiplier.
6. Two things occur: (a) The receiver of information on the bus becomes confused, since it cannot interpret the bus information, and (b) The three-state buffers eventually burn out.

## Unit 20

1. Accumulator input/output is a term associated with 8080-based microcomputer systems. The I/O instructions are IN and OUT and the data transfer occurs between the I/O device and the accumulator within the 8080 chip.
2. In microcomputer output, the objective is to "catch" data that is being output for only a short interval of time. The capture of data is accomplished using an 8-bit latch that is enabled during the short interval of time.

   In microcomputer input, the objective is to input stable TTL data into the microprocessor chip during the interval of time when the external input device has control over the data bus. At all other times, the input device should not influence the data bus. Such objectives are accomplished using a three-state 8-bit buffer.
3. The drive capability of an output pin on a microprocessor chip may be low, of the order 1.9 mA or even less. The fan-in of a typical TTL input is 1.6 mA, so it is not possible to connect more than one standard TTL input to a single microprocessor chip output. The solution, of course, is to use TTL chips that have lower fan-ins, i.e., the 74L or 74LS series, or else to use a driver between the microprocessor chip and output devices.
4. In the absence of a DAA instruction, when two 8-bit numbers are added using an ADD, ADC, ADI, or similar instruction, the sum of the two numbers is in binary and the augend and addend are assumed to be binary quantities. When a DAA instruction immediately follows an ADD, ADC or ADI instruction, the augend, addend and sum must all be considered to be 2-digit packed-BCD numbers. In other words, you can perform pure binary addition or pure BCD addition depending upon whether the DAA instruction is absent or present.

## Unit 21

1. Memory-mapped input/output is a term associated with 6800, 8080A, and other 8-bit microcomputer systems. The I/O instructions are memory-reference instructions and the data transfer occurs, in the case of the 8080A chip, between the I/O de-

vice and any of the general-purpose registers within the 8080A chip.

2. In accumulator I/O, the control signals are $\overline{\text{IN}}$ and $\overline{\text{OUT}}$, whereas, in memory-mapped I/O, the control signals are $\overline{\text{MEMR}}$ and $\overline{\text{MEMW}}$. Accumulator I/O employs only two 8080A instructions, IN and OUT. Memory-mapped I/O employs any memory-reference instructions, e.g., MOV r,M, MOV M,r, MVI M, STAX rp, LDAX rp, ANA M, ORA M, ADD M, and others. In accumulator I/O, data transfer occurs between the I/O device and the accumulator register. In memory-mapped I/O, data transfer occurs between the I/O device and any of the general-purpose registers, such as B, C, D, E, H, L, or the accumulator.

3. All sixteen bits of the address bus are decoded using a suitable decoder network so that each memory-mapped I/O device is uniquely identified and cannot be confused with a memory location in read/write memory or EPROM or accidentally addressed when program execution goes awry.

4. To prevent the accidental addressing of a memory-mapped I/O device when program execution goes awry. To clearly distinguish between a memory location and a memory-mapped I/O device.

5. A device select pulse is generated when you apply the 8-bit device code from the address bus and either $\overline{\text{IN}}$ or $\overline{\text{OUT}}$, which are control signals, to a suitable decoder circuit. An address select pulse is generated when you apply the 16-bit address bus and either $\overline{\text{MEMR}}$ or $\overline{\text{MEMW}}$, which are memory-reference control signals, to a suitable decoder circuit.

6. No, we do not agree. In most cases, 256 different input and 256 different output devices or device select pulses are more than adequate for a microcomputer system that includes an 8080A chip. A better reason for using memory-mapped I/O techniques is to permit direct data transfer between the I/O device and all of the general-purpose registers within the 8080A chip. In addition, the contents of a *memory-mapped input port* can be added to, subtracted from, compared with, or logically operate on the contents of the accumulator.

## Unit 22

1. A data logger is an instrument that automatically scans data produced by another instrument or process, and records readings of the data for future use.

2. Basically, you must determine how many bits of memory are required to store the data, whether such storage is short term or long term, and how many data points per second are to be logged. It is not difficult to exceed the capabilities of an 8080A-based

microcomputer (without DMA, or direct memory access) in high-speed data logging applications.
3. You would input the ASCII character into the accumulator and then compare (CMP) the accumulator contents with the specific ASCII code of the desired character. The ASCII code could be stored in registers B, C, D, E, H, or L, or in a memory location. When the desired character is detected, the zero flag goes to logic 1.
4. Time delays can be software or hardware generated. A software time-delay loop can generate delays as short as twenty to thirty microseconds and as long as hours or even days. Hardware methods of generating time delays include the use of a real time clock, probably operating at 60 Hz, that interrupts program execution, a crystal-based real time clock, or a programmable interval timer.

## Unit 23

1. Flags may be switches, flip-flops, counters, shift registers, or memories. Generally, any bistable device can qualify as a flag. Flags are generally used to indicate that some condition has changed and to synchronize flow and control operations in computer systems.
2. All the conditional instructions, i.e., jumps, calls, and returns, are useful, although jumps are the most frequently used. Other instructions such as rotate, AND, OR, etc., are useful, but they do not detect flag conditions.
3. Speed. Interrupts are generally sensed within microseconds, while polled flags can take much, much longer, depending upon software.
4. Interrupts can be single line, multilevel, and vectored. See Fig. 23-7 and the associated text.
5. The restart instructions, RST X, are generally the most useful. They are single-byte calls that, when executed, cause a return address to be pushed onto the stack. The computer then vectors to the address of the subroutine specified by the restart instruction. RST X calls a subroutine at 000 0X0. A return instruction must be used to end the subroutine.
6. It includes PUSH, POP, EI, and RET instructions in addition to the interrupt service software routine. See Fig. 23-10 and the associated text.
7. A priority interrupt is one in which there exists a preset priority for the order in which interrupts are serviced by the computer. Priority may be set up in hardware or software.
8. Determination of timing and priority are the big problems. These subjects have been covered in detail near the end of Unit 23.

# Index

## A

Abbreviations, 104-105
Absolute
　decoder, 70, 74, 76
　decoding, 38, 39, 44
Accumulator, 45-46, 73-74, 81, 91, 95, 113-114, 120-123, 133-134, 146, 151, 201, 206, 261, 268
　data, 196
　I/O, 191, 195, 202, 207, 213, 215, 227, 232, 239
　input, 320, 322
　instructions, 192, 203-204
　techniques, 264, 270
　vs memory-mapped I/O, 228-229
　rotate instructions, 82
Ac loads, 42
AD7522, 261, 296, 302-304
Addend, 158-159, 222-224
Addition, 120-122
Address
　binary, 102
　buffer, 91
　bus, 22, 24, 29, 36-37, 46, 59, 70, 151, 152, 228, 230
　　byte, 44
　　device code byte, 76
　byte, 25, 82, 307
　memory, 93, 230
　select pulses, 195, 238
Addressable memory location, 93
Addresses, symbolic, 315
Addressing
　memory, 81, 83
　memory I/O, 238

Addressing—cont
　modes, 102-103
Altman, Lawrence, 16
ALU, 93, *see also* Arithmetic/logic unit
Analog-to-digital converter, 304
Analytical instruments, 14
Arithmetic
　group, 100, 101, 115-128
　instruction, 222
　/logic units, 11, 24, 93
　operations, 23, 93, 95-96, 121-122, 127-128, 135, 139-140, 259
ASCII
　byte, 268
　character, 86, 267-269, 270, 278-281
　code, 161, 223, 268, 269, 281, 311
　listing, 90
Assembler format, 105
Assemblers, 17
Assembly language, 83, 101
　instruction, 90
　program, 87, 96
　programming, 82, 83, 96
　summaries, 90
Augend, 158-159, 222-224
Automation, factory, 13
Auxiliary Carry flag, 104, 105, 115, 116, 122, 123-124, 128, 222, 310

## B

Baker, R., 90
BCD, 122-123, 124, 155, 157-158, 161, 207

BCD—cont
  codes, 42, 223
  synchronous counters, 35
  up/down counters, 35
Bidirectional
  data bus, 25, 26, 33-34, 44, 48, 49, 51, 55, 65, 194-195, 196, 201, 202, 229, 237, 238, 282-284
  I/O port, 270, 288
  memory-mapped I/O using an 8216 chip, 270, 285-288
Binary
  address, 102
  code, 42
  coded decimal, 122
  counter, 181, 185
  full adder, 34
  mutliplication, 88
  number, 93, 94, 102, 107, 108, 155, 161-162
  processors, 223
  synchronous counters, 35
  up/down counters, 35
  word, 273
Bit(s), 93-94, 263, 291
  -by-bit logical operation, 128, 129, 130
  carry, 120-121
  condition, 83
  manipulation, flag, 316
  pattern, 48, 61-62, 104-105, 131
Branch
  group, 100, 101, 134-145
  instruction, 103, 134
Buffer(s), 11, 16, 175, 176, 313
  address, 91
  circuits, 191, 202-203
  three-state, 29, 34, 95, 177-179, 180, 181, 188-189, 201, 280
  Tri-State®, 173
Buffer/latch
  chip, 321, 356, 361
  data bus, 91
Bus, 173-190
  address, 22, 29, 36-37, 46, 59, 70, 76, 151, 152, 228, 230
  bidirectional data, 33-34, 194-195, 196, 229, 237, 238, 282-284
  buffer/latch, 91

Bus—cont
  control, 23, 25-26, 29, 36-37, 45-46
  data, 22, 45-46, 174, 284
  lines, 24
  monitor, 48-55, 58, 62-65, 211, 229, 270, 282-285
  single-line, 176, 181, 183
  systems, 176-177, 185
  unidirectional, 24
Buses, multiline, 177
Busing, three-state, 11, 173, 174-175, 181, 190, 195
Byte(s), 25, 51, 54, 90, 94, 96, 102, 265-266, 275
  absolute address, 307
  address bus, 44, 76, 82
  ASCII, 268
  data, 192, 223-224
  device code, 46, 67-68, 76
  device select address, 38-39
  instruction, 45, 56, 59, 67, 93, 161, 267, 268, 278, 281, 291, 320, 321, 352, 362, 377
  masked, 259
  timing, 342

## C

Card readers, 29
Carry
  bit, 120-121
  flag, 104, 105, 115, 116, 122-123, 128, 134, 139, 222, 268, 310
Cascaded 7490 decade counters, 272
Cassette tape, 262
Central processing unit, 23, 91-92, 100, *see also* CPU
Chart recorders, 263-264
Circuit(s)
  bus monitor, 49-52, 54, 270
  data logging, 261, 270
  decoder, 64
  decoding, 37-39, 42
  device select pulse, 31, 37-39
  input/output, 238-240, 270
  interface, 24, 48, 49, 270, 311
  latch, 191, 267
  latch/buffer, 181, 188-189
  output latch, 196-200
  single-step, 48
  three-state buffer input, 202-203

**402**

Clock(s), 12, 27
  circuitry, 17
  cycles, 265
  inputs, 35, 36
  outboard, 33, 62
  pulses, 32, 33, 63
  real time, 269
  subroutine, 86-87
  two-phase, 94
Clocked logic system, 32
Code
  ASCII, 161, 268, 269, 281, 311
  BCD, 42
  binary, 42
  device, 36, 42, 46, 47-48, 151, 152
  8-bit identification, 36
  hexadecimal, 81, 161, 305
  hex or octal instruction, 90
  instruction, 45-46, 81, 93-95, 101, 164
  listing, octal, 90
  machine, 101
  object, 101
  octal, 44, 81, 143, 280
Communications, time-sharing, 6
Comparators, 42-44, 74-75, 304-306
Compilers, 17
Computer(s), 15, 16
  digital, 15, 24, 25, 27
  games, 89
  hierarchies, 21
  interfacing, 28
  -on-a-chip, 12, 13, 17
  program, 14
  programming, 82
Condition
  bits, 83
  flags, 81, 103-104, 105, 106, 115, 124, 127, 128, 145
Conditional instructions, 81, 101, 134, 137-142
Control, 22-23
  bus, 23, 25-26, 29, 36-37, 45-46
  logic, CPU, 100
  pulses, 24, 25, 151
  register, 289, 291
  signal(s), 25, 26, 39, 44, 45, 50, 53, 62, 66, 71, 270, 283
  solvent-level, 313-316
  word, 292, 293

Controller, 11, 18
Conversion table, octal/hex, 389-390
Converter, digital-to-analog, 270-271, 296, 299, 302, 304
Core memory, 262
Counter(s), 12, 16, 29
  BCD, 35
  binary, 35, 181
  decode, 181
  program, 81, 83, 91, 93, 95, 134, 135
  7490, 63, 64, 66, 72, 75, 78, 264, 282
  74193 up/down, 199-200
CPU, 23, 24, 26, 29, 91-92, 94, 122, *see also* Central processing unit
Cross-assembler, 83
Crt displays, 29
Cycles, machine, 37, 47, 48, 59, 66-67, 71, 81, 82, 106, 237, 265
  FETCH, 45-46, 58
  INPUT, 46, 58, 63
  MEMORY READ, 56
  OUTPUT, 46, 57, 63
  STACK WRITE, 57

**D**

DAA, 122-124
  instruction, 102, 122, 124, 207, 208, 218-226; *see also* Decimal Adjust Accumulator
  operations, 124
Data, 102, 103
  accumulator, 196
  acquisition
    and control systems, 13
    rates, 263, 266
  bus, 45-46, 47, 49, 91, 174, 284
    bidirectional, 22, 26, 33-34, 51, 55, 65, 151, 152, 194-195, 201, 229, 237, 238, 282, 283
  buffer/latch, 91
  bytes, 25, 192, 206, 223-224
  formats, instruction and, 102
  input program, 278
  keyboard, 217
  logger, 261, 263, 271, 276, 308
  logging, 261-267, 270-278, 308
  multibyte numeric, 102-103

Data—cont
  output program, 278
  processors, 11, 15
  "raw," 263
  selector/multiplexer, 34, 35
  storage, 17, 262-263
  transfer, 29, 205
    group, 100, 101, 106-115
    operations, 23, 81
    synchronization pulses, 28-29
Decode counter
  circuit, 275, 276
  7490, 73-74, 78, 181, 185, 272
Decimal Adjust Accumulator; see
    DAA instruction
Decoder(s), 12, 35
  absolute, 70, 74, 76
  chip, 32
    7442, 184-185
    74154, 64, 66
  circuit, 64, 215, 230, 272
  /demultiplexer, 74154, 35, 36
  8205, 201
  instruction, 93-95
Decoding, 37-39, 42, 44
Decrement, 104
Decrementing, 125
Demultiplexers, 35
Diagnostic program, memory, 87
Digital
  bus, 174
  codes, 12
  computer, 15, 24, 25, 27
  controller, 18
  information, 22, 174, 176
  logic, 32
  register, 310
  servo, 35
  signals, 42
  tachometer, 35
  test instruments, 14
  -to-analog converter, 270-271, 296-301, 299, 302, 304
DM8220 9-bit parity generator/
    checker, 35
Double precision arithmetic
    operations, 121-122
D-type latch(es), 284
  7475, 181, 188-189, 196-197, 207-213, 238

D-type latch(es)—cont
  74100, 196

### E

Eadie, Donald, 15
8080
  /8080A programming information
    sources, 81, 82-90
  instruction set, 90, 100-106, 164-170
  microcomputer, 21-24
    data logging with an, 261-264
  microprocessor registers, 90-95
  mnemonic instructions, 96-100
8080A
  -based microcomputers, 191, 270, 271, 309
  central processing unit (CPU), 91-92
  input/output instructions, 44-45
  instruction set, 81-172, 192
  microcomputer, 25, 26, 28, 265, 356
  microprocessor, 17, 21, 24, 25-26, 93, 95-96, 100, 196, 310
  programming, 213
8095 three-state buffer, 179, 202, 207-213, 245, 280, 336
8111-2 static read/write memory, 201
8205 decoder, 201
8212
  I/O port, 201
  latch/buffer chip, 196, 202, 238, 240, 321, 356, 361
8255 interface chip, 202, 270, 288-296
Eikrem, Lynwood O., 14
Enable, 35, 36
  and disable interrupt, 320-321
  input, latch, 55, 63, 65
  interrupt, 142, 152-153, 321
Encoders, 35
Exclusive masking, 131-132

### F

4-bit magnitude comparators, 42
555
  monostable multivibrator, 34
  timer, 33
Fan-in, 44, 201, 288
Fan-out, 175, 201, 288

FETCH machine cycle, 45-46, 94
Flag(s), 117, 118, 119, 128, 135, 222, 309-382
  bit manipulation, 316
  carry, 268
  condition, 81, 103-104, 105, 106, 115, 124, 127, 128, 145
  external, 337-341
  flip-flops, 312, 314, 318, 326
  internal control, 145
  keyboard, 337
  register, 93, 95
  response time, 336, 341-344
  servicing software, 309
  status, 90
  Zero, 268
Flip-flops, 12, 20, 28, 33, 35, 36, 152-153, 312-314, 351, 379
  7474, 344
  7476, 42, 49, 78
  flag, 352
  interrupt enable, 356, 357
  outputs, 201
  positive-edge-triggered, 199
Floppy disk, 262
Flow chart, 311, 315
Format(s)
  description, 105-106
  instruction and data, 102

### G

Gansler, M. H., 155
Gate(s), 12, 20, 28, 33, 35, 44
  AND, 34
  circuit, 32
  Exclusive-OR, 34
  NAND, 42, 76
  NOR, 39, 70, 75
  OR, 34
  pulse, 32
Group
  arithmetic, 115-128
  branch, 134-145
  data transfer, 106-115
  logical, 128-132
  stack, I/O and machine control, 145-154
Gundestrup, Niels S., 159

### H

Hardware
  back-up, 335
  devices, electronic and mechanical, 32
  priority interrupts, 328-330
  substitution of software for, 31, 32-35
  vs software, 17
Hardwired logic systems, 21
Hex
  instruction code, 90
  /octal conversion table, 389-390
Hexadecimal, 87
  -ASCII listing, 90
  code, 81, 161, 225, 305
  instruction code, 164
  listing, 90, 162-163
Hierarchical systems, 13
Hierarchies, computer, 21
Hogan, C. Lester, 12, 14
Honea, Charles R., 13

### I

Inclusive masking, 131-132
Indicator, numeric, 48, 270
Input
  device select pulses, 38, 39, 69, 70, 201
  instructions, 60
  latch enable, 55, 63, 65, 270
  microcomputer, 201-202
  -output
    circuit, memory-mapped, 245
    operation, 192-194, 213
    ports, 250, 252, 253
  /output
    circuits, 238-240, 270
    data, 191
    device(s), 16, 23, 24, 29, 95, 192, 310, 317
    instructions, 44-45
    interface circuits, 227
    operation, 191
    program, 204
  port, 208-210, 292-294
  program, parallel, 296
  strobe pulses, 59

Instruction(s)
  8080 mnemonic, 96-100
  assembly language, 90
  byte, 25, 45, 47, 56, 59, 67, 161, 267, 268, 278, 281, 291, 320, 321, 352, 362, 377
  characteristics, 109-110
  code(s), 45-46, 93-95, 100, 165-170
    hex or octal, 90
    hexadecimal, 164
    octal, 164
  coded, 100
  conditional, 81, 101
  cycle, 45
  decoder, 93-95
  formats, 102
  IN and OUT, 36-37, 45-47, 64, 191, 195, 237
  input and output, 44-45, 46, 60
  I/O, 44-45
  interrupt, 103
  macro, 83
  memory-mapped I/O, 234-237
  memory-reference, 47, 64, 235-236, 250, 252
  microcomputer, 24, 32-33
  multiple byte, 102
  processor, 154
  programming language, 101
  register, 47, 81, 93-95, 101
  register pair, 47-48
  rotate, 132-134
  set, 17, 20-21, 96, 100-101, 154
    8080, 101-106, 164-170
    8080A, 81-172, 192
    summary, 47
  single-byte, 170, 318, 320, 362
  three-byte, 135, 170
  two-byte, 45, 170
  unconditional, 81
Instruments, test, 14
Integrated-circuit chips, 34-36
Intel program library, 83-88, 155, 159, 161
Interface, 68, 91, 278, 288
  chip, programmable peripheral, 180, 270, 288-296
  circuit(s), 24, 48, 49, 191, 202, 211, 227, 270, 311
  system, microcomputer, 242

Interfacing, 11, 12, 28, 29, 32, 82, 201
  applications, 38
  digital-to-analog converter, 270, 296-301
  keyboard, 310-313
  mechanical switch, 344-346
Interrupt(s), 25-26, 309-382
  acknowledge, 23, 25, 26, 320, 322
  circuitry, 17
  -driven keyboard interface, 321-325
  enable
    and disable, 320-321
    flip-flop, 356, 357
    instruction, 377-378
  flag, 321, 323
  hardware priority, 328-330
  instruction, 103
    port, 309, 321
    register, 321, 355-360, 362
  modes, 318
  multilevel, 318
  operation, 317
  priority, 325-328
  pulser, 357-358, 360, 367
  service routine, 87, 309
  simultaneous, 337
  single line, 318
  software, 320, 331-336, 366, 380
  time-oriented, 336
  timing problems, 309, 337
  types of, 317-318
  vectored, 318
  what is an?, 317-318
Interval timer, programmable, 269-270
I/O, 192, 228-229
  accumulator, 191, 192, 195, 203-204, 270, 288
  address select pulses, generating memory-mapped, 230-232
  and machine control group, stack, 145-154
  bidirectional memory-mapped, 270, 285-288
  decoder circuit, 215
  device(s), 23, 24, 29, 32, 36, 45-46, 101, 153, 191, 228, 232, 317
  group 100, 101-102
  instructions, 44-45
  memory-mapped, 194, 232-237, 288-296

I/O—cont
  port(s), 16, 257, 289, 291
    8212, 201
    bidirectional, 270, 288
    memory-mapped, 245
    programs, 191
    read, 23, 25, 26
    write, 23, 25, 26

## K

KEX, 58-59, 62
  program, 48, 56, 218
  software, 346, 348
  time-delay subroutine, 342
Keyboard
  data, 217
  EXECUTIVE (KEX) EPROM, 50, 53, 299
  flag, 337, 346, 347
  input subroutine KBRD, 347
  interface, interrupt-driven, 321-325
  interfacing a, 310-313
Kleffman, Donald V., 14

## L

Latch(es), 12, 16, 29, 34, 35, 36
  8-bit, 34
  7475 D-type, 51, 181, 188-189, 245
  74175, 197-199
  /buffer, 179-181, 188, 196, 202, 238, 240
  circuit(s), 191, 196-200, 267
  enable input, 55, 63, 65, 270
  general-purpose, 200
  output, 95, 201
  STROBE input, 50
Latched 7490 counter, 282
Linear ramp output, 306
Linking addresses, 323
Liquid-level sensing, 337, 348-352
Lloyd, Sheldon G., 13
LM311 comparator, 304, 305, 306
Logging
  slow data points, 265-266, 276-278
  sixty-four 8-bit data points, 264-265
Logic
  CPU control, 100
  operations, 11-12, 23
  states, 33, 34
  switches, 12
  synchronous, 27-28

Logic—cont
  systems, 18, 21, 32
Logical
  group, 100, 101, 128-132
  operation(s), 93, 95-96, 101, 104, 127-130, 135, 139-140, 259
Loops, time-delay, 32-33, 79, 265, 269
LSI, 16-17, 34

## M

Machine
  code summaries, 90, 101
  control group, 100, 101-102
  cycle(s), 37, 48, 56-58, 59, 63-64, 66-67, 71, 81, 82, 106
  memory read and memory write, 237
  instructions, 101
  level programming, 82
Macro instructions, 83
Magnetic tape, 29, 262, 263
Marley, Richard, 14
Magnitude comparator, 7485 4-bit, 35
Mask pattern, 329
Masked byte, 259
Masking, 131-132
Master clock, 27
Math programs, 87-88
Memory, 13, 14, 16, 23, 24, 29, 35, 47, 48, 54, 71, 94-95, 102, 115, 305
  8111-2 static read/write, 201
  address(es), 93, 114, 161, 230
    select pulses, 227
  addressing, 81, 83
  core, 262
  devices, external, 234, 235
  diagnostic program, 87
  I/O, 194, 195, 227, 232, 240, 257
    addressing, 238
    device, 236, 294
    techniques, 244, 299
  location(s), 24, 26, 45-46, 103, 113, 129, 135, 227, 232, 235-237, 265, 267, 273, 277, 305, 353
    addressable, 93
    read/write, 256, 257
  -mapped
    input-output circuit, 245
    I/O, 191, 227, 228-229, 285-296

Memory—cont
-mapped
address select pulses, generating, 230-232
bidirectional, 270, 285-288
characteristics of, 229
instructions, 234-237
interface circuits, 257
port, 245
techniques, 259
use of Address Bit A-15, 232-234
using the 8255 chip, 294-296
vs accumulator I/O, 228-229
read, 23, 25, 26
and memory write machine cycles, 237
/write, 262, 263, 266, 267, 268, 274, 275, 307, 345
-reference instruction, 47, 64, 230, 235-236, 250, 252
write, 23, 25, 26
Microcomputer(s), 16, 18, 20-21
8080, 21-24
8080A, 25, 26, 28, 191, 265, 270-271, 346
-generated synchronizing signals, 31
input, 192, 201-202, 238
/output, 191, 240
instructions, 24, 32-33
interface, 202, 242
interfacing, 12
output, 191-192, 195-196, 199-200
program(s), 32, 45-46, 67
programming, 82, 89
Microprocessor(s), 12-13, 15-17
8080, 21
8080A, 24, 25-26, 93, 95-96, 100, 114, 135, 149, 174, 181, 196, 310
registers, 8080, 90-95
vs microcomputer, 15-17
Minicomputers, 13, 20, 21
MMD-1 microcomputer, keyboard characteristics of the, 346-348
Mnemonic(s), 96-99, 101
code, 164
instructions, 8080, 96-100
Modes, addressing, 102-103

Modulo-N-divider, 35
Monitor circuit, bus, 48, 49, 50, 51, 52, 211, 270, 282, 284
Monostable multivibrator, 12, 33, 34
MOS circuitry, 16, 17
MPU technology, 13-15
MSI integrated circuit chips, 34
Multibyte
instruction, 336
numeric data, 102-103
Multilevel interrupt, 318
Multiline busses, 177
Multiple byte instructions, 102
Multiplexers, 35
Multivibrator, 12, 33-34

## N

NAND gate, 7430 8-input, 42, 76
Nibble, 81, 122-123
NOP instruction, 153-154
NOR gate, 7402 2-input, 39, 70, 75
Number generation, pseudo-random, 35
Numerical listing, octal/hexadecimal, 90

## O

Object code, 101
Octal
code, 44, 81, 90, 143, 225, 280
/hex conversion table, 389-390
/hexadecimal, 90
instruction code, 90, 164
OUT
control, 44, 50, 53, 66, 151
instructions, 36-37, 45-46, 47, 64
Outboards, 33, 62
Output
circuits, microcomputer, 191, 192
device, 95
select pulses, 38, 39, 69, 270
drive capability, 191, 201
from a data logger, 266-267
instruction, 46, 60
latch circuits, 95, 196-200
linear ramp, 306
lines, 94
machine cycle, 46
microcomputer, 195-196
open-collector, 176

Output—cont
  port, 208-210, 292-294
  strobe pulses, 59
  three-state, 174, 176, 180, 181

**P**

PACE microcomputer programs, 34-35
Packed-BCD, 206, 208, 220, 223-224, 225, 273
Paper tape, 262-263
Parallel
  bit-by-bit logical operation, 128, 129, 130
  input program, 296
Parity
  flag, 104, 105, 115, 116, 128, 139, 310
  generator/checker, DM8220, 35
Pattern, bit, 48, 61-62, 104-105, 131
Peripheral
  equipment, 16, 17
  interface chip, programmable, 270, 288-296
Perlowski, Andrew A., 13
Petritz, Richard L., 14
Polled operation, 316-317
Polling, 316-317
Ports, I/O, 289, 291
Power dissipation, 44
Priority
  encoders, 35
  interrupt timing, 337, 375-378
  interrupts, 325-330, 337, 369-374
Process industries, 13-14
Processor
  CMOS-on-sapphire, 16
  data, 11, 15
  instructions, 154
  status word, 146-147
Program(s), 17, 26, 100, 101
  8080A microcomputer, 45-46
  addressable registers, 91
  assembly language, 87, 96
  computer, 14
  counter, 81, 83, 91, 93, 95, 134, 135
    register, 91, 95, 105, 137, 141, 192

Program(s)—cont
  data logging, 264-267, 270
  debugging, 153
  input/output, 204
  I/O, 191
  KEX, 48, 56
  library, Intel, 83-88
  math, 87-88
  memory diagnostic, 87
  microcomputer, 67
  PACE microcomputer, 34-35
  parallel input, 296
  software, 315
    debugging, 336
  source, 101
  storage, 17
  text storage, 86
Programmable
  interface chips, 180
  interval timer, 269-270
  I/O lines, 17
  peripheral interface chip, 8255, 202, 270, 288-296
  registers, 180
Programming, 82-101
  assembly language, 82, 83, 96
  experiments, 81
  8080/8080A, 81-90
  languages, 101
  machine level, 82
  microcomputer, 82, 89
Pulse(s), 27-28
  address select, 195
  clock, 32, 33, 63
  control, 24, 25, 62, 66
  data transfer synchronization, 28-29
  device select, 28-29, 31-80, 191, 195, 196, 202, 213, 270, 283
  gate, 32
  input strobe, 59
  negative, 32, 66
  output strobe, 59
  positive, 66
  single-step, 211
  strobe, 29, 32, 48
  synchronization, 36, 37, 195, 230, 282
  timing, 94
  width, 26, 32
Pulser(s), 12, 33, 62, 186, 344, 356

## R

Random
- -access memory, 102
- logic, 20

Rates, data-acquisition, 263, 266
"Raw" data, 263
Read-only memory, 102
Read/write memory, 102, 262, 263, 266-268, 274, 275, 277, 307, 345
Real-time clock, 269, 328
Recorders, chart, 263-264
References, 383-384
Register(s), 81, 90-95, 101, 102, 115, 128, 181
- 8080 microprocessor, 90-95
- accumulator, 45-46
- contents, 106, 107, 120
- control, 289, 291
- digital, 310
- flag, 93, 95
- general-purpose, 81, 91, 93, 95, 114, 115, 119, 128, 129, 149, 192, 194
- instruction, 81, 93-95
- internal, 191, 192, 234
- interrupt instruction, 355-360, 362
- operate instructions, 101
- pair(s), 81, 96, 103, 104-105, 125, 254
- program addressable, 91
- program counter, 91, 95, 105, 137, 141, 192
- programmable, 180
- sense, 313, 315, 318
- shift, 11
- stack pointer, 91, 95, 105, 137
- temporary, 93
- word, 269

Relay(s)
- optically-isolated solid-state, 42
- solid-state, 49, 78, 79, 313

Riley, Wallace B., 21
Rotate instructions, 132-134

## S

16-bit operations, 89
74H21 dual 4-input AND gate, 34
74L42 decoder, 39-40
74L85 low-power comparator, 44
74L154 decoder chip, 32
74LS05 inverters, 40
74S85 high-power comparator, 44
7402 2-input NOR gate, 39, 70, 75
7408 quad 2-input AND gate, 34
7409 quad 2-input AND gate, 34
7411 triple 3-input AND gate, 34
7432 quad 2-input OR gate, 34
7442 decoder chip, 36, 184-185
7430 8-input NAND gate, 42, 49
7474 flip-flop, 36, 336, 344
7475
- 4-bit D-type latch chip, 196-197, 207-213
- D-type latches, 181, 188-189, 238
- latch, 36, 39, 51, 245

7476 flip-flop, 42, 49, 78
7483 4-bit binary full adder, 34
7485 comparator, 35, 42-44, 48, 74-75
7486 quad 2-input Exclusive-OR gate, 34
7490
- counter, 48, 63, 64, 66, 72, 75, 78, 264, 282
- decade counter, 36, 73-74, 181, 185, 272

7493 binary counter, 36, 181, 185
74100 8-bit D-type latch chip, 196
74121 monostable multivibrator, 34
74125 three-state buffer, 177-178, 181, 188-189
74126 quad three-state buffer, 176, 178-179, 181, 185-186
74150 data selector/multiplexer, 34
74151 data selector/multiplexer, 35
74154
- decoder chip, 31, 64, 66
- decoder/demultiplexer, 35, 36
- decoders, 37, 38, 39

74175 4-bit latch chip, 197-199
74184 BCD-to-binary converter, 35
74185 binary-to-BCD converter, 35
74190 up/down BCD counter, 35
74191 up/down binary counter, 35
74193 up/down counter, 199-200
74198 8-bit shift register, 29, 197-199, 238
Scallon, Nicholas P., 14

Sense register, 313, 315, 318
Servo, digital, 35
Shift registers, 11, 29, 35
   74198, 197-199, 238
Short-term data storage, 262-263
Sign flag, 104, 105, 115, 116, 128, 139-140, 310
Single-bit synchronization pulse, 230
Single-byte instructions, 170, 318, 320, 362
Stack
   I/O, and machine control group, 100, 101-102, 145-154
   pointer, 81, 83, 91, 93, 95, 102, 145-146, 148, 149, 159, 192, 353
   register, 91, 95, 105,137, 145-146, 151
Static read/write memory, 8111-2, 201
Status
   pulses, 25
   signal, 312
   word, processor, 146-147
Strobe
   input, 50
   pulses, 29, 32, 48, 59
Strobing integrated-circuit chips, 35-36
SUB flag, 124, 148, *see also* Subtract flag
Subroutine(s), 83
   clock, 86-87
   instruction, 137, 148-149
   KBRD, listing of, 217-219
   time-delay, 58, 266, 270
   TIMOUT, listing of, 219
   TIMEOUT, listing of, 343-344
Subtract flag, 124, *see* SUB flag
Subtraction operations, 120-122
Symbolic addresses, 315
Symbols, 104-105
Synchronization pulse(s), 36, 37, 195, 230, 282
   data transfer, 28-29
Synchronizing signals, microcomputer-generated, 31
System(s)
   bus, 176
   data acquisition and control, 13
   hierarchical, 13
   logic, 18

**T**

Tachometer, digital, 35
Tape, 262-263
Teletypewriters, 29
Temporary registers, 93
Three-byte instruction, 135, 150, 159, 170
Three-state
   buffer, 29, 34, 95, 173, 177-179, 180, 181, 188-189, 201, 202, 336
   8095, 179, 202, 207-213, 245
   74125, 177-178, 181
   74126 quad, 176, 178-179, 181
   chips, 280
   circuits, 191
   input circuits, 202-203
   bus systems, 185, 195
   bussing, 11, 173, 174-175, 181, 190
   devices, 173-190
   latch/buffer, 179-180
   output, 174, 176, 180, 181
TIL311 numeric indicator, 48, 49, 50, 52
Time-delay(s)
   loop(s), 32-33, 79, 265, 269
   other methods of generating, 269-270
   routine, 299, 301-302, 342
   software, 33
   subroutine, 58, 266, 270, 342
TIMEOUT
   listing, 299, 301-302, 342, 343-344
Timer
   555, 33
   interval, 269-270
Timesharing communications, 87
Timing
   bytes, 342
   problems, 309
   pulses, 94
Transfer group, data, 106-115
Triple precision arithmetic operations, 121-122
TRI-STATE
   buffer, 173
   concept, 174-175
   devices, 180-181
TTL subfamilies, characteristics of, 201

Two's complement arithmetic, 115

## U

Unconditional instructions, 81, 134, 137
Unidirectional bus, 24
Up/down counter, 74193, 199-200

## V

Vectored interrupt, 318

## W

Wadsworth, Nat, 89
Wait loop, 50, 53, 270
Wampler, Galen W., 14
Webb, Sidney, 12
Width, pulse, 32
Word
   control, 292, 293
   length, 93
   processor status, 146-147
   register, 269

## Z

Zero flag, 103, 105, 115, 116, 128, 139, 140, 142, 268, 310

# NOTES

**NOTES**